D1827084

International Series in Operations Research & Management Science

Volume 172

For further volumes:
www.springer.com/series/6161

ManMohan S. Sodhi • Christopher S. Tang

Managing
Supply Chain Risk

 Springer

ManMohan S. Sodhi
Cass Business School
City University
London, UK

Christopher S. Tang
Anderson School of Management
University of California, Los Angeles
Los Angeles, CA, USA

ISSN 0884-8289   International Series in Operations Research & Management Science
ISBN 978-1-4614-3237-1   e-ISBN 978-1-4614-3238-8
DOI 10.1007/978-1-4614-3238-8
Springer New York Dordrecht Heidelberg London

Library of Congress Control Number: 2012933784

Printed on acid-free paper

Springer is part of Springer Science+Business Media (www.springer.com)

To

Katarzyna Zechenter,

Amelka Sodhi

and

Richard Paegelow

Preface

At the core of the layer-upon-layer creation of every pearl—perfect or not—is an irritating tiny grain of sand. While we cannot claim this book is a valuable gem, we can say for sure that the effort has painstakingly entailed layer-upon-layer of concepts drawn from the surrounding business world: the annoying grit for us has been the idea of trying to understand supply chain risk conceptually. We have been working in the field of supply chain risk since 2004 at least and have co-authored many articles with each other and with others. And our work, rather than remove the grain of sand, simply built up the layers around this grit as we understood supply chain risk a little more each time.

Certainly over time there have been excellent edited books with different authors picking up different aspects of supply chain risk. The same applied to special issues of well-regarded scholarly journals. But as we reviewed these and our own work, we felt there was a need to present supply-chain risk as a whole and that would be possible only with a book.

Once we decided we should write a book, the next question was for whom. Supply chain risk is a nascent field, going back only to the early 2000s. This does not mean there was no risk in supply chains before then, but that supply chain risk began to be identified as a domain of research and practice only around that time. For us, it meant we should definitely target scholars looking to start work in supply chain risk.

However, just like supply chain management, supply chain risk first became a domain of practice than of research. For us, this meant we should definitely target managers if the book is to have any validity: if it does not pass muster with practitioners then perhaps it is not really about supply chain risk. Hence, we decided we would target not only scholars but, importantly, practitioners and senior managers working in risk management or in supply chain chain management.

Finally, the question we faced was content and presentation. As academics, we do have access to colleagues who work in risk or in supply chain management. We could draw on risk literature from accounting, actuarial sciences, or finance. However, we did not seem to be able to find overlaps between these different views other than perhaps the mathematics of probability. From our own experience, we felt that this mathematical approach would be too limiting this early in the development of supply chain risk. Instead, as we listened to managers, it became clear that managing drew on enterprise risk management and business continuity. However, we did not want to forget the "supply chain" aspect of supply chain risk either and therefore wanted to bring as much of the supply chain management literature into our purview so researchers could build upon what is already well established in the supply chain literature pertaining to risk.

To present all this, we decided we would present the book in four parts, layered upon each other: (A) introducing supply chain risk with four steps of supply chain risk management, borrowing from enterprise risk management, (B) presenting broad mitigation approaches in practice tohighlight concepts,drawing on industry examples and cases to emphasize practice, (C) detailing modelling-based approaches to mitigate supply chain risk, drawing on the supply chain literature, and (D) concluding by including the viewpoints of many researchers as well as by listing areas for further research.

This is what we have here—it's been a rather long journey for us and I hope both researchers and practitioners will find this book a useful first step whether for research or for setting up supply chain risk management in companies.

ManMohan S. Sodhi
London
Christopher S. Tang
Los Angeles

Foreword

Risks are inherent in every aspect of business, and the ability of manage risks is one important aspect that differentiates successful business leaders from others. In a supply chain, risks can be found in almost every stage. From product development to product launch, from procurement to manufacturing, and from distribution to after-sales support and product end-of-life, supply chains are vulnerable to risks. What makes such risks even more complex and challenging to manage is that the risks are often not within the direct control of the business executive. They can exist outside of the company boundary, and yet the impact to the company can be huge.

This book is the most up-to-date and comprehensive treatment of the subject. There are several distinct features that also distinguish this book from others. First is the comprehensiveness of the treatment of risks. As mentioned earlier, supply chain risks require an end-to-end coverage, and this book has that end to end coverage. The chapters on new product development and on product recall, for example, show how we need to contain the risks right at the beginning and at the end of the supply chain. Second, the book is rich in current examples. These are what managers and executives have to worry about in the current business environment. The authors have provided illustrations of risky situations and case studies that are most relevant and current. Third, this book ties observations in practice to methodologies and research. The rich case examples motivated the approaches and methodologies used to mitigate risks, and in the course of doing so, the authors also provided insights on existing and new research opportunities. As a result, this book is highly relevant to both practitioners and academics. Fourth, the book is also written with management lessons on how risks can be mitigated, and how risks can be contained once disruptions have occurred. As such, it is also a book for management to gain insights and to develop management skills.

It is not easy to write a book that can appeal to and be highly useful to both practitioners and academics. I applaud the authors for having successfully achieved such an admirable goal. What they have been able to do is admirable, and I welcome the addition of this book as a significant contribution to our understanding of supply chain management.

Hau L. Lee
Thoma Professor of Operations, Information and Technology
Director of the Stanford Global Supply Chain Management Forum
Graduate School of Business, Stanford University

Endorsements

"Supply Chain Risk Management is an issue that many companies face and yet few companies know how to deal with it in a systematic and pragmatic manner. While avoiding and reducing supply chain risks are certainly preferable, developing ways to restore and stabilize supply chain operations rapidly after a major disruption is critical for managing global supply chains. Sodhi and Tang present important concepts, frameworks, strategies, and analyses that are essential for managing supply chain risks. Not only does this book suggest some practical ways to work with different partners to manage the risks that are present in a global supply chain, it creates a framework that would enable practitioners to engage researchers to work on this important area."
—Thomas A. Debrowski, Executive Vice President, Worldwide Operations, Mattel, Inc.

"When a firm outsources its operations to external suppliers, the firm is vulnerable to major and rare disruptions that can occur at any link in the global supply chain. Because these disruptions rarely occur, few firms take commensurable actions to identify, assess, mitigate and respond to various types of supply chain risks. By introducing frameworks and concepts along with several case studies and a review of academic literature, Sodhi and Tang treat this important subject with practical relevance and academic rigor. This book will bring practitioners and researchers to develop effective and efficient ways to manage supply chain risks."
—Marshall L. Fisher, UPS Professor, Professor of Operations and Information Management and Co-Director of Fishman-Davidson Center for Service and Operations Management, The Wharton School, University of Pennsylvania

"This book ties observations in practice to methodologies and research. The rich case examples motivated the approaches and methodologies used to mitigate risks, and in the course of doing so, Sodhi and Tang provided insights on existing and new research opportunities. As a result, this book is highly relevant to both practitioners and academics. Also, the book is also written with management lessons on how risks can be mitigated, and how risks can be contained once disruptions have

occurred. As such, it is also a book for management to gain insights and to develop management skills."
—Hau L. Lee, Thoma Professor of Operations, Information and Technology and Director of the Stanford Global Supply Chain Management Forum, Graduate School of Business, Stanford University

"As companies have extended their supply chains globally and as the face increasing resource issues, they face a number of new risk challenges. While there are various case studies written about supply chain risks, this book gives a comprehensive treatment of the subject with clarity. The concepts and frameworks developed by Sodhi and Tang in this book would create awareness of this important and yet not well understood subject, and strategies described in this book would stimulate practitioners to develop a holistic approach for identifying, assessing, mitigating, and responding to different types of supply chain risks."
—Nick Wildgoose, Global Supply Chain Proposition Manager, Zurich Insurance

Acknowledgements

As we noted in the Preface, this book has been built layer-upon-layer. This would not have been possible without drawing upon our own past publications with each other and with other co-authors. We adapted our published work with Sunil Chopra (Chapter 4), with Seongha Lee (Chapter 6), with Joshua Zimmerman (Chapter 11), with Brian Tomlin (Chapter 14) and with Byung-Gak Son (Chapter 16) and we acknowledge their intellectual input.

We would like to thank Fred Hillier, the series editor, and Camille Price for their continuous support and encouragement over the past few years. Mirko Janc has typeset the book really well and we thank him for that and for his suggested improvements.

Last but not least, we would like to thank our families—Katarzyna and Amelka, and Richard—for their support over the years it took to put this book together. This book is dedicated to them.

Contents

Part I
Introduction—Identifying, Assessing, Mitigating and Responding to Supply Chain Risk

Chapter 1
Supply Chain Risk Management

Abstract Recognizing various natural and man-made disasters have caused major supply chain disruptions over the last two decades, this chapter illustrates the vulnerability of many global supply chains and provides justifications for companies to develop a systemic approach to managing supply chain risks. Viewing supply chain risk management as comprising four steps—identifying risks, accessing risks, mitigating risks, and responding to risks—this chapter highlights the overall structure and key objectives of the book.

1.1 Introduction

The two decades straddling the turn of the millennium, 1990-2000 and 2000-2010, have witnessed many varied natural and man-made disasters causing havoc with the supply chains of many global companies. Likewise, the supply chains of companies have wrought disasters on society and the ecosystem. And, more often, problems in one link of the supply chain have caused unmitigated disaster to another link, resulting in large financial and non-financial damage. Indeed, as supply chains have grown globally with companies from the west as well as those from emerging economies seeking new markets and low-cost sources worldwide, natural, man-made or even self-inflicted disasters disrupt these far-flung supply chains of today much more so than they did in the distant past.

Natural and man-made disasters have disrupted supply chains of many companies. Such disasters include terrorist attacks, wars, earthquakes, economic crises in North America, Australia and Europe (2007–08), devaluation of currencies in Indonesia, South Korea, Thailand and other Asian countries (1997–98), SARS virus outbreak starting in China (November 2002 to mid-2004), the Indian Ocean tsunami affecting Indonesia but also India, Sri Lanka, and Thailand (2004), strikes in France by public sector workers and others against the proposed increase of retirement age (October 2010), and computer virus attacks like the Stuxnet attack on an Iranian nuclear plant by a foreign government (September 2010). Then there was the

M.S. Sodhi, C.S. Tang, *Managing Supply Chain Risk*,
International Series in Operations Research & Management Science 172,
DOI 10.1007/978-1-4614-3238-8_1, © Springer Science+Business Media, LLC 2012

petroleum leak off the coast of Louisiana from a deep-sea drilling platform (summer of 2010) and the leakage of toxic waste in Hungary (October 2010).

Disruptions can come from within the supply chain itself. Supply problems caused the delay of the Sony PlayStation 3 (2006) resulting in Sony not only losing revenues but also long-term market share. Dell had to recall 4 million laptop computer batteries made by Sony owing to some problems with the batteries. In 2001, mobile-phone manufacturer Ericsson lost 400 million Euros after their supplier's semiconductor plant caught on fire upon being hit by lightning; eventually, Ericsson ceded the mobile-phone business to a new entity, Sony Ericsson. A supplier's insolvency resulted in UK auto-maker Land Rover having to lay off 1400 workers. The impact of major disruptions need not be financial alone: in 2007, over 100 brands of tainted pet food contaminated with the toxic chemical melamine killed thousands of dogs and cats in the United States.

Besides such newsworthy disruptions, there are many mostly small, but sometimes large, hiccups and bumps in the day-to-day challenge of matching supply and demand, although this challenge could persist for the long-term. There may be delayed supplier deliveries or delayed product launches. Losses could result from customer dissatisfaction caused by late delivery or otherwise less-than-perfect order fulfilment in an environment increasing supplier lead times. There could be quality problems, whether or not serious enough to require recalls, in the face of shortening product lifecycles and increasing customization. Thus, the supply chain's ability to match supply to demand comes under threat with possibly disastrous consequences for the company as small losses accumulate over time despite the absence of any highly visible disaster.

How should companies make sense of these different kinds of risks stemming from different sources? While such risk incidents generally get categorized under the rather big umbrella of *supply chain risk*, neither the term nor its related terminology is well defined or scoped. Nor is there any consistent approach to managing these or indeed who should manage these risks: should it be the supply chain management people or should it be the enterprise risk management personnel or indeed cross-company with teams of people from the company and from its business partners? And where do we begin dealing with the kinds of risks above?

1.2 This Book

This book seeks to provide a starting point for companies seeking to develop an approach to managing supply chain risk and for young researchers wishing to start out in this field. There is already a body of literature building up. On one hand are the edited books targeting researchers with compilations of diverse views of different researchers: Brindley (2004); Kersten and Blecker (2006); Handfield and McCormack (2008); Zsidisin and Ritchie (2008); Wu and Blackhurst (2009); and Khan and Zsidisin (2010) along with special issues of operations management journals (cf. Seshadri and Subrahmanyam, 2005). On the other hand are books targeting practitioners: Sheffi (2005 a and b); Waters (2007); Kaye (2008) and Lynch (2009).

In contrast, we seek to bridge practice and research by providing a strong and common foundation for both by providing frameworks for supply chain risk management and research. This foundation is layered, each successive part of the book building upon the previous: we introduce the basics of the process of managing supply chain risk in Part I and develop and illustrate broad approaches to mitigating supply chain risk in Part II. The chapters in these two parts provide a common starting point for practitioners as well as for researchers. Next, we lay the foundations of mathematical modelling through broad reviews of the research literature in Part III and, finally, we provide researchers' perspectives and topics for future research in Part IV. While the chapters in the third part are primarily for researchers and technically trained practitioners, those in the fourth part can benefit both practitioners and researchers in helping them understand the challenges and opportunities that lie ahead for supply chain risk management.

A more detailed view of the book is as follows: In **Part I**, successive chapters (Chapters 2–5) deal with different aspects of the process of risk management: identifying risks, assessing risks, mitigating risks, and responding to risk incidents. In addition, Chapter 6 discusses how global companies manage supply chain risk with an emphasis on the supply chain's "connectedness" using the example of Samsung Electronics.

In **Part II**, the first two chapters (Chapters 7 and 8) deal with strategic and tactical approaches for mitigating supply chain risks. The remaining chapters in this part, Chapters 9–12, are applications of these broad strategic/tactical areas to specific domains: long-term demand uncertainty and consequent risk entailed in capacity decisions (Chapter 9), outsourcing (Chapter 10), supply management for new product development (Chapter 11) and product recall (Chapter 12).

Part III provides three detailed reviews of the research literature to get the researcher (or practitioner) started: Chapter 13 introduces risk models for supply chain Management, Chapter 14 provides different models to value flexibility in making the supply chain more robust, and finally Chapter 15 explores the application of stochastic programming for planning under demand uncertainty.

Finally, **Part IV** provides researchers' perspectives on supply chain risk management and where the existing research literature is compared against the perceived need for research in Chapter 16. Practitioners will also find this chapter useful to understand how the researchers view supply chain risk management as a basis for engaging with them. The final Chapter 17 outlines topics for further research for researchers; practitioners can also see areas where they can cooperate with researchers to create a better understanding of supply chain risk management.

1.3 The Supply Chain Management Context

To discuss supply chain *risk* management, we should start with *supply chain management* and state the basics. A *supply chain* is a network of organizations possibly including suppliers, manufacturers, logistics providers, wholesalers/distributors, and retailers that aims to produce and deliver products or services for the end customer. *Supply chain management* is the management of material, information and financial flows through the supply chain. It includes the coordination and collaboration of processes and activities across different functions such as marketing, sales, production, product design, procurement, logistics, finance, and information technology within the supply chain.

Supply chain management includes tracking and seeking to improve *operational performance metrics* including:

- Time-related measures such as product development cycle time, time to market, production lead time, replenishment/delivery lead time, cash-to-cash cycle;
- Cost-related measures such as product development cost, material cost, production cost, labor cost, inventory cost, shipping and handling cost, sales and marketing cost, fixed and overhead cost; and
- Customer-satisfaction-related measures such as product availability, product and service quality, reliability, after sales support, and total cost of ownership.

In addition, there are *strategic performance metrics* such as profits, market share, revenue growth, return on assets (ROA), and share-price performance.

Throughout the 1990s and 2000s, many firms strived to improve their strategic performance measures by implementing various supply chain initiatives that led to supply chains becoming more global, and thus not only longer but also more complex. These initiatives were intended

- To increase revenue (e.g., more product variety, more-frequent new product introductions, more sales channels/markets),
- To reduce cost (e.g., supply base reduction, online sourcing including e-markets and online auctions, offshore manufacturing, Just-in-Time inventory systems, vendor managed inventory), and
- To reduce assets (e.g., outsourced manufacturing, information technology and logistics).

These initiatives have led to complex supply chains. According to an industry study conducted by AMR Research in 2006, over 42% of the companies manage more than five different supply chains because of the need to produce multiple products for multiple markets.

1.4 The Need for Supply Chain Risk Management

Supply chains today are vulnerable to disruptions with large unanticipated conse-quences of seemingly contained events (c.f., Craighead et al., 2007). Three under-lying reasons are: (1) these supply chains have more points of possible disruption than they did in the past; (2) being longer, these supply chains have less visibil-ity, which causes slow decision-making and response in case of a disruption; and (3) local "fixes" create problems in other parts of the supply chain. In a 2004 study, Computer Sciences Corporation found 60% of the firms in their sample reporting that their supply chains are vulnerable to disruptions. At the same time, according to two independent studies, one by the Center for Research on the Epidemiology of Disasters (www.cred.be) and the other by the world's largest re-insurer Munich Re (www.munichre.com), historical data indicates that the total number of natural and man-made disasters has risen dramatically in the new millennium. Moreover, Munich Re has reported that the average cost of disasters has increased by a factor of 10 since the 1960s.

Based on anecdotal observations, disruptions can also leave supply chains dam-aged for a long time. For example, out of the 350 businesses operating in the World Trade Center before the 1993 bombing of the World Trade Center, 150 were out of business a year later (Eskew, 2004). Indeed, these disruptions can have a negative impact in the long-term on a firm's market value and strategic performance mea-sures. Hendricks and Singhal (2005a) found that companies suffering from supply chain disruptions experienced 33-40% lower stock returns relative to their indus-try benchmarks over a 3-year time period that starts one year before and ends 2 years after the disruption announcement date, based on a sample of 827 disruption announcements made over a 10-year period.

1.5 Apprehension without Action

Despite the headline news of disasters and their known detrimental effects, most firms find it difficult to justify investment for mitigating supply chain disruptions that occur rarely or that have never occurred in the past. Indeed, Rice and Caniato (2003) and Zsidisin et al. (2000; 2004) comment that most companies invest little time or resources in managing supply chain risks even though they carry out supply chain risk assessments.

Two surveys capture this perplexing dichotomy. According to a study conducted by Computer Sciences Corporation in 2003, 43% of 142 companies, ranging from consumer goods to healthcare, reported that their supply chains are vulnerable to disruptions, but 55% of the companies had no documented contingency plans (c.f., Poirier and Quinn, 2003). According to another survey conducted by CFO Research Services, 38% of 247 companies acknowledged that they had too much unmanaged supply chain risk (c.f., Eskew, 2004).

To our knowledge, there is no specific study that examines the paradox of why firms perceive serious supply chain risk and yet do not take commensurable actions. Possible reasons could be:

- Firms underestimate the risk in the absence of accurate supply chain risk assessment;
- Firms are not familiar with ways to manage supply chain risks (Closs and McGarrell, 2004); and
- With inaccurate estimates of the likelihood of the occurrence of a major disruption, many firms find it difficult to perform cost/benefit or return-on-investment analysis to justify certain risk reduction programs or contingency plans (Rice and Caniato, 2003; Zsidisin et al., 2000)

The last reason appears the most compelling when we consider Total Quality Management (TQM) as an analogy. Anecdotal and empirical evidence confirmed that TQM can provide significant value to the firms and their customers (c.f., Hendricks and Singhal, 1996). However, fewer than 10% of the Fortune 1000 have well-developed TQM programs, according to a survey study conducted by Rigby (2001). This motivated Repenning and Sterman (2001) to conduct over a dozen of in-depth case studies in industries including automobiles, chemicals, oil, and semiconductors. Their analysis suggested that most firms do not invest in improvement programs like TQM because "nobody gets credit for fixing problems that never happened." We propose that the reason why so many firms do not invest in programs to reduce supply chain risk is the same as it was with TQM.

1.6 Robust Strategies for Supply Chain Management

Disasters and other dramatic incidents can take away the focus from the *supply chain* aspect of supply chain risk. Many executives and researchers started paying more attention to responding to severe disruptions after the September 11, 2001 attacks (c.f., Sheffi 2005 a and b, and Zsidisin and Ritchie 2008). However, it is important to manage risks in the larger context of coordinating supply and demand, the eventual goal of supply chain management.

As such, companies need to devise "robust" supply chain strategies that possess the following properties: On one hand, under normal circumstances such a strategy would enable the company to manage the supply and demand fluctuations efficiently in the typical supply chain management context (see section below). On the other hand, such a strategy would help the company sustain operations at some basic level during a disruption and restore operations soon after.

To help develop robust strategies, Lee (2004) proposed "Triple-A" principles: Alignment, Adaptability, and Agility referring to long-, medium-, and short-term perspectives, respectively. Let us discuss each of these three principles in turn.

1. **Aligning** interests among supply chain partners can reduce supply chain risks. For example, to mitigate its supply cost risk, Intercon Japan first developed a

second supplier Nagoya Steel in addition to Asahi Metal. Then Intercon Japan developed incentives and penalties for these two suppliers: a supplier would get a higher share of the business for offering a lower supply cost (Mishina and Flaherty, 1988). To align interests among multiple parties, trust and a long-term perspective are necessary. Lee (2004) provides other examples about how the Alignment principle can be used to reduce supply chain risk.

2. **Adaptiveness** means being able to respond to the changing demand and supply. Li & Fung (www.lifung.com), the largest trading company in Hong Kong for durable goods such as textiles and toys, has established a supply network of over 15,000 suppliers around the globe. This supply network enables Li & Fung to adapt to market conditions quickly. Consider the case when Indonesia Rupiah devalued by more than 50% in 1997. Many Indonesian suppliers were unable to deliver their orders to their U.S. customers because they were unable to pay for imported materials; however, Li & Fung adapted to the situation quickly by shifting some production to other suppliers in Asia and by providing financial assistance to those affected Indonesian suppliers to ensure business continuity. With an adaptive supply network, Li & Fung was able to serve their customers in a cost-effective and time-efficient manner. The reader is referred to St. George (1998) for details. According to Lee (2004), adaptability can be achieved by using intermediaries such as Li & Fung, by creating flexible product designs such as Xilinx's programmable integrated circuits, or by monitoring new markets.

3. **Agility** enables a firm to reduce the impact of short-term changes in demand or supply. For example, to reduce the overstock and under-stock costs of different versions of DeskJet printers, HP redesigned its DeskJet printers by delaying the point of product differentiation. Specifically, HP first manufactures and ships generic printers to the distribution centers in different regions. These generic printers are then customized for different country-specific markets at each distribution center. The generic printers are produced according to a make-to-stock system, while the country-specific printers are customized in a make-to-order manner. This "postponement" strategy has enabled HP to respond to the demand changes quickly and effectively. The reader is referred to Lee and Tang (1997) for a detailed description of various mechanisms for delayed product differentiation such as modular design, standardization, commonality, etc., and to Feitzinger and Lee (1997) for a detailed description of successful implementations of various postponement strategies at HP.

Besides these three principles, firms can also build-in redundancies throughout the supply chain to reduce the impact of undesirable risk events associated with risks related to supply, process, and demand. For example, extra inventory, extra back-up production capacity, extra back-up suppliers, etc. are potential redundancies that would enable firms to make supply meet demand. However, redundancies are usually expensive because they are put to use only when certain unanticipated events occur (Sheffi, 2005 a and b). Also, redundancies disguise inefficiencies in the supply chain, which could inhibit the achievement of a lean supply chain.

1.7 Disciplinary Roots of Supply Chain Risk Management

While it is tempting to think of supply chain *risk* management as simply extending supply chain management to include risk, it is really a multi-disciplinary area with research and practice that draws on at least three fields: ***supply chain management***, ***enterprise risk management***, and ***crisis management***. These domains in turn draw on a broad range of the academic literature in areas such as organizational behaviour, psychology, decision analysis, empirical analysis, stochastic modeling and mathematical programming.

Based on our analysis of the surveys of researchers presented in Part IV (Chapter 16), there is a general view that supply-chain risk management overlaps with both supply chain management and enterprise risk management. There are other viewpoints too, of course, for instance, that of supply-chain risk management as a subset of crisis management and business continuity. It may well be that identifying problems lies in enterprise risk management but solution approaches and their deployment lie in supply chain management and business continuity.

Each of these related areas has its own primary purpose that is different from that of supply-chain risk management. Enterprise risk management tends to focus on risk disclosure in financial reporting of a company so as to comply with such regulations as the Sarbanes-Oxley Act in the US or the KonTraG requirements in Germany. Likewise, supply chain management is primarily concerned with ways to improve the operational performance of a supply chain under "normal" circumstances. Crisis management focuses on the survival of the organization or even that of society and may seem distant from worries stemming from a delayed delivery of parts at a plant. Still, we need to draw on these related domains to structure supply-chain risk management research and practice as this book attempts to do.

While rooted in these disciplines, supply-chain risk management is not fully a part of any of these. It is not merely a part of supply chain management although good supply chain management needs to include principles of risk management. Likewise, it is not part of enterprise risk management, although to do effective enterprise risk management, a company needs to excel in supply-chain risk management. Finally, supply-chain risk management is not part of crisis management and business continuity, although if the supply chain experiences a crisis, then any solution to crisis management has to incorporate supply-chain risk management. Thus, supply-chain risk management is important not just to supply chain professionals, and given its growing importance to companies, it is beginning to be recognized as a discipline in its own right.

Next, we view supply-chain risk management as comprising the four steps in enterprise risk management:

1. Identifying risks (Chapter 2),
2. Assessing these risks (Chapter 3),
3. Mitigating these risks (Chapters 4), and
4. Responding to incidents through communication, coordination and other means (Chapter 5).

The first three entail activities that take place *before* the occurrence of an incident that generates negative (and mostly) unanticipated consequences (Chapters 2, 3, and 4). The fourth one applies to actions taken *during* and *after* the occurrence of an incident – the focus here is on time, which is also called *time-based risk management* (Chapter 5); of course, the planning of such actions can and should take place before risk events occur. In addition, we discuss how global companies mitigate risk (Chapter 6) with a focus on Samsung Electronics.

Chapter 2
Risk Identification

Abstract As a first step to managing supply chain risks, this chapter focuses on ways to *identify* and categorize risks. First, as a way to distinguish causes from effects, drivers from consequences, risks whose consequences are apparent soon after an incident from those whose consequences are spread over months, and risks that have already materialized from incidents that might happen in the future, we present a "butterfly" model that conceptually separates underlying causes, actual events and consequences. Second, keeping in mind who should manage supply chain risk, we categorize risks in a way that follows the supply chain organization. Most companies have different groups facing the supply side (purchasing), those working in internal processes (manufacturing, storage and internal distribution), and those facing the demand side (distribution and sales). Therefore, we categorize risk motivated by the supply chain management organization: supply risks, process risks, and demand risks.

2.1 Introduction

There are many ways to identify and categorize risks, and each organization has its own way for developing its *risk register*: a list of identified risks with their importance rating. But why should we categorize risks (or supply chain risks) at all? Possible reasons include helping us to understand the distinctions among these risks and to prioritize different risk mitigation investment decisions. In this chapter, we provide two different ways for categorizing supply chain risks by suggesting answers to (1) *how* we should manage any particular category of risks and (2) *who* should manage this category of risks.

Many *risk categorization* methods for identifying risks do not suggest *how* to address risks. There are many methods for charting different types of risk using taxonomy-based risk identification (c.f., Carr et al., 1993). Juttner et al. (2003) provide a classification based on the sources of risk: environmental risk sources, network risk sources, and organizational risk sources. Chopra and Sodhi (2004) dis-

M.S. Sodhi, C.S. Tang, *Managing Supply Chain Risk*,   13
International Series in Operations Research & Management Science 172,
DOI 10.1007/978-1-4614-3238-8_2, © Springer Science+Business Media, LLC 2012

cuss delays and disruptions, Sodhi and Lee (2007) present supply-related, demand-related and contextual risks and Tang and Tomlin (2008) discuss supply risk, process risks, and demand risks.

There are also different ways of *risk identification* including identifying critical uncertainties in scenario planning (c.f., Garvin and Levesque, 2006). In identifying risks, supply chain researchers have used the terms *uncertainty* and *risk* interchangeably although economics researchers have attempted to narrow *risk* to only those situations where possible outcomes can be assumed to follow a known probability distribution. In fact, there are *upside risks* that can help an organization to achieve desired objectives, *downside risks* that can inhibit achieving these objectives, and *other risks* that create uncertainty about outcomes—Hopkin (2010) calls these opportunity risks, hazard risks, and control risks; respectively.

The diversity in risk identification and categorization is not simply because each organization has its unique circumstances or that different risk researchers come from different backgrounds. We believe that the real problem is that the vocabulary of risk, including that of supply chain risk, still needs to be fine tuned in practice. As such, we need to develop a conceptual view of risk in general and supply chain risk in particular so that everyone, not only in the organization but also in supplier and customer organizations in the supply chain, uses the vocabulary consistently. Such a shared vocabulary should also help us in answering *how* to manage different categories of supply chain risks.

We also need to categorize risks in a way that allows us to match any category to a specific entity within the organization for risk management, i.e., to enable us to answer the question of *who* should manage a particular risk category. For supply chain risk, this is not easy because any risk incident could start anywhere in the supply chain and end up causing huge impact somewhere else. In this chapter, we provide a categorization based on the perspective of a supply chain organization within the company, giving examples of mitigation efforts. This risk categorization maps each category to that part of the organization that is in the best position to prevent the risk incidents or at least contain the impact of risk incidents in this category. We also categorize risks whose prevention requires a centralized approach under a category that maps to the corporate entity.

As such, we take a twofold view of categorizing and identifying risks. First, we present a conceptual view of risk in general and supply chain risk in particular. The intent is to develop a shared vocabulary for risk within the organization (or supply chain) and to give an indication of *how* each risk category can be managed. Second, we present a tangible categorization of supply chain risks that suggests *who*, i.e., which part of the supply chain organization or the corporate entity itself, should seek to prevent the risk incidents or at least contain the impact of these incidents in each risk category. We also give examples of how companies manage the different categories of supply chain risks, although we discuss various risk mitigation strategies in Chapter 4.

2.2 A Conceptual View of Risk for Identification and Categorization

To develop a shared vocabulary for risk or supply chain risk, we need to sharpen the existing vocabulary first. We should be able to distinguish causes from effects, drivers from consequences, risks whose consequences are apparent in seconds and minutes from those whose consequences are spread over months, and risks that have already materialized from incidents that could happen in the future—in common parlance, all are confusingly termed "risks". We first present a "butterfly" depiction of risk that conceptually separates underlying causes, actual events and consequences.[1] Next, we present two categories of risks based on their impact: disruptions and delays (Chopra and Sodhi, 2004). Finally, we present a view of risk that is typical of supply-chain risk in contrast to enterprise risk: network risk associated with local-and-global causes and local-and-global effects.

2.2.1 A Butterfly Depiction of Supply Chain Risk

It is helpful to think of the commonly used term "risk" as issues ranging from underlying *causes* to actual *risk events* to *impacts*. A useful point of delineation is the occurrence of a risk event or incident. Causes are *before* the risk event while the impact is felt *after* the occurrence of such an event. Likewise, prevention efforts are *before* the event and the response efforts are made *afterward*. The location of the causes, the event and the consequences can be quite different especially in a supply chain: causes for a risk event may lie far from where the incident actually occurs as we shall discuss in Section 2.3 and the impact of any incident may be felt far beyond the location of this event.

We find it helpful to use a "butterfly" depiction of supply chain risk (Figure 2.1) atop an *x*-axis representing time as well as the relative location in the supply chain. The risk event is the body (thorax and abdomen) of the butterfly. The left-hand wing represents underlying causes and prevention efforts that could lead to this event, while the right wing depicts the post-event impact and response efforts. The left wing can be extended to represent causes upstream in the supply chain and the right wing extended to depict impacts further downstream in the supply chain relative to the time and location of the risk incident. Any preparation for quick response to risk events has to be done before the response but such preparation could be before or after the event (Figure 2.1).

There are several benefits of the butterfly depiction as regards how to manage any risk category. The observable risk event separates causes and effects, adding clarity to what is *the* risk. The timeline reflects prevention, response, and preparation-to-response efforts in time so we can plan these for any particular risk category.

---

[1] This idea was given to us by Paul Hopkin. The butterfly has been used to depict risks in many ways other than the way we present here as well.

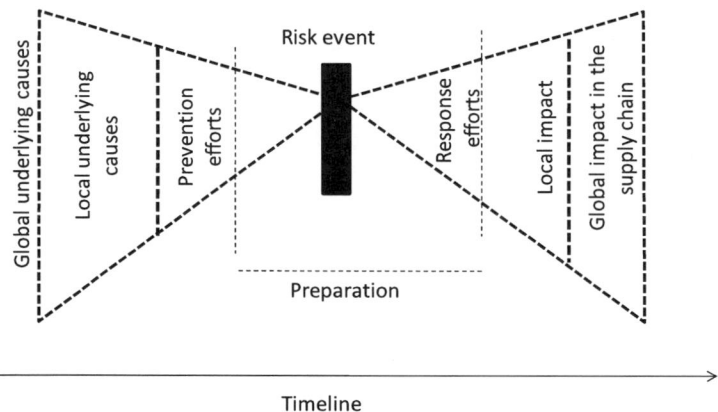

Fig. 2.1 A butterfly depiction of supply chain risk

Although the butterfly depiction is quite clear in representing risk as a spectrum around a single event, we know from practice that one risk event can trigger another risk event, building up negative consequences even when each event in itself may have contained consequences. For example, consider the Union Carbide gas leak in Bhopal (India) in 1984 that caused the deaths and blindness of tens of thousands of people.[2] This disaster was a result of a combination of missteps with one risk event

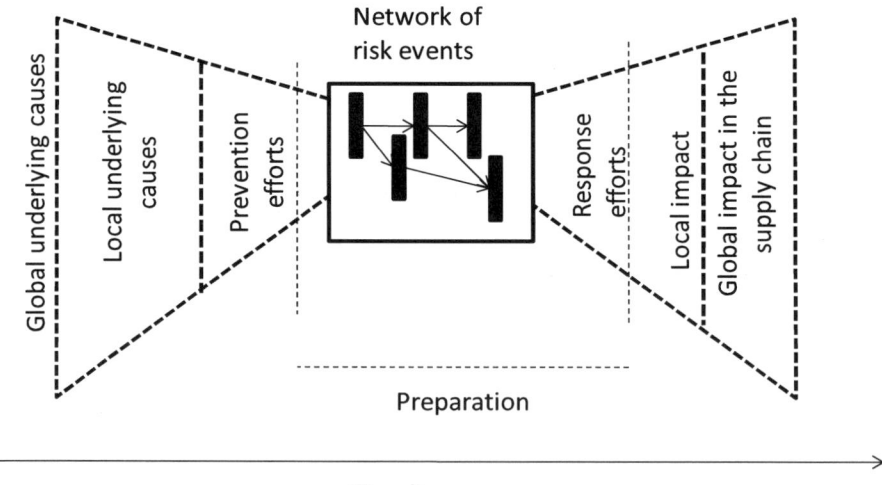

Fig. 2.2 A butterfly depiction of risk with a network of risk events

---

[2] There is an interesting organizational follow-up. Dow Chemical agreed to buy Union Carbide in 1999, completing the deal in 2001. Dow's stated policy is that "the company has neither a connection to nor legal liability for the tragic events of the 1984 gas release." As a subsidiary of

triggering another, eventually resulting in enormous negative impact. As with any perfect storm, it is not clear where or how to respond in such a network of events when this network has already triggered but the depiction should at least help to highlight the need to prevent the triggering of such a "network" of events. We can modify the butterfly depiction by replacing the single event, i.e., the body of the butterfly, by a network of events (Figure 2.2).

The butterfly depiction is quite clear when the event has large consequences; however, we do have events in the supply chain whose consequences are small individually but may still be large in the aggregate. For instant, a stockout for an individual order at a warehouse may have only a small consequence, but over time, repeated stockouts can lead to large accumulated losses hitting the company's bottom line. And, customers experiencing such repeated stockouts may choose to take their business elsewhere. We need to use the history of past events to prevent future events, or at least make them much less likely to occur. The timeline in the butterfly diagram can then be depicted as circular to show that the event is repeated over time, slowly accumulating the negative losses (Figure 2.3).

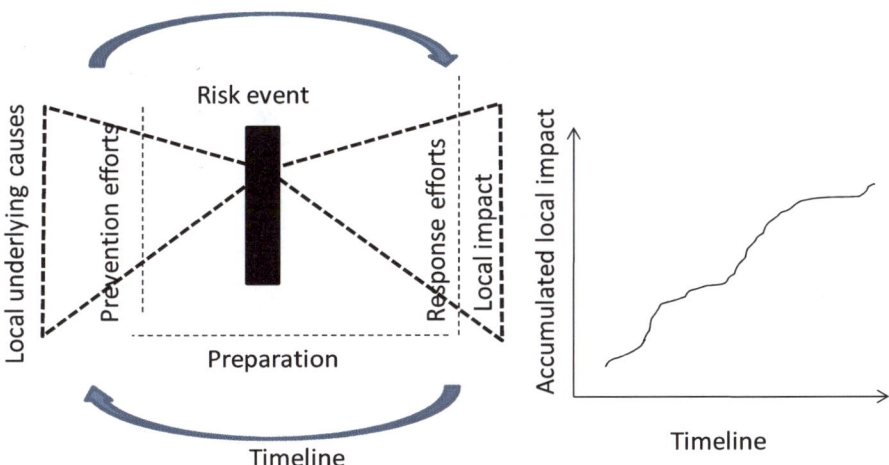

Fig. 2.3 A butterfly depiction of repeated risk events—note the circular timeline, both causes and effects being primarily local, and the gradually accumulating losses

Dow Chemical after the takeover, Union Carbide maintains there are no outstanding legal claims on it. Dow Chemical also says its subsidiary, Dow India, has no influence over Union Carbide to provide any help to victims. See http://news.bbc.co.uk/1/hi/programmes/bhopal/4023447.stm.

2.2.2 Delays and Disruptions: "Normal" and "Abnormal" Risks

There are two fundamentally different types of supply chain risks—delays and disruptions. These correspond to the "normal" risks experienced in supply chain management with the small and somewhat expected fluctuations of matching supply to
demand, and to the "abnormal" risks of huge and unexpected mismatch of supply to
demand due to a big failure with either the supply or the demand collapsing.

Delays in material flows often occur when a supplier, through overly high utilization or other causes of inflexibility, cannot respond to changes in demand. Other
culprits include poor-quality output at supplier plants or at *their* suppliers' plants,
high levels of handling or inspections during border crossings, or changing transportation modes during shipping.

Disruptions to material flows anywhere in the supply-chain are unpredictable and
rare, but often quite damaging. Natural disasters, labor strikes, fires and terrorism
can all halt the flow of materials. Consider the following examples. Following a fire
in February 1997 at an parts factory owned by Aisin Seiki, a key supplier, Toyota
was forced to temporarily shut down production at most of its Japanese plants.[3] The
1994 Kobe earthquake in Japan left many small companies like Kelly Micro Systems of Irvine, California without any supply of parts.[4] The California dockworkers
strike in 2002 produced shortages of high-demand retail items.[5] The bankruptcy in
2001 of UPF-Thompson, sole chassis supplier to Land Rover, caused major problems for the auto maker.[6] Immediately after the attacks of September 11, 2001,
U.S. auto manufacturers ran short of parts; transport trucks had been delayed at the
Canadian border.[7] Supply disruptions can also increase prices, as the Midwest discovered painfully in August, 2001, regional gasoline process skyrocketed following
a refinery fire at the height of summer demand.[8]

Many executives and researchers started paying more attention to handling major
disruptions after the September 11 event (c.f., Chopra and Sodhi, 2004; Kleindorfer
and Saad, 2005; Rice et al., 2003; Sheffi, 2001; 2005 a and b). However, it remains critical for firms to develop ways to manage supply chain risks arising from
delays, the problems of coordinating supply and demand, because these risks are
ever-present everywhere in the supply chain and can have an enormous accumulated impact.

Classification of risks as delays or disruptions suggests different answers to the
question of *how* to deal with such risks. Companies can avoid or be prepared for
delays by appropriately placing and sizing their capacity and inventory reserves at
the lowest cost. If material-flow delays are frequent, companies can plan mitigation

---

[3] "Fire hits parts supply network at Toyota," *Financial Times*, February 4, 1997

[4] "Damage to chip makers puts sourcing in spotlight," *Journal of Commerce*, January 30, 1995

[5] "Retailers scramble to keep stores stocked," *Wall Street Journal,* October 21, 2002

[6] "Making it safe to rely on a single partner," *Financial Times*, April 1, 2002

[7] "Inventory controls reexamined: Attacks reveal vulnerability of Just-in-time," *Plain Dealer*, October 29, 2001

[8] "Prices jump at gasoline pumps in the area," *St. Louis Post-Dispatch*, August 29, 2001

strategies based on historical information. One simple solution is to have excess flexible capacity in existing plants. Toyota, for example, carries excess capacity at its plants at the shop floor by way of team leaders who can work on any station in the assembly line, thus reducing the need for spare station-specific workers to cover absences. This ensures that daily production goals are met even if minor problems occur along the assembly line.

Another solution is positioning capacity and inventory depending on the cost of the products. Telecom equipment maker Cisco has *capacity* to assemble higher-value items in the United States so it can respond quickly to orders from up-market American customers. In contrast, the company holds an *inventory* of lower-value, high-demand items produced in low-cost (but not very responsive) locations overseas. So by matching its approach to product value, Cisco keeps supplier-delay risks low with low inventory costs.

Yet another solution is to combine inventory with different transport modes. PC manufacturer Dell holds very little inventory of components in North America or Europe for computer sales in the west, instead, it keeps most of the component inventory in Asia as part of its "postponement" strategy. By doing so, Dell "pools" its inventory and ships high-value components by air from the Far East to U.S. and Europe as needed, and ships less expensive components regularly by sea at low cost. In this way, Dell minimizes delay-related risk as well as inventory-related costs.

Disruptions require different approaches. Companies can counter disruptions in material flow by building inventory or by having redundant suppliers; after all, it is unlikely that all suppliers would be disrupted at the same time. However, holding inventory in this context can get very costly because holding and obsolescence costs incur continually, while the inventory would be used only in the rare event of a disruption. In essence, the company pays (and pays) for reserves that probably will never get tapped. Still, building inventory *does* make sense if the company can predict the disruption with reasonable confidence. In 2002, many retailers selectively built up inventories after learning of the impending California dockworker strike. When the strike disrupted supplies, damage to retailers was minimal. Stockpiling inventory as a hedge against disruption also makes sense for commodity products with low holding costs and no danger of obsolescence. The large petroleum reserve kept by the United States is an example of this strategy.

However, if the products have *high* holding costs and/or a *high rate of obsolescence*, it may be better to use redundant suppliers rather than to hold inventory against the threat of disruption. Motorola, for example, buys many of its handset components from multiple vendors, depending on the components' volume. Doing so prepares the company against disruptions without building up fast-depreciating inventory. It lowers the cost of redundancy by using multiple suppliers for high-volume products and single-sourcing for low-volume products. By doing so, the company maintains economies of scale at its suppliers to get good prices.

2.2.3 Local-and-Global Drivers and Local-and-Global Consequences

We need to distinguish the occurrence of uncertain events from their consequences: the uncertain event is the risk, while the eventual consequences (and to whom) depend on the actions taken by different parties after the event occurs. For example, from the viewpoint of many manufacturers, the September 11 event was unforeseen and hence it was an uncertain event. However, it was the subsequent suspension of air transportation that was directly responsible for causing supply delays that disrupted the manufacturing operations of many U.S. manufacturers.

Consequences may be local or global depending on whether the impact of the risk incident is limited to a particular location within the supply chain or whether the entire supply chain is affected. *Global consequences* impact the entire supply chain. For example, due to a recent surge in gasoline prices, Ford's supply chain for SUV manufacturing came to a halt in 2008 as consumers switched to compact cars; and Mattel suspended its production after recalling over 20 millions of toys in 2007. *Local consequences* impact a particular market or a particular site. For example, Ford closed five plants in the United States for several days after all air traffic was suspended after the September 11 event.

However, where an incident occurs may or may not be connected to where the consequences are felt. Thus, assessing supply chain risk entails understanding where a risk incident can occur as distinct from where the consequences might be. We can categorize supply-chain risks as being related to incidents that occur globally, spanning the entire supply chain, or those that occur locally at a particular supply chain entity only.

Global risks are defined in the context of the global environment within which the supply chain operates. The corresponding uncertainties pertain to social or political instabilities, credit crunch crises, and commodity price increases.

Local risks are defined in the context of specific supply chain entities. Risk events could be natural disasters, labour union strikes, supplier bankruptcies, contaminated production processes, or loss of intellectual property rights at a specific supply chain entity. Local risks could also stem from the damaging behaviour of particular supply chain partners.

We can thus categorize supply chain risks by the "location of occurrence" and by "location of consequence" (Table 2.1) to set the stage for developing mitigation strategies. Identifying the point of occurrence of each type of risk and its consequences would create shared awareness of different types of risks and their potential impact on different supply chain parties. This process can enable supply chain partners to better define their roles and responsibilities and to generate support for collaborative efforts for mitigating risks for all parties. For example, after suffering from a *global* $2 billion loss in sales caused by a *local* fire that disrupted the production of microchips at a supplier's plant, Ericsson worked with its supply chain partners to develop proactive plans should an incident of such time recur (c.f., Norrman and Jansson 2004).

Table 2.1 Supply chain risk drivers and consequences (also see Chapter 6)

		Consequences (Region of possible eventual impact)	
		Local (impact to a particular supply chain entity or market)	*Global (impact to the entire supply chain or multiple markets)*
Supply Chain Risk Driver (Point of possible occurrence)	*Local (originating at a supply-chain entity or in a particular market)*	Local risks stemming from supply, demand, or failure match supply and demand.	Risks originating in a plant or region whose consequences eventually impact the entire supply chain or markets.
	Global (originating supply-chain-wide or globally).	Global risks that affect a particular supply chain entity.	Global risks that can affect the entire supply chain as a whole.

Clearly, the affected supply chain entity is responsible for reducing the local risk when the consequences are also local. Some of this is very much part of supply chain management, e.g., maintaining inventory to reduce the impact of unexpected delays in supplies.

At the other extreme, for global risk drivers that have global consequences, collaborative efforts from the entire supply chain are needed at the corporate levels of companies. For example, to improve supply chain security, U.S. Customs established the Customs-Trade Partnership against Terrorism (C-TPAT) certification program in 2002. This certification program requires all supply chain partners to comply with a standard operating procedure to ensure security. To entice companies to comply with the best security practices, C-TPAT certified companies are allowed to clear customs faster with less inspection (c.f., Tang 2006a).

To handle local risk with global consequences—some would argue this should be the focus of managing supply chain risk—we need to understand the risk management efforts both at the point of occurrence and at the affected locations. For example, after recognizing the global consequence of a world-wide recall of lead-tainted toys associated with the local risk of a particular sub-supplier using lead-tainted paints in the production of these toys, Mattel announced a safety check system to prevent the manufacture of toys with noncompliant levels of lead in paint (c.f., Pyke and Tang, 2008).

Finally, situations entailing global risks with local consequences are usually the result of corporate policy applied inappropriately or incorrectly to a local situation. For instance, there may be local negative consequences of a corporate procurement policy. This can be mitigated by communication between the local facilities with the corporate entity.

2.3 Risk Categorization Motivated by the Supply Chain Organization

We now present a categorization that suggests *who* should prevent risk events or at least contain the impact of risk events for a particular risk category. Given the importance of supply chain management, companies already have their respective organizations in place to carry out the supply-chain activities that entailed. Therefore, it makes sense to categorize risks in a way that follows the supply chain organization. This way, different entities within the supply chain organization can determine who are best positioned to prevent or contain the consequences of certain types of risks events and who should be responsible for managing the corresponding risk categories. Most companies have different groups facing the supply side (i.e., purchasing), those working in internal processes (manufacturing, storage and internal distribution), and those facing the demand side (distribution and sales). Therefore, we can consider three types of risk motivated by the supply chain management organization:

- **Supply risks**
- **Process risks**
- **Demand risks**

There are also

- **Corporate-level risks**

for which the entity charged with risk management is the corporate entity with a remit over the entire company (Table 2.2).

Table 2.2 Supply chain risk categorization motivated by the supply chain organization

Supply risks	Process risks	Demand risks	Corporate-level risks
• Supplier failure • Supply commitment • Supply cost	• Design • Yield • Inventory • Capacity	• Forecasting • Change in technology or in consumer preference • Receivable	• Financial • Supply chain visibility • Political/Social • IT systems • Intellectual property • Exchange rate

Prevention in risk management (and hence motivating this categorization) is important, because one risk event can trigger another of the same or of a different type. As indicated above, Toyota's accelerator-related problem resulted in loss of reputation and decrease in demand along with political risks in the US.

2.3.1 Supply Risks

Supply risks pertain to risk events on the supply side that include supplier defaults or other unexpected changes in supply cost, delivery, quality or reliability. Outsourcing risks fall in this category and their importance is growing as more manufacturers reduce the number of direct suppliers and source globally. Indeed, many U.S. manufacturers reduced the number of direct suppliers throughout the 1980's and 1990's to reduce the cost of managing multiple suppliers and to foster better supplier relationships. Some companies even pushed for sole sourcing. While managing a smaller number of suppliers is more efficient, a smaller supplier base can expose the company to greater risk. Consider the following types of risks on the supply side:

Supplier failure. This is a well-documented risk with many examples such as the bankruptcy in 2001 of UPF-Thompson, sole chassis supplier to Land Rover, which caused major problems for the auto maker. During the 2008-09 economic crisis including the credit squeeze, automakers BMW and Daimler along with UK defence company VT Group told their top suppliers to come to the company for financial help as a way to reduce their supply risks.

Supply Commitment. If the buying organization has to commit to long-term purchases from its supplier without the option of revising the quantities, it can have the risk of having unmet demand or excess inventory over time. For instance, Canon is the sole supplier of the engines for the HP LaserJet printers. Hewlett-Packard has to place its order six months in advance and is not allowed to change the order quantity once the order is placed. This arrangement limits the company's ability to react to changes in demand (Lee, 2004).

Supply Cost. This refers to unanticipated increases in acquisition costs resulting from supplier price hikes or from fluctuating exchange rates. Price increases are more likely when a company uses only one supply source. When Intercon Japan's connector manufacturer sourced a special type of bronze from a single metal supplier (Asahi Metal), it had little bargaining power. Consequently, Intercon Japan experienced significant price increases from Asahi Metal (Tang, 1999). We consider exchange-rate risk in a separate category at the corporate level as managing it centralized risk mitigation, for instance, sourcing from a country with a different currency can be hedged on the demand side by selling to the same country (see Chapter 8). Still, as an example consider the fact that in the mid-2000s, the weakening of the dollar drove up not only oil prices in dollars for U.S. companies but also the costs of imports from Europe.

Price increases by suppliers can be blunted by signing long-term contracts, by having redundant suppliers, or, in rare instances, by holding inventory. But long-term purchasing can badly damage profits if prices for the contracted goods fall. For example, obligations signed by California during the peak of its electricity crisis in 2001 forced the state to pay 800% more than the 2002 market price.[9] Also, long-term contracts with little quantity flexibility, as already indicated above as supply

---

[9] "California may have new energy deals," *New York Times*, August 8, 2002

commitment risk, can lead to a company not being able to match its supply to a changing demand.

Contracting with redundant suppliers can work if companies can maintain economies of scale. Global giants like Toyota seek out local economies of scale by single-sourcing at the plant level, but use different suppliers globally. That way, even though a company might be the sole-supplier to a particular Toyota plant(s), it must keep prices down to compete for business across the entire Toyota network. Alternatively, some firms use multiple, redundant suppliers, even if it means sacrificing some economies of scale. Cisco, for one, claims to have four or five more suppliers than it needs. The company keeps the resulting higher costs in check by monitoring and benchmarking suppliers against each other.[10] A good example of using inventory to counter the threat of price increases is the U.S. strategic oil reserve policy. Meant primarily meant to prevent oil supply disruption, the reserve also has been used on occasion to keep down prices. Another instance comes from U.S. based International Paper Company. To keep prices of raw material down, the company sources raw materials from independent forest owners, as well as from its own forests with trees as raw-material inventory.

2.3.2 Process Risks

These risks pertain to risks within the organization's internal supply chain, typically pertaining to design, manufacturing and distribution. Consider the following categories of supply-chain risks:

Design. Despite significant efforts in implementing Total Quality Management (TQM), Lean Manufacturing and Six Sigma, many companies are still facing risks from products produced as a result of faulty design or manufacturing. Toyota's recall of cars in late 2009 and early 2010 owing to "sticky" accelerators has hurt the company's reputation, demand and stock price. In 2007, Mattel recalled over 17 million toys designed internally that were unsafe for small children owing to small loose magnets. (This risk incident was different from the one where toys Mattel had to recall because of a supplier using lead paint.)

Yield. If the manufacturing yield at a plant is uncertain, it can result in the company not being able to match its supply to its demand. Yield problems in 2004 at IBM's plant in East Fishkill, New York contributed to the $150 million first-quarter loss by its microelectronics division (c.f., Krazit 2004). The lower-than-expected yields reduced the plant's effective capacity and limited IBM's ability to meet customer demand.

Inventory. Excess inventory hurts financial performance. That was the case in late 2000, when the PC industry carried roughly 12 weeks of inventory. The combination

---

[10] Randy Pond, Senior VP, Cisco, November 2002.

of excess inventory and falling prices hurt many companies such as Compaq.[11] The extent of the risk stemming from inventory depends on (1) the value of the product, (2) its rate of obsolescence, and (3) uncertainty of demand or of supply. Holding excess inventory for products with high value or short life cycles can get too expensive but it can work well for low-value commodity products with low obsolescence rates. Naturally, the larger the product variety, the greater a company's exposure to inventory risk.

Three proven approaches can help managers can mitigate inventory risk: (1) pooling inventory, (2) creating common components across products, and (3) postponing or delaying till the receipt of orders the last stage of production from which emerges product variety. Online bookseller Amazon.com serves all its customers in the United States with inventory housed in a handful of warehouses, while book retailer Borders supplies its customers with inventory in several hundred stores. Besides pooling inventory, each Amazon warehouse pools demand over a large geographical area, leading to more stable forecasts and lower total inventory. The strategy helps Amazon achieve 14 inventory turns per year, versus two at Borders.[12]

The paint industry illustrates well how to leverage component commonality and postponement to manage product variety. Traditionally, manufacturers held paint inventory in a rainbow of different colors. Today, paint inventory is held as a common base, which is then is mixed to exact color specifications after the customer orders. This simple but powerful change has significantly lowered paint inventory at retail stores. Apparel maker Benetton also practices pooling and postponement. An inventory of un-dyed sweaters gets stockpiled in one location. The garments are dyed after orders have been received. This pooling of demand across geographical areas and across colors helps Benetton reduce inventory risk while meeting customer demand more effectively.[13]

Companies can also minimize inventory risk by working with a highly responsive supplier, especially for high-value, short life-cycle products. Excess capacity can also lower the amount of inventory required. By running plants at 80 percent utilization, Toyota can handle demand variation without having to hold inventory.

Capacity. Inadequate capacity means a company may be unable to meet its demand and thus suffer from unmet demand. To avoid this, companies can err on the side of having excess capacity. However, building excess capacity is usually a strategic choice as it may take much longer to ramp capacity up or down compared to changing inventory levels and may cost a lot more. Moreover, excess capacity hurts financial performance in terms of providing lower returns on investment and on investment. That was the case in 2002–03, when many semiconductor firms had to run at 50 percent capacity because of soft demand.

Managers can lower excess capacity risk by making existing capacity more *flexible*. Flexibility is a form of pooling that allows use of the same capacity for a variety

---

[11] "PC prices fall with demand," *USA Today*, December 13, 2000

[12] Borders filed for bankruptcy in the US in February 2011.

[13] Benetton (A), *Harvard Business School* case 9-685-014.

of products. For example, Hino Trucks plants employ multiple assembly lines, the number of workers on each line determining the line speed. This flexibility not only lets Hino change production on any line by moving its capacity of workers in response to fluctuating demand, but it also keeps the excess capacity of workers that Hino would have to carry much lower than a situation where workers would be line-specific.

Toyota decreases risks from idle capacity by ensuring that each plant is flexible enough to supply more than one market. Demand fluctuations can be satisfied from a variety of plants, which decreases total capacity required. The company carries the idea of flexibility down to the shop floor, where team leaders can work on any station in the assembly line, reducing the need for spare station-specific workers to cover absences.[14]

Lastly, a company can minimize excess capacity by serving geographically scattered customers from the same location. Italian automaker Ferrari, for example, minimizes total production capacity by centralizing production of all cars in a single plant. The arrangement also provides Ferrari with economies of scale, even though Ferrari produces much fewer cars than the big auto companies.

2.3.3 Demand Risks

The uncertain nature of product demand is one of the supply chain risks that all companies need to face with uncertainty surrounding volume and product mix. To increase revenue, many firms sell their products in multiple countries. To satisfy country-specific requirements such as power supply and language driver, Hewlett-Packard (HP) has to develop multiple versions for each model of their DeskJet printers. Each version serves a particular geographical region (Asia-Pacific, Europe, or Americas). Due to uncertain demand in each region, HP faced the problem of over-stocking certain printers in one region and under-stocking certain printers in other regions (Kopczak and Lee, 1993). For companies that sell multiple products, not only is the total demand volume unpredictable but also the demand mix, i.e., the individual demand for each of the product variants. Demand risk therefore encompasses uncertainties in both volume and mix.

Forecasting. Forecast risk stems from the mismatch between a company's forecast and actual demand. If the forecast turns out to have been too low, then there may not be enough products available to sell. If forecast turns out to have been too high, the weak demand will result in excess inventories and price-markdowns. Long lead times for production (hence a farther forecast horizon), seasonality of demand, high product variety, and short product life cycles all increase forecast error. Also, errors tend to be larger when a few customers make larger purchases as opposed to many customers making smaller purchases.

---

[14] "Toyota Motor Manufacturing USA Inc.," Harvard Business School Case 9-693-019.

Forecast errors also result from *information distortion* within the supply-chain. In late 2003, for example, product shortages in western Europe led Nokia customers to order more than they needed, so they would be able to meet demand in case Nokia began rationing or allocations. These exaggerated figures distorted Nokia's reading of the market, causing the company to inaccurately forecast sales.[15] Other causes of information distortion include: promotions and incentives that lead to forward buying; batching of purchases, which leads to higher volatility in orders; and lack of knowledge of end-customer demand at upstream locations.

Distortion increases in the supply-chain as you get further away from the end consumer, a phenomenon known as the *bullwhip effect* (Lee et al., 1997). Companies can reduce the sting of the bullwhip effect, though, by adjusting pricing and incentives to decrease variation in orders. Increasing the visibility of demand information across the supply-chain also helps. Continuous Replenishment Programs (CRP), and Collaborative Planning, Forecasting, and Replenishment (CPFR) and other supply-chain initiatives also can soften the bullwhip effect.

The impact of the resulting forecast errors can be lessened by selectively holding inventory or by building responsive production and delivery capacity. Holding inventory is appropriate for commodity products with relatively low holding costs; responsive delivery is better for expensive products with short lifecycles (and corresponding large forecast errors). Motorola practices responsive delivery each day when it flies in phones from China in response to demand by customer Nextel. Instead of stocking parts for uncertain demand, Dell also flies in high-value items from Asian suppliers on an as-needed basis as mentioned before.

Change in technology or in consumer preference. Closely tied to forecast risk is the longer-term trends of changes in technology introduced by competitors and change in consumer preference whether tied to the change in technology or to something else. Such changes not only undermine a company's demand but also render capacity investment highly optimistic, thus hitting goals on return on investment. The electronics industry is constantly buffeted by new technology or designs and Apple's 2010 introduction of the iPad left competitors such as Amazon's Kindle scrambling for comparable offerings or having to offer price reduction. Large investment in new technologies may require collaboration with competitors to reduce the risk of another technology or a different set of standards leaving a company high-and-dry. The question is not just about the availability of technology but which technologies consumers will adopt. This means constant research and development not only of products but also of consumers in different market segments. One approach is to research and monitor existing and potential customers not only through sales but also through online forums and social networking sites.

---

[15] "Nokia feels the squeeze from shortage," *Off the Record Research*, November 13, 2003

2.3.4 Corporate-Level Risks

There are also risks to the entire supply chain—supply side, within the organization and demand side—and hence to the enterprise itself. As such, companies need centralized action coordinated with internal and external entities in the supply chain.

Financial risk. When a company is expanding markets, it is rare that the focus will be on profitability or other financial measures. Instead, the goal is typically market share, "presence", name recognition, and other objective and subjective measures. Another issue related to conglomerates is shareholder interest translating to cash injections to keep some companies within the conglomerate afloat in difficult times. Add to this a currency crisis and credit markets drying up, we have a situation where the company may suddenly face bankruptcy. Moreover, in a conglomerate or in a supply chain, one company going bankrupt could have a domino effect on other members of the conglomerate or on suppliers and other supply chain partners.

Maintaining liquidity in an industry with short product life-cycles and high obsolescence rate can be a challenge. Over-investing in inventory can lead to big losses or even bankruptcy. This was the case for Hayes, the company that developed the standard for modems at the start of the Internet revolution.

There is also risk from receivables as inability to collect on receivables can torpedo the performance of any company. In 2002, Sears Roebuck's credit division reported unexpected losses caused by delinquent cardholders.[16] As a result, Sears stock plummeted more than 30% in one day. The company learned the hard way that filtering customers for creditworthiness is a very prudent and powerful way to reduce receivables risk. Filtering customers for creditworthiness is necessary to reduce receivables risk, but as with Sears, aggressive sales growth goals can result in sloppy credit checks.

Another approach is to spread the risk across more customers. McMaster Carr, a maintenance-materials supplier with hundreds of thousands of customers, enjoys a much lower receivables risk than a competitor selling to a single, large customer. The Achilles' Heel here is a widespread economic shock that harms the creditworthiness of all customers, a fate that befell Cisco during the dotcom bust in 2001 and, many suppliers in the 2008-09 economic downturn in western countries.

For companies selling globally, exchange rate fluctuations and movements can create receivables risk tied to not only the ability to collect against invoices in the future but also the exchange rate. As such, many companies sell to distributors or large retailers in many parts of the world using only cash (rather than 30-day terms) and that too in standard currencies: US dollar, euro, or pound sterling.

Finally, receivables risk can stem from errors in processing invoices and in invoicing and payment delays. Samsung Electronics uses proof-of-delivery technology to reduce receivables risk by reducing errors and by cutting down on the amount of time between delivery and invoicing. This time could be as much as 15 days if it were paper-based. Instead, a driver scans packages upon delivery as proof of re-

---

[16] "Sears earnings will be hurt by credit unit," *The New York Times*, October 18, 2002

ceipt. The information is uploaded to the SAP system, thereby triggering the invoice (Sodhi and Lee, 2007).

Supply chain visibility. As the number of partners increases in a global supply chain, the level of visibility and control can be reduced significantly. For instance, according to a study conducted by AMR Research in 2006, supply chain visibility is relatively low: few companies have either future demand or current inventory information from downstream partners and more than half the companies take more than two weeks to sense changes in actual demand. The low visibility level and the low control level reduce the "confidence" of each supply chain partner regarding the replenishment lead time/order status quoted by upstream partners and demand forecasts provided by downstream partners. Such a low confidence level can cause the entire supply chain enters a "risk spiral" so that each supply chain partner either "inflates" their order or "disguises" their on-hand inventory (Christopher and Lee, 2004). The confidence level deteriorates further as every partner starts gaming the system, and hence, the "risk spiral" continues. To break this vicious cycle, supply chain visibility both within the organization and with suppliers as well as customers by way of timely communication, and coordinated corrective actions are needed to improve the confidence level of each internal or external supply chain partner.

Political/social risks. A global supply chain is subjected to social/political risks when multiple countries are involved. For example, Airbus, a four-nation consortium, incurred an opportunity loss of 4.8 billion euros due to a two-year delay in launching the super-jumbo A380. In addition to technical problems associated with the wiring system, political issues among the four countries with manufacturing plants to make these planes are thought to be a contributing reason. Airbus' parent, EADS, struggled in the mid-2000s to develop a restructuring plan to replace political haggling with industrial logic (Gumbel, 2006).

IT systems risk. The more a company connects its systems into an efficient network, the greater the threat that a failure anywhere can cause failure everywhere. A breakdown of information infrastructure can devastate today's highly networked environments. In 2002, the fast-spreading "Love Bug" computer-virus infection shut down email systems at the Pentagon, at NASA and at the Ford Motor Company among others causing billions of dollars in estimated damages.[17] In 2010, the Stuxnet virus (actually "worm") that takes control over equipment caused worldwide industrial alarm as a foreign government succeeded in infecting the Iranian nuclear plant in Bushehr.[18] The same year Google also discovered it was under "cyber-attack".

The banking industry has long recognized systems risk as a major threat to its business systems. In 1988, the Basel Committee on Banking Supervision warned about the growing reliance on globally integrated systems. "The greater use of more

---

[17] "FBI hunts love bug source: Damage from e-mail source cuts across USA and worldwide," *USA Today*, May 5, 2000

[18] Stuxnet worm causes worldwide alarm, *Financial Times*, Sep. 23, 2010

highly automated technology has the potential to transform risks from manual processing errors to system failure risks," the committee wrote.[19]

A defense against systems failure is use of robust backup systems and well-designed, well-communicated recovery processes that duplicate all data and transactions. Such approaches helped securities firms recover quickly and convincingly following the World Trade Center attacks in 2001.[20] Still, these are relatively simple ways that may not help with the kind of attack made on the Iranian facilities on industrial equipment. As such, companies (and governments) are struggling with cyber-security even as many of the same governments are investing in developing technology for cyber-attacks.

Intellectual property. This risk of loss of intellectual property has grown rapidly, as supply chains become less vertically integrated and more global, and companies outsource to contract manufacturers who are used by their competitors or who can turn into competitors. Intellectual property risk has long-term implications on a company's profitability. While outsourcing or offshoring to low-cost countries does lower the cost of goods sold, the company can become more vulnerable to loss of its intellectual property.

For example, even though the reform of the Intellectual Property protection law made progress after China's WTO entry in 2001,[21] infringements can still occur in China. For instance, multinational firms are not necessarily protected legally when their Chinese suppliers start producing unauthorized products using virtually identical design and materials. When the relationship between New Balance shoes and Qiuzhi Footwear, one of its Chinese suppliers, went sour, the supplier started producing "New Barlun" shoes using a logo that resembled the New Balance's "N". New Balance filed a lawsuit in China in 2002 (Chandler and Fung, 2006) eventually winning it in 2006.[22] However, it can still be difficult to protect intellectual property and to eliminate the risk of near-counterfeits under certain licensing or contractual agreements.

Companies can mitigate intellectual property risk by bringing, or keeping, some production in-house or, at least, under direct company control. This is one reason why Motorola owns some of the testing equipment at its supplier locations. Another way managers can decrease risk is by limiting the flow of new intellectual property into countries with weak legal controls protecting it. Companies like Cisco, which outsources all manufacturing, also lower risk by creating business processes that no single manufacturer can replicate the entire product (or the process that goes into making it). Electronics manufacturer Sharp even repairs equipment itself, thus preventing any possibility, accidental or otherwise, that its vendors will share propri-

---

[19] See the BIS website for the Basel Committee's communications, in particular, "Sound practices for the management and supervision of operational risk".

[20] "Backup systems pass trying test," *Washington Post*, September 27, 2001

[21] See Judicial Protection of IPR in China, http://www.chinaiprlaw.com/english/news/news5.htm, accessed 4th Nov. 2010.

[22] See New Balance's press release http://www.newbalance.com/public-relations/library/2006/new-balance-wins-landmark-lawsuit-in-china-against-counterfeit-brand/ accessed 4th Nov. 2010.

etary information with Sharp competitors. The company goes so far as to reprogram various computer-aided machines used by its vendors without sharing the information.

Exchange-rate. This type of risk, mentioned earlier with supply costs, can be countered by creating financial hedges, balancing cost and revenue flows by region, and building flexible global capacity. Toyota's manufacturing strategy, to cite one good example, allows each plant to serve the local market and at least one other market across the world. This flexibility lets Toyota shift production if exchange rates change appreciably. Another way to reduce exchange-rate risk is to source, to the extent possible, from the same currency region where sales are being made. Limiting global sales to a few currencies—the US dollar and the euro, for instance—reduces the exchange-rate risk to only these currencies and the company's home currency if different. In general, the nature of global supply chains is such that sourcing and selling may be in very different currencies. As such, Samsung Electronics uses futures at the country or region level to help stabilize the operational effect of currency fluctuations (Sodhi and Lee, 2007).

However, things can get more complicated than simply hedging the rates between two countries. For a country-based subsidiary of a global company, one issue is the exchange rate at which profits can be sent back to (or at least reported in) the parent company's country. This subsidiary may source parts from other countries and may distribute and sell products in yet other countries. As such, exchange rates and even the question of what base currency to use become complex issues for companies with global supply chains.

While daily exchange rate movements are of operational interest, long-term trends like the decline of the US dollar in the mid 2000's relative to other major currencies or the decline of the pound against the euro from 1.5 to 1.1 euros/pound over the two-year starting January 1, 2007 period have strategic implications for the supply chain. Britain's decision to not join the eurozone impacts companies who are contemplating investment in the UK or in the rest of Europe. The various currencies of the emerging economies of Central and Eastern Europe also provide a challenge as regards exchange rate given their importance as "near-shore" locations for manufacturing for western consumption.

Environmental risk and compliance cost. Environmental standards in the UK and the rest of the EU are quite high compared to other European countries and to the US. But these higher standards also mean higher costs. Industries producing waste through manufacturing or through the end-consumers disposing the products at the end of the life face the question of who should pay the cost of doing so: local or central government, the company, end-consumers, or eventually of course, future generations if we do not do anything at all. Many companies have taken proactive steps in anticipation of requirements to be placed by the EU. Companies can also conduct R&D to reduce waste, energy consumption in manufacturing and/or consumer use, or otherwise reduce environment impact. Toyota in Europe has used its famed Toyota Production System to cut down water and energy consumption per car in manufacturing and to eliminate waste going to landfill.[23] Samsung Europe set up a new environment team in 2003 comprising specialists to work with government regulators and other electronics manufacturers (Sodhi and Lee 2007).

Regulation compliance. While compliance on accounting and in particular on transfer pricing is not a supply chain issue per se, it does have implications on reputation risk as consumers may not want to buy products from a company with a reputation of not having transparent transactions.

2.4 Summary

We described two fundamental ways of identifying and categorizing supply chain risks motivated by the questions of how to manage these risks and who should manage these risks. The first way required an abstract view of risk, the intent being to help a company to distinguish between causes and effects and to suggest prevention and response efforts respectively. To this end, we provided three different ways to view risk: (1) a butterfly depiction of supply chain risk to distinguish between causes, events and consequences, (2) delays and disruptions to distinguish risk with small losses that accumulate to significant losses from risks that cause large losses right away, and (3) a network view to distinguish risks on a 2-by-2 framework of the origin of the risk drivers (local or global) from the region of impact (local or global).

The second way required categorizing risks in a way that follows the typical organization structure of supply chain management functions in any company. Doing so, we can help a company to figure out where risk events could best be prevented or at least responded to in a timely fashion. Specifically, we identified and categorized risks along the company's supply chain: supplier-related, process-related and demand-related risks. We also provided another category of risks that apply to the enterprise as a whole, which we called corporate-level risks.

---

[23] See http://www.toyota-europe.com/corporate/environment/360-approach/making-the-car.aspx

Chapter 3
Risk Assessment

Abstract This chapter discusses different ways to assess the risks in the different categories identified in the previous chapter. Risk assessment is a critical step because the results of the risk assessment can influence the decisions (prioritization, resource allocation) to be made by the organization and the commitment from its top management. In this chapter, we present a real risk assessment exercise to illustrate an approach for assessing supply chain risks. While this approach is not the only way to access supply chain risks, it does highlight the challenges and opportunities of risk assessment within the organization and its supply chain.

3.1 Introduction

Having identified different types of supply chain risks in the previous chapter, we now discuss different ways to *assess* these risks. Risk assessment is a critical process for helping top management to make informed prioritization and resource allocation decisions. To ensure our discussion is not purely conceptual, we provide a real example of a risk assessment exercise, using a cross-company survey, as an illustration. Although this illustration is not presented as a universal approach for assessing supply chain risks in all situations, it does highlight the challenges and opportunities of risk assessment within the organization and its supply chain.

Although the intention of risk assessment is to generate a shared sense of urgency about a subset of identified risks and to develop risk mitigation effort, there are many related purposes. First and foremost, risk assessment is the basis for allocating funds to different risk mitigation efforts against a backdrop of competing needs. We have to be careful in noting that the assessment of risk can be different from the assessment of the "perception" of risk (cf. Slovic, 1987): for instance, because any lives lost in a single airplane incident get much more attention than lives lost in car accidents, air travel is perceived to be riskier than car travel. As such, much investment goes in making air travel safer even though many more lives are lost in car travel each year. For supply chains, we do not have to worry about such a

M.S. Sodhi, C.S. Tang, *Managing Supply Chain Risk*,
International Series in Operations Research & Management Science 172,
DOI 10.1007/978-1-4614-3238-8_3, © Springer Science+Business Media, LLC 2012

distinction, but we should keep in mind that there are competing needs; e.g., investments in inventory and capacity—what, where and how much (Chopra and Sodhi, 2004). In a supply chain, there is also the question of who should make such an investment, the organization, its suppliers or its customers.

Secondly, the risk assessment exercise can help management focus on specific areas of vulnerability whether within the organization's four walls or in its extended supply chain. In the event when the risk assessment exercise is conducted as a result of a recent risk incident or as a result of pressure from an insurance company, risk assessment may entail estimating the unmitigated risk (we discuss this later in the example), the company's or supply chain partners' efforts in prevention, mitigation or response, and the level of insurance.

Thirdly, companies also use risk assessment program to meet legal or regulatory requirements such as the Sarbanes-Oxley Act of 2002 in the U.S. and KonTraG in Germany that require companies to inform their shareholders of the company's risk profile and related measures being taken.[1] These risks are those for the company's extended supply chain.

Finally, companies use risk assessment to develop contingency plans by attempting to understand the nature of threats and other risks to help counter these better.

3.2 What Risks to Assess

We discussed risks associated with delays and disruptions in the previous chapter, which we referred to as "normal" risks and "abnormal" risks respectively. Frequent fluctuations of process yields, material costs, currency exchange rates, and product demands can be viewed as normal risks and these would be relatively straightforward to quantify, given historical data about past events. In our view, *normal* risks are typically managed routinely as part of supply chain management rather than of supply chain risk management; this implies that a supply chain risk assessment exercise will typically exclude such risks.

On the other hand, rare events such as natural/man-made disasters, contaminated products, and system failures are hard to assess, given the paucity of data. Even if a particular type of risk incident has never occurred in the past, it could still occur in the future. Risk assessment of such risks is part of *business continuity* effort in many companies. Many researchers too feel that such risks should be the domain of *supply chain risk management* (see Chapter 16), leaving the day-to-day risks of "normally" varying supply and demand to *supply chain management*.

---

[1] Sarbanes-Oxley Act of 2002 requires U.S. companies to inform shareholders of their risk profile and their approach to manage risk. The reader is referred to http://www.aicpa.org/info/ sarbanes_oxley_summary.htm for details. KonTraG is a German law implemented in 1998 that is analogous to the Sarbanes-Oxley Act. The reader is referred to http://www.germanlawjournal .com/print.php?id=622 for a detailed description of KonTraG.

3.3 How to Assess Risks

Most companies use different methods, ranging from formal quantitative models to informal qualitative plans, to assess supply chain risks (Rice and Caniato, 2003; Zsidisin et al., 2000; 2004). Depending on the purpose, various tools can help prioritize risks in: (a) *relative* sense—keep in mind the need to allocate risk management efforts and budget; or (b) *absolute* sense—to reflect the importance of a particular risk category in monetary terms, say, for the purpose of deciding whether or not to transfer the risk to an insurance company. Thus, relative rating of risk can help create a shared view within the organization whereas an absolute rating could go further in helping resolve whether or not a risk is worth the cost of risk mitigation or insurance.

Companies typically assess supply chain risks or the risks associated with business continuity in general by carrying out a ***risk mapping*** exercise. This exercise entails rating, for each identified risk category, two aspects of risk incidents: (1) the ***likelihood*** or ***frequency*** of incidents in this risk category to occur, say, in a year; and (2) the "typical" ***impact*** or associated consequences, usually economic, should an incident in this risk category actually occur.

In the *relative* risk view, companies assess likelihood of incidence and impact per incident in relative terms on a 1–5 scale (or similar), the aim being to focus attention on certain categories of risk. For each risk, the values of these two variables are presented as a two-dimensional table called a ***heat map***: a risk category for which both the likelihood and the impact per incidence are high is in the red zone, while a risk category for which both are low is in the green zone; the remaining risk categories are yellow or amber. This helps create a shared view of risks that are 'red hot' and therefore requiring greater attention than the others.

In the *absolute* risk view, likelihood is sought to be specified as the probability of occurrence over a period of time, say, a year, and impact of that occurrence in terms of monetary losses, say \$10–12 million or 50% of annual income. In probabilistic terms, the likelihood is depicted as the rate of events per year (say, 0.01 events per year or one event in a century) as we would specify for a Poisson distribution and the impact is the (expected) value of the loss per event.

The importance rating, or sometimes the ***risk intensity***, is a combination of likelihood and impact. In relative terms, we obtain a basis for "colouring" the risk red through green—when both likelihood and impact. In absolute terms, the risk intensity is expected loss over a year (or other period of time) so we must ensure that the units for impact and likelihood are consistent. The two numbers can be combined as a single number obtained through multiplication or addition.[2] Either way we get an importance rating or risk intensity: in the chart as the colour of different

---

[2] Whether it should be a sum or a product or something else depends on how the scores are reported—the rationale is to get a sense of the expected total loss per year based on the expected number of incidents per year and the expected loss per incident. If both numbers were on a logarithmic scale (e.g., earthquake impact on a Richter scale is logarithmic), then it would make sense to add them as discussed in the survey reported later in this chapter.

zones ranging from red for high-likelihood-high-impact to green for low-likelihood-low-impact.

However, there are many challenges surrounding the rating of likelihood and impact. Likelihood has to be based on probability and probability itself can have one of (at least) two interpretations: (1) an objective or frequency-interpretation or (2) a subjective or belief-interpretation.[3] If taking the frequency view of likelihood, there is history for "normal" risks (in the sense we discussed in the previous chapter) that could reflect the frequency of occurrence. However, such risks are the domain of normal supply chain management. However, for "abnormal" risks that are the focus on supply-chain risk assessment, there is little history to take a frequency view. If we take the belief view of probability, then there is the issue of whose beliefs should be taken and how we should combine the beliefs of many.

Understanding impact, i.e., the loss per event, is even more challenging. Although we may rate impact per incident for any risk category as a single number, it is often multi-dimensional and those dimensions cannot all be mapped into a single monetary value. In many instances, impact may comprise short-term financial losses by way of reduction in income, long-term losses by way of a drop in share prices, the loss of human lives, and loss in the value of the brand. Even for each of such impacts, we don't have a single number, we really have to think about the losses being a probability distribution but as we have already seen, probability itself is subject to frequency or belief interpretation—so we have to think of which single number we want. And even if we could devise a probability distribution, we would need one number, say, the expected value or value-at-risk (VaR) (cf. Sodhi 2005).

Yet another challenge is related to looseness of vocabulary of 'likelihood' being understood and used differently within an organization. Likelihood may be viewed by some as occurring in the next 12 months, by others as occurring any time in the future at all. The likelihood of 0.001 depicted as a risk incident occurring once in a thousand years is not the same as the *probability* of such an event occurring over the coming year—the numerical difference is small when the risk event is extremely rare but these are conceptually different nonetheless. Such differences in understanding may show up in spreadsheet models within the organization or when communicating within the extended supply chain.

With all these challenges, it may sometimes be useful to look at the actual annual loss accruing from incidents in broad category of risks rather than the impact and likelihood of risk incidents in these categories. (Recall our discussion of risk intensity earlier.) For large categories of risks, an organization may have some history of total annual losses or at least some agreement as to the range of annual total losses due to events in the particular risk category. Then simulation models could be developed to match such losses with minimal starting information; however, the results may be extremely sensitive to the starting assumptions and should be used with caution. For instance, given the arrival process of a major disruption with certain probability, Deleris et al. (2004) develop a simulation of a stochastic process

---

[3] There is also a logical view of probability but that only underscores the point that there is more than one view.

to estimate the probability distribution of supply chain losses caused by the disruptions.

3.4 Example of a Cross-Company Survey: What Risks and How to Assess

Now we discuss an example of a supply-chain risk assessment that we carried out with senior risk officers of several companies.[4] Our purpose here is threefold: (1) to show one way to conduct a supply chain risk assessment exercise as well as to generate some useful charts; (2) to show how to combine the information across the people surveyed within the company's extended supply chain (even though in this case we have actually used a survey of independent companies with one senior manager, the Chief Risk Officer or equivalent, responding from each company); and (3) to introduce the concept of *residual risk* as a way to prioritize mitigation effort.

Residual risk is the level of risk that remains after taking the company's prevention and responsiveness efforts into account. Thus residual risk entails assessing the unmitigated risk assuming no prevention or responsiveness in place and then adjusting it for prevention and responsiveness. If a company were in an industry where its plants manufacture an inflammable chemical, then unmitigated risk of fire would be quite high but if the company is good at preventing fires to begin with and responding quickly if and when a fire does start, then its residual risk would be quite low.

In 2010, we surveyed Chief Risk Officers (CRO) of companies from different (and multiple) industry sectors to see how they viewed risks and risk management in their own organizations. We also wanted to see whether there is any relation between insurance coverage and residual risk. Our questionnaire sought to understand the risk in absolute terms while still depending on subjective responses on a 10-point scale.

3.4.1 Our Sample

The sample comprised major UK-based companies mostly in the FTSE 250 list, from different sectors, with many companies covering more than one sector. We started with 41 companies but dropped two companies due to inadequate data for the analysis presented here. The top three sectors in our sample are: (1) retail and distribution; (2) industrial and manufacturing; and (3) food and drink. Other sectors include chemicals, construction, financial institutions, media/telecoms/IT, utilities, and transportation with four or more CROs responding. The companies are mostly large, with three-fourths that had revenues exceeding £1 billion (approxi-

---

[4] We are grateful to Paul Hopkin for his help with this survey.

mately US$1.6 billion) in 2009. Two-thirds of the responding CROs have a remit over more than 10,000 personnel spread over many locations in multiple countries; nearly 90% are responsible for more than 1,000 personnel. More than half the respondents have responsibility spanning multiple continents and nearly all have a remit across multiple locations. Details can be obtained from the authors.

Our first (non-demographic) question was about these CRO's overall satisfaction with the current state of risk management in their organization on a 1–10 scale, 1 being 'completely dissatisfied with all aspects at all locations—couldn't get any worse' and 10 being 'completely satisfied and confident about all aspects of risk management at all locations—couldn't get any better'. Nine-tenths of the managers are quite satisfied with risk management at their company with scores at or above 6, although there is room for improvement as none of them rated risk management in their organization as 10 out of 10 (Fig. 3.1).

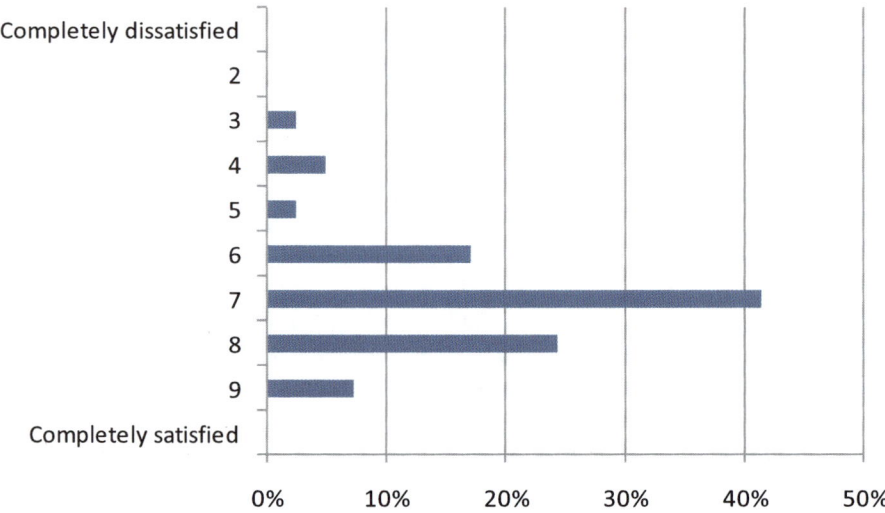

Fig. 3.1 Number of risk managers responding to their satisfaction with the current state of risk management in their organizations

3.4.2 What Risks Were Assessed

Next we asked these CROs to assess for their companies specific risk categories that are tied to single identifiable incidents. We emphasized that we were seeking ratings of risk assuming no prevention or response controls in place to get an estimate of the "threat" of such incidents.

1. Fire/explosion

2. Flood, earthquakes and other natural catastrophes
3. Incidents involving loss of life
4. Manufacturing or distribution failure within the supply chain
5. Failure of a key supplier disrupting the supply chain
6. Unintentional IT/communications failure
7. Hacking/wilful damage to IT/communication infrastructure
8. Strike/people disruption
9. Product recall/unexpected warranty costs
10. Escalation in cost of supplies/commodities
11. Environment/pollution incident
12. Fraud perpetrated on the company/theft or loss of tangible assets
13. Company involved in offering bribes/perpetrating fraud/price fixing
14. Corporate social responsibility issues
15. Litigation losses-class action lawsuit, patent infringement, etc.
16. Other (please list. . .)

In the 'other' category, some respondents provided categories such as reputation, leakage of sensitive commercial information to competitors, risks related to developing strategy and its planning and execution. One respondent provided his/her company's categorization of all risks. The risk categories that we provided is hardly universal: while more companies consider these risk events at the category/sub-category level than otherwise; there is wide variation in categorizing risks (Fig. 3.2). This means that it would be difficult for companies to benchmark their risks even with other companies. This further underscores the point made in the previous chapter about the challenges of risk identification and the vocabulary pertaining to risk.

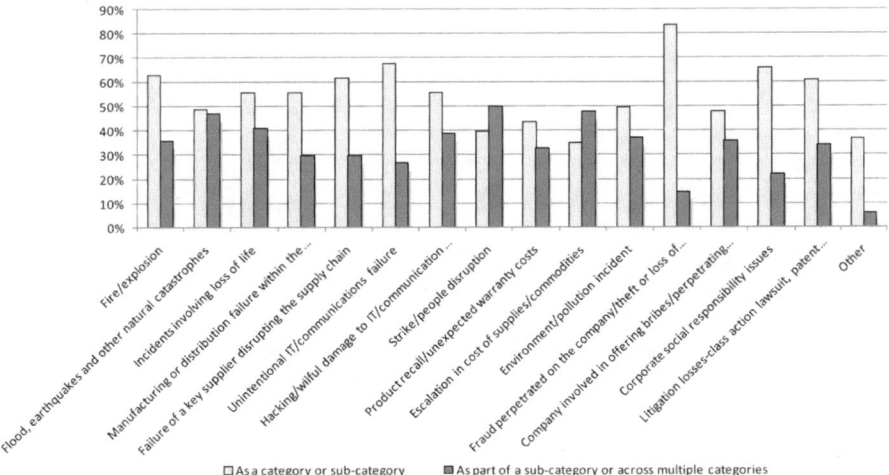

Fig. 3.2 Percentage of all respondents who deal with this risk at the category/sub-category level or at a lower level in their company's risk categorization

3.4.3 How We Assessed These Risks

We first asked the respondents the FREQUENCY of risk incidents assuming there were no prevention controls. This rating really depends on the environment, whether internal or external, that engenders risk incidents in the first place. Of course, many of these risks will be blocked in reality as organizations will have some prevention controls in place but the idea is to separate threats from the possible damage due to incidents (refer to the butterfly diagram in the previous chapter). The ten-point scale we used for the respondents was as follows:

1	Not relevant or never	6	Once a year
2	Once in ten thousand years or longer	7	Once a quarter
3	Once in a thousand years	8	Once a month
4	Once in a hundred years	9	Once a week
5	Once in a decade	10	Daily or more often

Note that this scale is (almost) logarithmic in that a frequency rated 4 is ten times a frequency rated 3, etc.

Likewise, we asked the respondents the potential IMPACT per incident associated with different risks, i.e., without any response controls (or insurance) in place. In reality, companies do have response controls so the purpose of this rating is really to understand the "threat" level in the internal or external environment. We used a 10-point scale of possible responses as shown in the table below:

1	Not relevant or never	6	Wipe out 10% of a year's earnings
2	Not significant	7	Wipe out a quarter of a year's earnings
3	Small but still noticeable	8	Wipe out half a year's earnings
4	Wipe out a small percentage of a year's earnings	9	Wipe out a year's earnings
5	Wipe out 5% of a year's earnings	10	Threaten the very existence of the company

Each score is about double that of the next lower score so this scale is also (approximately) logarithmic. From the data from all respondents on frequency and impact of potential incidents, we were able to see how different CROs perceived the threat levels to their respective companies in different risk categories. Once again, these threats do not take their organization's prevention or response (or insurance) into account (Fig. 3.3).

The same analysis, if done with multiple respondents within the same organization can help focus attention on specific risk categories more than others, a necessary step for allocating resourcing to prevention and response to different risk categories. Considering the frequency and impact together allows us to consider risk intensity. Our scales are *logarithmic* so the expected loss, and thus risk intensity, is indicated by a *sum* of the frequency and impact scores rather than their product, i.e.

$$\text{Risk Intensity} = \text{Frequency} + \text{Impact}$$

We can therefore divide up the graph for each risk category in different zones of risk intensity by using lines that represent different sums: we chose the particular lines only to visually separate different responses (Fig. 3.3). As depicted in Fig. 3.3, the top line corresponds to the case when the risk intensity score = 14, the middle line has risk intensity score = 10, and the bottom line has risk intensity score = 6. If all respondents were from the same company, a more visual way of presenting this information is would be a "smoothened" two-dimensional histogram to focus on areas of concentration. Within a company, such a chart can help build consensus on the seriousness of a particular (unmitigated) risk category (Fig. 3.4). As shown in Fig. 3.4, there is a general consensus about the risk intensity associated with "Strike/people disruption" but not "Fire/Explosion." In that case, more data collection or discussion would be needed to examine the risk intensity about "Fire/Explosion" further.

Next we asked about PREVENTION CONTROLS, i.e., for risk events that can be prevented from occurring in the first place, how good were the existing prevention controls in preventing such risk incidents. The purpose of this rating is to obtain the residual frequency by adjusting the (unmitigated) frequency in the environment by the (perceived) efficacy of prevention controls. Here we used a linear 10-point scale:

1	Not relevant or non-existent	6	Effective in preventing nearly 60% such incidents
2	Barely effective	7	Effective in preventing nearly 70% such incidents
3	Effective in preventing nearly 30% such incidents	8	Effective in preventing nearly 80% such incidents
4	Effective in preventing nearly 40% such incidents	9	Effective in preventing nearly 90% such incidents
5	Effective in preventing nearly half such incidents	10	Fully effective in preventing all such incidents

Likewise, for RESPONSIVENESS, we asked the respondents about their organization's ability to respond to risk events that had already started unfolding. Their response would indicate how good their existing response controls were in detecting and communicating the event to the right people within their organization to mitigate the impact of such a risk event. Again, we used a 1-10 point scale as follows:

1	Not relevant or non-existent	6	Effective in reducing the potential impact by 60%
2	Barely effective in decreasing the impact if a risk incident were to occur	7	Effective in reducing the potential impact by 70%
3	Effective in reducing the potential impact by 30%	8	Effective in reducing the potential impact by 80%
4	Effective in reducing the potential impact by 40%	9	Effective in reducing the potential impact by 90%
5	Effective in reducing the potential impact by half	10	Fully effective in avoiding all of the potential impact even if a risk incident were to occur

Now we can talk about residual risk, i.e., the risk after prevention and responsiveness. To calculate residual risk, we started with the "unmitigated" threat level using

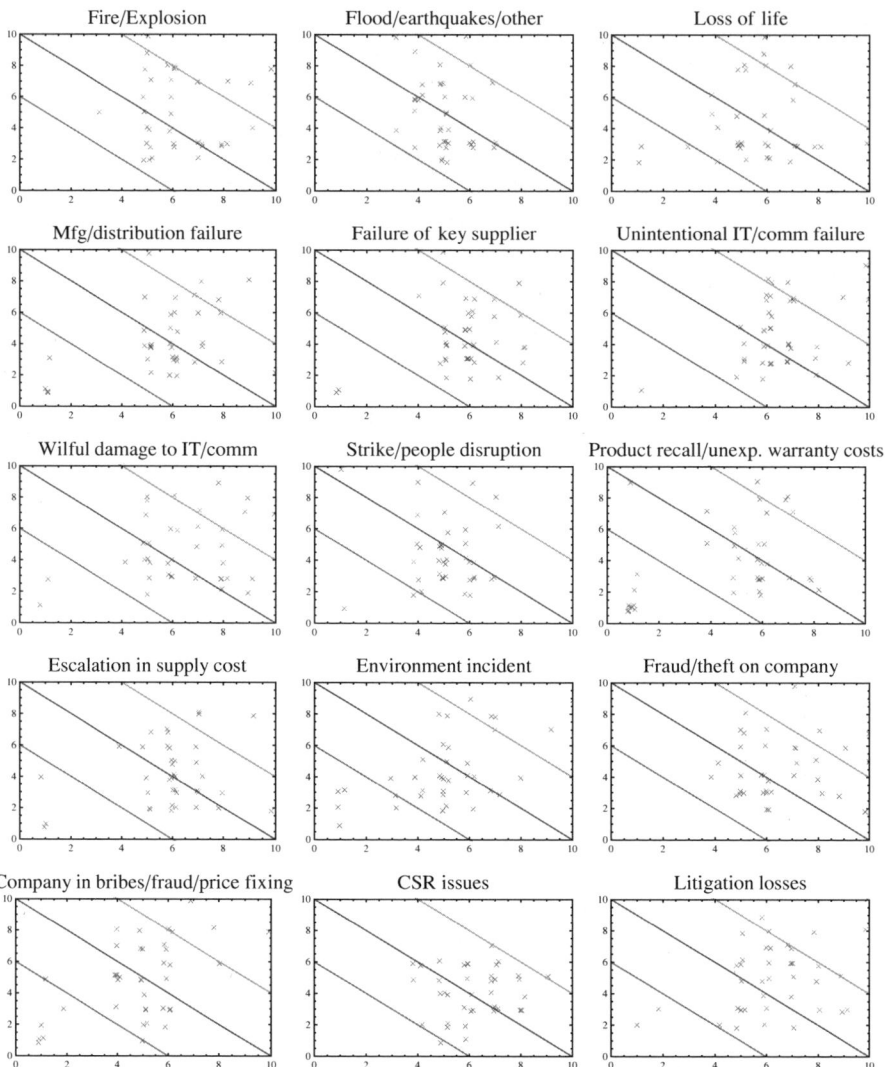

Fig. 3.3 Graphs showing impact vs. frequency of the threat level, i.e., the unmitigated risk, for different risk categories as perceived by chief risk officer of various companies with the *x*-axis being the *frequency* of incidents in a risk category and the *y*-axis being the *impact* per incident. The back-diagonal lines divide the graph into different zones of *risk intensity*, the top right corner represents the risk officers who perceive this risk category as a major threat if left unmitigated and the bottom left corner as only minor threats.

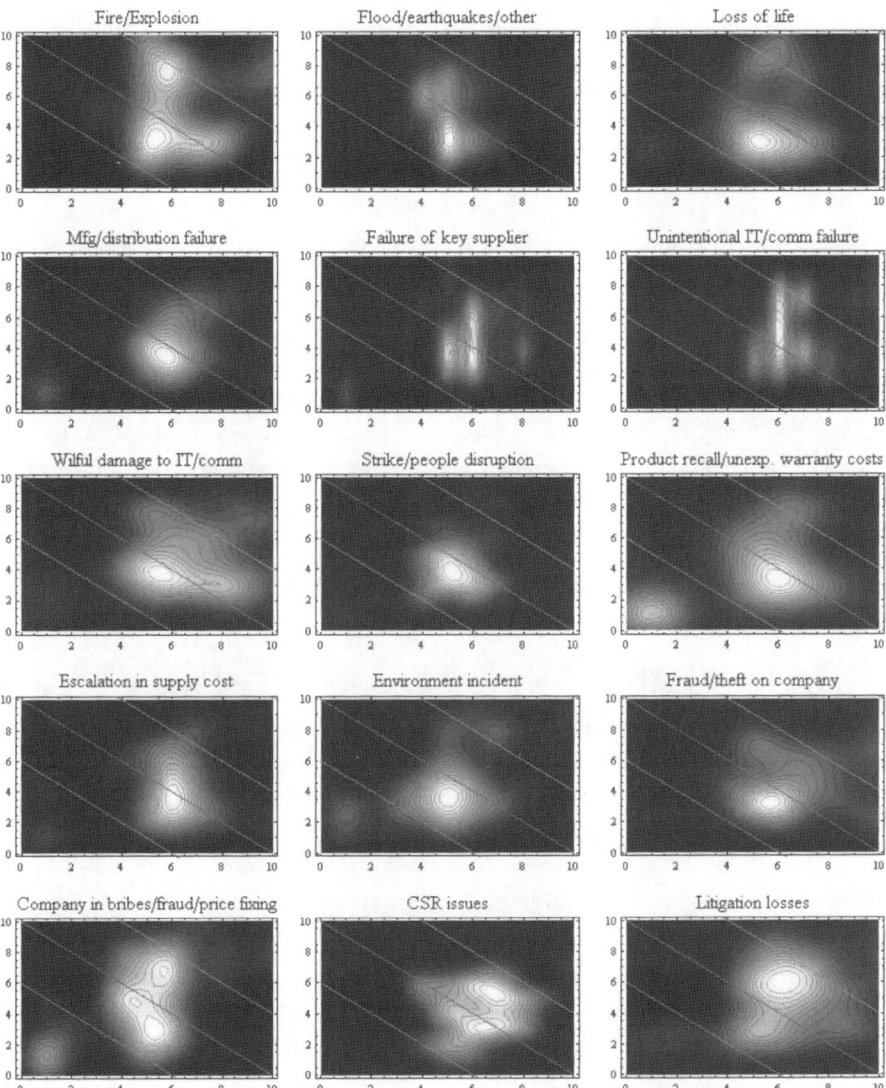

Fig. 3.4 A smoothened two-dimensional histogram of unmitigated impact vs. frequency for different risk categories as perceived by multiple respondents; light colour corresponds to more respondents (see also Fig. 3.3)

the (unmitigated) frequency score and (unmitigated) impact score presented earlier, and then adjusted it down with respondent's view of their organization's prevention controls and the responsiveness to an incident. As a simple calculation,[5] we take

$$\text{Residual frequency} = \text{Frequency} \times (1 - \text{Prevention Controls}/10) + 1$$
$$\text{Residual impact} = \text{Impact} \times (1 - \text{Responsiveness}/10) + 1$$

As shown in Fig. 3.5, if *prevention controls* were rated at 10, then *residual frequency* would drop to the lowest score of 1 regardless of how much the (unmitigated) frequency was. Also, because both (unmitigated) frequency score and prevention score are measured on a ten-point scale, it is easy to check that the residual frequency ranges from 1 to 10. Likewise, if *responsiveness* were 10, *residual impact* would drop to 1. Also, both residual frequency and residual impact are higher when the prevention controls score and the responsiveness score are lower, respectively. The purpose of this simple, even simplistic, calculation is to show which risk categories should be focus of the company's efforts in terms of better prevention, better responsiveness or even transferring the residual risk through insurance.

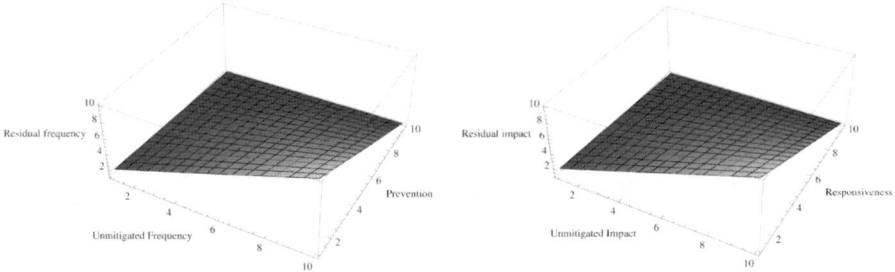

Fig. 3.5 Residual frequency and residual impact

Thus, relative to (unmitigated) frequency and impact scores as shown in Fig. 3.3, the residual frequency and residual impact scores, adjusted for prevention and responsiveness respectively, are lower as shown in Fig. 3.6. However, it is readily seen that for some respondents (in this survey, some companies) even the residual risk remains high—this could explain why satisfaction for some respondents with their companies was not high although we do not test this here (Fig. 3.1).

We can also view this data as a smoothened histogram across the two dimensions of residual frequency and residual impact per incident to see the areas of risk that still remain as a consensus among the different respondents—this would be meaningful within the company's extended supply chain although not in the present context (Fig. 3.7).

Now consider insurance. It would seem logical to assume companies would seek to transfer the residual risk, if high, to insurance, i.e., the higher the residual risk, the

---

[5] This linear correction to a logarithmic quantity is not in the spirit of expected loss but it is simple to use and effective in visualizing the efficacy of prevention and response controls.

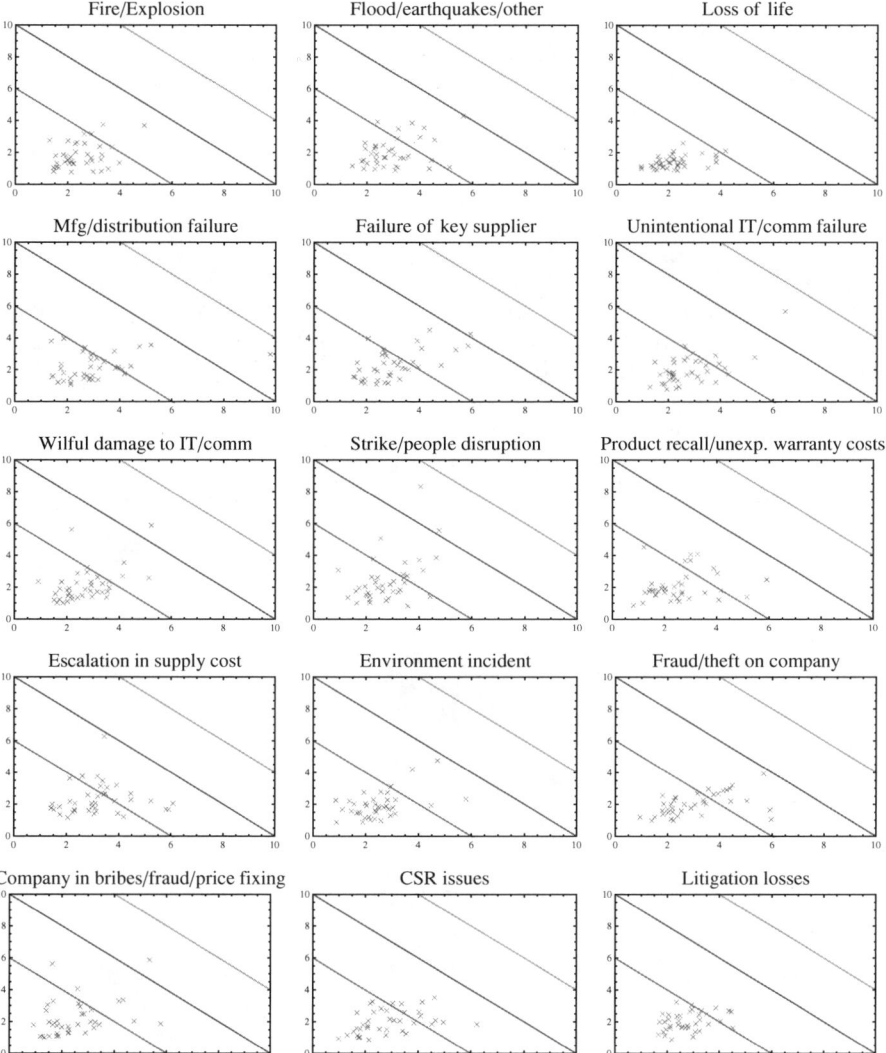

Fig. 3.6 Residual threat level through prevention controls and responsiveness, with the *x*-axis being the *residual frequency* and the *y*-axis being the *residual impact* per incident (compare with the unadjusted or unmitigated numbers in Fig. 3.3)

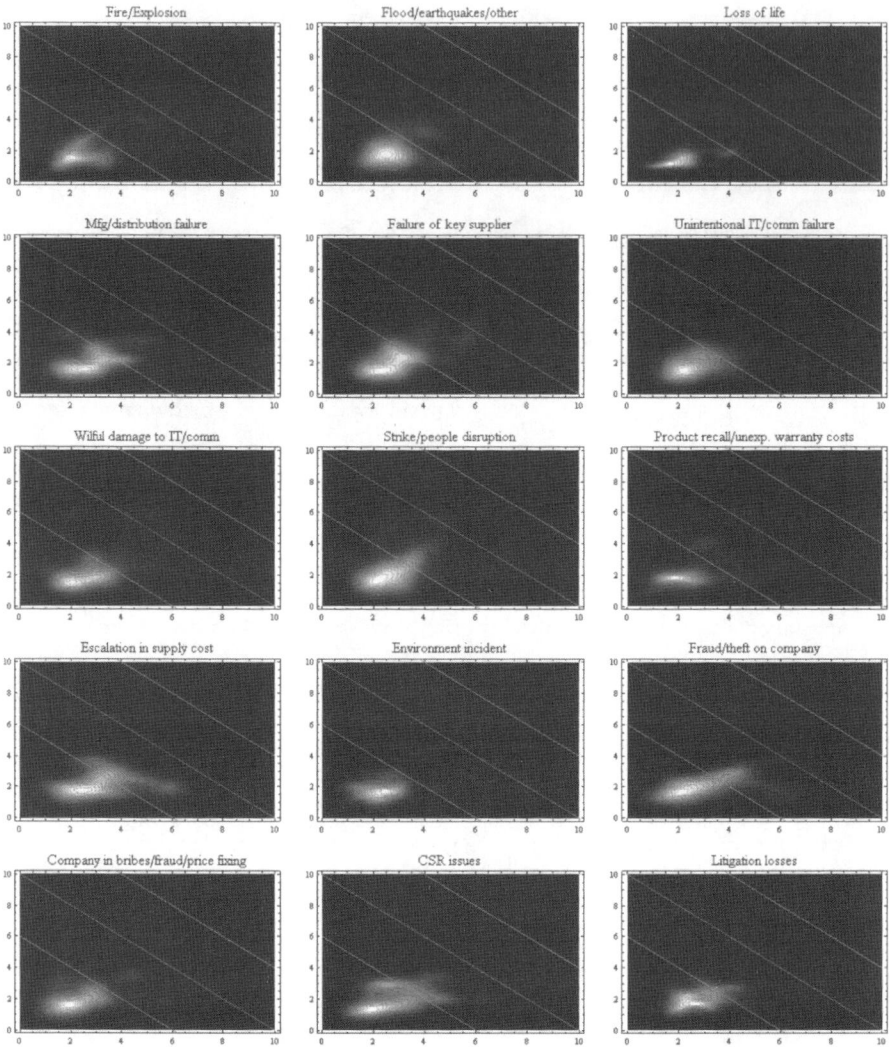

Fig. 3.7 Smoothened histogram in two dimensions of residual impact per incident versus residual frequency of risk incidents (compare with Fig. 3.4)

higher we may expect insurance levels to be. Therefore, we asked the risk officers about the insurance coverage for the different categories of risks. Our question in the survey was: If a risk event of the following type were to occur, how much of the eventual loss they expected their current level of insurance would cover as per the following linear 10-point scale:

1	Not relevant or no coverage at all	6	Around half
2	Around 10%	7	Around 60%
3	Around 20%	8	Around 70%
4	Around 30%	9	Around 80%
5	Around 40%	10	More than 90%

We compared insurance coverage to the residual risk intensity. As an example, we assume

$$\text{Residual risk intensity} = \text{Residual frequency} + \text{Residual impact}$$

(Alternatively, one can consider defining unmitigated risk intensity in terms of the product of those two unmitigated measures instead of the sum, or consider using the logarithmic scale as discussed earlier.) When the unmitigated risk intensity is measured in terms of the sum of those two unmitigated measures, we plot the insurance coverage against the unmitigated risk intensity (Fig. 3.8). The results were surprising in that we did not find a clear pattern, something we raised with respondents and other managers.

3.4.4 Discussion

While our risk assessment method is based on CROs' perceptions (with any entailed biases or other measurement errors) and some of the calculations and adjustments are simplistic, it goes a long way in addressing the reasons for doing risk mapping as we outlined in the Introduction of this chapter. Despite limitations, such a survey method and the corresponding results could be used to help a firm develop shared understanding on what risks need further attention. Risk management personnel could also explain the value of risk mitigation efforts and investments by separating unmitigated risk from residual risk. Some of the assumptions behind risks that have been considered as non-threatening may need to be re-evaluated in light of such findings. Thus, such an approach can be used to create shared awareness of the threats to the organization and the capability of the organization to mitigate them through prevention and responsiveness and possibly developing a consistent approach to insurance based on a shared rationale rather than selective fear.

When we showed these results to risk managers of companies that included many of the respondent companies, we learned that insurance companies do not offer the products they want or that they offered products eo cover risks that were already well managed within their organization. Another feedback was that losses could exceed annual earnings. Finally, the managers cautioned us against drawing too many conclusions as different companies had different risks they faced and the number of companies in any particular sector in our sample is quite small. On the other hand, many companies in an organization's supply chain may be from different sectors so doing supply-chain risk assessment across sectors may be quite important.

This is one reason why we combined mix purely subjective and purely objective ratings. Doing so allows us not only to use different companies from different sec-

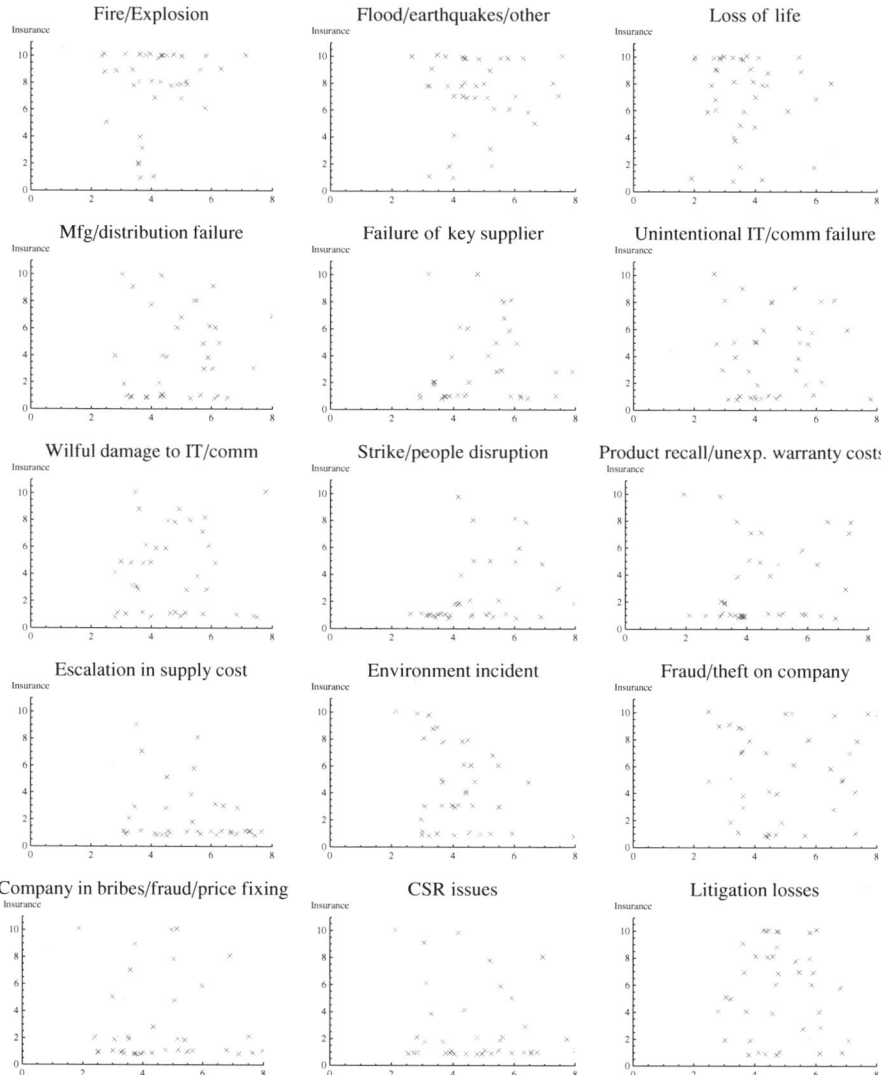

Fig. 3.8 Insurance levels versus intensity of residual risk (sum of residual frequency and residual impact)

tors but also to draw more information from the risk assessment exercise that would be possible only with objective data. At the same time, the complexity of generating information this way is not any more than using the purely subjective 1–4 type of ratings in common use.

Our purpose in showing this example is that this is one way to do risk assessment within an organization or across its supply chain even though we conducted this particular survey across many companies. Hopefully this can be used as a basis for more refined data collection or for creating a common basis for risk discussion within a company or between a company and its suppliers and customers.

3.5 Summary

We started this chapter by discussing the purposes of supply chain risk assessment. As with risk identification in the previous chapter, we have to be guided by the decisions that that the organization needs to make. Such decisions will typically be about focusing limited resources on fighting different risks. As such, the main purpose of risk assessment should be to develop a shared sense of urgency about a subset of identified risks and catalyze risk mitigation effort including devising a consistent approach to insurance. This urgency should also translate into investment in inventory and capacity—what, where and how much. The organization can use risk assessment to develop contingency plans by attempting to understand the nature of threats and other risks to help counter these better.

A risk incident that did a lot of damage in recent history or pressure from an insurance company can spur risk assessment. But rather than focus on a horse that has already bolted the stable, risk assessment can help focus on the horses, the stable and the business processes of managing the stable. We presented an example of how residual risk can be measured through surveys within or across companies. As companies also use risk assessment program to meet legal or regulatory requirements, a survey like this one done periodically within the organization and across the supply chain can be part of the formal (and documented) process that reflects the mitigation efforts of the company.

Chapter 4
Mitigating Risks

Abstract To mitigate supply chain risks identified by the managers, this chapter describes three general risk mitigation strategies: (1) alignment of supply chain partners incentives to reduce the behavioral risks within the supply chain, (2) flexibility to reduce not only demand risks but also supply and process risks, and (3) building "buffers" or redundancies. We also discuss ways for managers to adapt these general risk-mitigation strategies to the circumstances of their particular company.

4.1 Introduction

In this chapter, we describe strategies to mitigate the supply chain risks we identified in Chapter 2. We also discuss how to tailor these strategies for a given company's context.

A notable example of supply chain risk is a fire that erupted at a Philips' chip manufacturing plant that impacted mobile-phone manufacturers Ericsson and Nokia. The two companies responded differently with dramatically different results. On March 17, 2000, a lightning bolt hit a power line in Albuquerque, New Mexico. The bolt caused a massive surge in the surrounding electrical grid, which in turn started a fire at a local Philips plant, damaging millions of radio frequency microchips. Nokia, a Scandinavian mobile-phone manufacturer and a major customer of this Philips plant, was ready to deal with such an unpredictable event, however. Almost immediately, the company began switching its chip orders to other Philips plants, as well as to other Japanese and American suppliers. Thanks to its multiple-supplier strategy and responsiveness, Nokia's production suffered little during the crisis.

In contrast, Ericsson, another mobile-phone customer of the Philips plant, employed a single-sourcing policy. As a result, when the Philips plant shut down after the fire, Ericsson had no other source of microchips and its production of mobile phones was disrupted for many months. Ultimately, the disaster and company de-

M.S. Sodhi, C.S. Tang, *Managing Supply Chain Risk*,
International Series in Operations Research & Management Science 172,
DOI 10.1007/978-1-4614-3238-8_4, © Springer Science+Business Media, LLC 2012

pendence on a sole supplier cost Ericsson $400 million in lost sales after the first quarter alone.[1]

These two dramatically different outcomes from this risk incident demonstrate the importance of proactively managing supply-chain risk. Ericsson has since implemented new processes and tools for managing supply-chain risk proactively (Norrman and Jansson, 2004).

This risk incident raises a key question: how can companies avoid risk incidents or reduce the impact of risk incents? Essentially, once a particular supply chain risk is identified, assessed, and selected for mitigation, it can be mitigated by decreasing its likelihood, reducing its potential consequences or both. Paulsson and Nilsson (2008) present 23 methods for mitigating risks; however, there are three basic approaches: *accept, avoid*, and *mitigate*; and it is the last of these that we focus on in this chapter. *Responding* to a risk incident that has already occurred, usually regarded as within the domain of business continuity, is the subject of Chapter 5.

Accepting the risk does not require doing anything other than the company bearing the entire consequence in case there is a risk incident or the company transferring part of the consequences to its insurance company or its supply chain partner. However, transferring risk through insurance or through financial instruments like swaps does not actually reduce the likelihood of the risk. Even if it reduces the impact of the risk to a certain extent, it may result in *moral hazard* whereby the company can become more risk-prone knowing that is can transfer some or all of the financial consequences. Likewise, liability insurance may offer financial compensations to customers who suffer from using unsafe products, but it does not reduce the damage to the reputation of the company nor the suffering of the people who used these products.

Avoiding risks entails efforts to prevent the occurrence of undesirable incidents. Such efforts entail the development of fail-safe systems, i.e., systems that cannot fail or that can trigger corrective actions to prevent failures, and the application of quality-based principles to ensure there is no failure in the detection of a risk incident with highly negative consequences. These approaches can be quite useful in security-related risks where preventing an incident from ever happening is the best approach. Lee and Wolfe (2003) illustrate how certain technologies, say, biometric systems for positive identification of personnel and smart container systems for monitoring internal temperature and pressure of each container, can be used to prevent containers being tampered with throughout the shipping process. The U.S. Homeland Security has developed the Container Security Initiative (CSI) that requires all containers to be pre-screened at the port of departure before they arrive at U.S. ports to reduce the likelihood of seaport terrorist attacks in the U.S.

While acceptance or avoidance are appropriate in some circumstances, companies tend to focus on developing approaches for mitigating supply chain risks, whether normal (delays) or abnormal (disruptions).

---

[1] "Can suppliers bring down your firm?" *Sunday Times* (London), November 23, 2003

4.2 Risk Mitigation Strategies

Risk mitigation entails efforts to reduce the impact of risk incidents in case such incidents do occur. Broad risk mitigation strategies in this category are (1) alignment of supply chain partners' incentives to reduce the behavioural risks within the supply chain, (2) flexibility to reduce not only demand risks but also supply and process risks, and (3) building "buffers" or redundancies. All three are useful to mitigate normal and abnormal risks in the supply chain. We describe the first two briefly below and the third approach in the next section.

Alignment. Besides long-term partnerships, there are other mechanisms to coordinate the interests of the different supply chain partners. Tang (2006b) reviewed different types of supply contracts to coordinate a supply chain so that all parties will act in the interest of the entire supply chain when dealing with demand risks including *wholesale price contracts, buyback contracts, and revenue sharing contracts*. A two-part tariff, i.e., a fixed cost and a per unit wholesale price, can be used to entice the downstream partner to order according to the optimal order quantity (as per the newsvendor solution) for the entire supply chain. A buy-back contract is a returns policy under which the manufacturer is required to buy back the retailer's excess inventory at a reduced price. Under certain conditions, doing so can achieve supply chain coordination. Under a revenue sharing contract, the retailer shares the revenue with the manufacturer and obtains a reduction in the wholesale price in return, achieving supply chain coordination. Narayanan and Raman (2004) have studied *risk sharing* and revenue sharing to align incentives across supply chain partners.

Flexibility. There are at least five different types of flexibility strategies corresponding to *multiple suppliers, flexible supply contracts, flexible manufacturing process, postponement* and *responsive pricing*. The ability to shift order quantities across suppliers can be a powerful mechanism for the manufacturer to hedge against supply risks. Under flexible supply contracts, the manufacturer is allowed to adjust the order quantity within a pre-specified range, say, a few percent of the order quantity. This helps to mitigate the impact associated with demand risks (c.f., Tsay and Lovejoy, 1999). The manufacturing process is flexible if different types of products can be manufactured in the same plant, enabling the manufacturer to reduce supply, process, or demand risks (c.f., Jordan and Graves, 1995). Postponement calls for delayed product differentiation by producing a generic product initially and then customizing it for different markets and customers later, thus allowing a company to respond to demand changes across multiple markets quickly (c.f., Lee and Tang (1997)). Responsive pricing is an effective tool to mitigate supply or demand risks by manipulating demand when the supply is inflexible. For example, as the supply of certain components from Taiwan was affected by an earthquake, Dell's response was to lower the price of certain products so as entice their online customers to "shift" their demands to other Dell computers that utilized components from other countries.

Similar to the above strategies are the "AAA" principles of *alignment*, *agility*, and *adaptability* that are intended to reduce the impact associated different types of supply chain risks (Lee 2004). *Alignment* calls for an aligned interests among supply chain partners so as to facilitate close communication, cooperation, and collaboration; *agility* entails flexibility and responsiveness; and *adaptability* requires close monitoring of the environment so that one can deploy a recovery plan in a timely manner, something we will discuss as part of responding to risk incidents in the next chapter.

4.3 Building Reserves for Redundancy

Firms can always build in some redundancies throughout the supply chain so as to reduce the cost implications of certain undesirable events associated with supply, process, and demand risks; Chopra and Sodhi (2004) refer to these as "reserves" against supply chain risk. For example, extra inventory, extra back-up production capacity, and extra back-up suppliers are "buffers" to absorb the impact against delays and disruptions in the supply chain. However, redundancies can be expensive when used against (rare) unanticipated events (Sheffi, 2005). Also, redundancies disguise inefficiencies in the supply chain, inhibiting the development of a lean supply chain. Flexibility overcomes these disadvantages: it can reduce the impact of the occurrence of certain unanticipated events and it can also be put to use with planned changes, for instance, to produce a greater variety of products.

To prevent the kind of heavy sales losses suffered by Ericsson after the Philips plant fire, managers must perform a delicate balancing act: keeping inventory, capacity and other elements at appropriate levels across the entire supply-chain in a dynamic, fast-changing environment. Dell, Toyota, Motorola and other leading manufacturers excel at identifying risks in their supply chains, and at creating powerful mitigation strategies that neutralize potentially negative effects. With a clear understanding of the types of supply chain risks, managers in many types industries can tailor effective risk-reduction approaches for their own companies.

As we discussed in Chapter 2, supply chain problems resulting from natural disasters, labor disputes, supplier bankruptcy, acts of war and terrorism and other causes can seriously disrupt or delay material, information and cash flows, any of which can damage sales, increase costs—or both. In that chapter, the categories we considered, at different levels, include delays, disruptions, forecast risks, systems risks, intellectual property risks, procurement risks, inventory risks, and capacity risks. Because each risk category has its own risk drivers, it is natural to develop ways to address these drivers (Table 4.1).

However, addressing these risk drivers and thus mitigating the corresponding risks is not the whole story. Managing supply chain risk is difficult. One big reason is that individual risks often connect with other risks. As a result, actions that mitigate one risk can end up exacerbating another. Consider a lean supply chain. While its bare-bones inventory *decreases* the impact of over-estimating customer

Table 4.1 Drivers of different types of risk

Risk Categories	Risk Drivers
Disruptions	• Natural disaster • Labor disputes • Supplier bankruptcy • War and terrorism • Dependency on a single source of supply as well as capacity and responsiveness of alternative suppliers
Delays	• High capacity utilization at supply source • Inflexibility of supply source • Poor quality or yield at supply source • Excessive handling due to border crossings or to change in transportation modes
Systems risk	• Information infrastructure breakdown • System integration or extensive systems networking • E-commerce
Forecast risk	• Inaccurate forecasts due to long lead times, seasonality, product variety, short lifecycles, small customer base • "Bullwhip effect" or information distortion due to sales promotions, incentives, lack of supply-chain visibility, and exaggeration of demand in times of product shortage
Intellectual property risk	• Vertical integration of supply chain • Global outsourcing and markets
Procurement risk	• Exchange rate risk • Fraction of procurement of a key component or raw material from a single source • Industry-wide capacity utilization • Long-term versus short-term contracts
Receivables risk	• Number of customers • Financial strength of customers
Inventory risk	• Rate of product obsolescence • Inventory holding cost • Product value • Demand and supply uncertainty
Capacity risk	• Cost of capacity • Capacity flexibility

demand, it simultaneously *increases* the impact of a supply chain disruption. Similarly, actions taken by any company in the supply-chain can increase risk for any other participating company (Table 4.2).

Table 4.2 Mitigation strategies can reduce some risks but may also exacerbate other risks

Mitigation strategy	Disruption	Delays	Forecast risk	Procurement risk	Receivables risk	Capacity risk	Inventory risk
Add capacity		↓↓		↓		↑↑	↓
Add inventory	↓	↓↓		↓		↓	↑↑
Have redundant suppliers	↓↓			↓		↑	↓
Increase responsiveness		↓↓	↓↓				↓↓
Increase flexibility		↓		↓		↓↓	↓
Aggregate or pool demand			↓↓			↓↓	↓↓
Increase capability		↓					↓
Have more customer accounts					↓		

Note: An up arrow ↑ or ↑↑ indicates how much the strategy increases risk when applied to a particular problem, with two arrows signifying greater risk. A down arrow ↓ indicates a decreased risk. Systems risk and intellectual property risks are not included. Adapted from Chopra and Sodhi (2004).

Building reserves for redundancy is useful for reducing the impact caused by disruptions and delays that can cause the affected organization(s) problems ranging from a minor to serious. A simple delay along the chain may create a temporary impact, whereas a sole supplier holding up a manufacturer to force a price increase represents a long-term risk. A machine breakdown may have a relatively minor impact for a manufacturing company with redundant capacity, whereas a war that disrupts shipping lanes can have a dramatic impact on a shipping company. Most companies develop plans to protect against the normal risks that are recurrent and low-impact in the supply chain; however, few companies develop plans to handle disruptions. For instance, a supplier with quality problems represents a common, recurrent disruption. Without much effort, the customer can demand improvement or find a substitute. In contrast, in regions where earthquakes are rare, preparedness may be weak or uneven to prevent major disruption.

One way to build a shared vision against both types of risks is to do a stress-testing of the supply chain. *Stress testing* is a group exercise that helps managers and their companies understand and prioritize supply chain risks. "What if" scenarios can help key players focus on the supply-chain one link at a time. This exercise offers an especially effective way to gain buy-in and shared ownership in project teams tackling supply-chain risk.

The first step in stress testing is to identify key suppliers, customers, plant capacity, distribution centers and shipping lanes. Next, the team surveys locations and amounts of inventory represented by components, work-in-process, and finished goods. Then managers probe each potential source of risk. This helps assess possible impacts as well as the level of preparedness within the supply-chain. Facilitators ask questions such as, "What might happen if a particular supplier could not deliver for a month?" or "What if a supplier raised prices by 20% at the termination of contract?" Questions pertaining to key customers might include, "What if demand went up or down by 20%?" or "What if a customer delayed cash payment by a month?" These and other questions related to various sources of risk are summarized in Table 4.3. When considering questions during stress testing, managers should realize that "20%" or "one month" are not sacred figures, but simply represent numbers large enough to be significant and small enough to be realistic. Also, it is wise to position stress testing as "thought experiment" to help the company prepare for unforeseen events rather than ignite a debate on the likelihood of such events. As such, it is useful to frequently remind people of the goal: preparing the supply-chain for unforeseen events and greatly lowering risk, at the lowest cost.

Through stress testing, managers should be able to identify risk-mitigation priorities for the near, the medium, and the long term. They will have identified product families at risk, as well as individual plants, shipping lanes, suppliers, or customers that could pose risks. Managers will also have a clear idea of what is at risk in terms of impact: sales, procurement costs, revenues, prices, or even reputation.

4.4 Tailoring Risk Management for any Given Company

With so many related risks and risk-mitigation approaches to consider, managers must do two things when they begin to construct a supply chain risk management strategy. First, they must create a shared, organization-wide understanding of supply chain risk. Then they must determine how to adapt general risk-mitigation approaches to the circumstances of their particular company. Managers can achieve the first through *stress testing* and the second through *tailoring their reserves to develop robust strategies*.

Leading companies mitigate risk by building various forms of "reserves" including inventory, capacity, redundant suppliers, and responsiveness. Managers must understand and evaluate the tradeoff between the risk and the cost of building a reserve to mitigate it—the research literature refers to this as *the newsvendor problem* (cf. Chopra and Miendl, 2004: 346-352).

Table 4.3 Stress-testing the supply chain by exploring what-if scenarios

	Supplier-related	Internal	Customer-related
Disruptions	• Supplier of a key part shuts down plant for a month Supplier capacity drops by 20% overnight	• Key plant shuts down unexpectedly for one month • Capacity at a key plant drops by 20% overnight	• Demand goes up by 20% for all products / a key product • Same questions with demand going *down* by 20%
Delays	• Delivery of of key parts or raw materials delayed by a month	• Distribution or production schedule delayed by a month	• Customer orders arrive later than expected by a month
Systems risk	• Supplier's order-entry system goes down for a week	• Key customer's procurement system inside your company goes down for a week • Company's inventory/accounts system goes down for a week	• Order entry system goes down for a week • Key customer's procurement system inside your company goes down for a week • Credit card information stolen from hacked e-commerce system
Information and material flows distortion	• Supplier rations supplies by 20 % • Supplier increases minimum order quantity by 20%	• To take advantage of volume discounts, company begins to order in quantities twice as large as usual, but half as frequently, which impacts supplier's ability to forecast	• To take advantage of volume discounts, key customer begins to order in batches that are twice as large as usual but less frequent, which impacts manufacturer's ability to forecast
Intellectual property risk	• Key supplier redesigns parts or develops its own product		
Procurement risks	• Supplier delays in processing returns by a month • Supplier forced to increase price of key components by 20% due to increase in material costs • Transportation costs go up 20% overnight	• Unforeseen cash squeeze, which causes a month-long delay in paying key suppliers	
Receivables risks			• Key customer withholds payments one month longer than usual • 20% of receivable payments delayed by one month

Leading companies deal with this range of supply chain risks by holding "reserves." Just as insurance company holds cash reserves to meet claims, top manufacturers hold supply chain reserves that include excess inventory, excess capacity and redundant suppliers. The big challenge for managers here: mitigate risk by smart positioning and sizing of supply chain reserves, without decreasing profits. So while stockpiling inventory may shield a company against delivery delays by suppliers, building reserves willy-nilly also drives up costs and hurts the bottom line. The managers' role here is akin to a stock portfolio manager: achieve the highest achievable profits (reward) for varying levels of supply-chain risk and do so efficiently (Fig. 4.1). This means the manager must seek additional profits for any level of risk protection and preparedness, or increase prevention and preparedness without reducing profits. Success at this task requires a good understanding of supply chain risks and remedies, both broad and tailored in the context of the manager's own company.

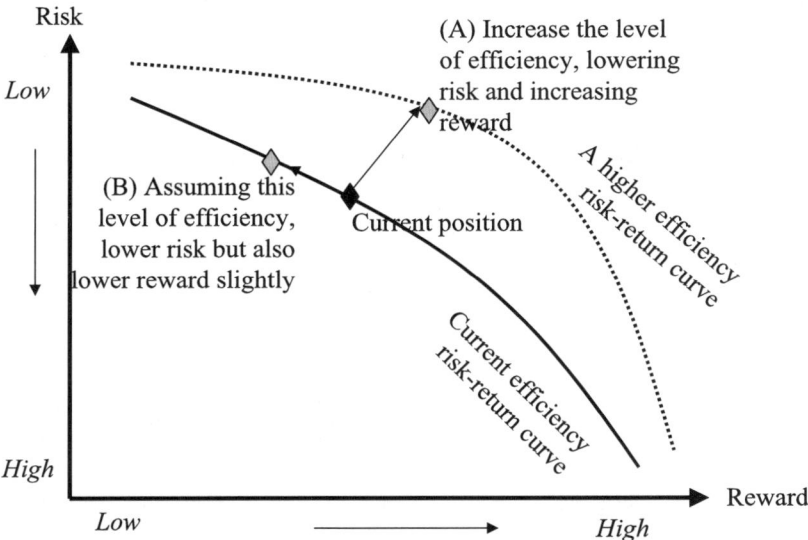

Fig. 4.1 Choosing supply-chain risk/reward tradeoffs—Choice (A) entails moving to a higher level of efficiency, reducing risk while increasing rewards. Choice (B) entails remaining at the current level of efficiency and therefore reducing risk, which also means reducing the reward. Adapted from Chopra and Sodhi (2004).

Just as a generic portfolio strategy needs to be tailored for any given portfolio, we need to tailor the portfolio of reserves for any given company's supply chain risk. There are four aspects to a company's situation that are useful to understand tailoring:

- Cost of a 'reserve' (capacity, inventory, etc.)
- Centralizing versus decentralizing the reserve in question

- Level of risk (high or low), and
- High volume stable demand (low risk) versus low volume uncertain demand (high risk) products.

Given these four aspects, three key relationships influence this optimal balance (see Fig. 4.2) as discussed below:

1. The *first* relationship is the increasing cost of risk reduction. This simply means that using inventory to cover a high level of demand risk costs much more than covering a low level of risk.
2. The *second* relationship shows that pooling forecast risk, receivables risk, or some other risk reduces the amount of reserve required for a given level of risk coverage. For example, the required level of inventory needed to ensure a given level of fill rate decreases when the demand forecasts at different locations are pooled.
3. The *third* relationship shows how the benefit of pooling grows with the level of risk covered: the benefit of pooling inventory is large only if the product has high forecast risk or a high inventory risk.

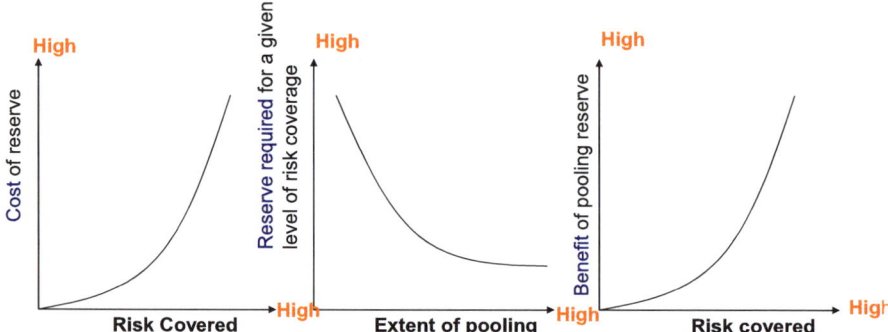

Fig. 4.2 Balancing supply-chain risk/reward relationships: In order to match the cost of building a reserve to the extent of the risk covered, managers must optimally balance three key relationships. (a) The **increasing cost of risk reduction** means covering a high level of demand risk using inventory is proportionally much more expensive than covering a low level of risk using inventory. (b) The **pooling forecast risk, receivables risk**, or other risk reduces the amount of reserve required for a given level of risk coverage so that the required level of inventory needed to mitigate forecast risk decreases as it is pooled. (c) The **benefit of pooling** grows with the level of risk covered so that the benefit of pooling inventory is large only if the product has high forecast or inventory risk. Adapted from Chopra and Sodhi (2004).

Managers can balance these relationships to tailor their response to risk with a surer grasp of extent and cost of reserve. The following rules of thumb can be applied to tailor risk-mitigation strategies: When the cost of building a reserve is low, reserves should be decentralized so that decentralized reserve would enable local entity to respond faster to risk incidents. When the cost for the reserve is high, reserves should be pooled so as to manage supply chain risk at affordable costs. If the level of risk is low, focus on reducing costs. If the risk is high, focus on risk mitigation (Fig. 4.3). By tailoring reserves for all risk-mitigation strategies, companies can maximize rewards for the same level of risk, or lower risks for the same reward (Fig. 4.3, see also Table 4.4).

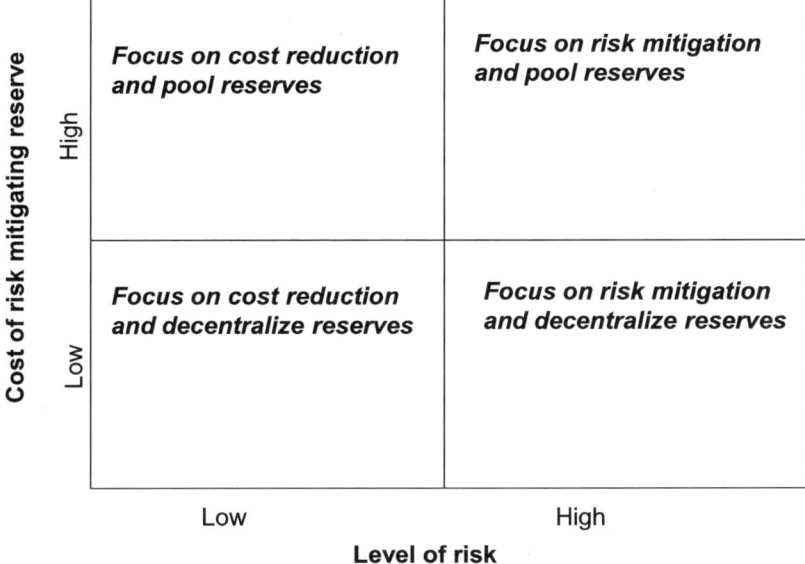

Fig. 4.3 Rules of thumb for tailored risk management

Besides the cost of building the reserve, companies must consider product volumes. Fast-moving standard products, with low margins and low forecast risk, call for different types of reserves than slow-moving special products with high margins and greater forecast risk. When planning capacity, managers should select a low-cost supplier for supplying high-volume stable demand (low-risk and low-margin) items. In contrast, a more responsive supplier is better suited for supplying low-volume uncertain demand (high-risk and high-margin) items. Case in point: Cisco tailors its response by manufacturing high-volume products with stable demand in specialized, inexpensive but not-so-responsive Chinese plants. Low-volume items with uncertain demand (and with high margins) get assembled in responsive, flexible (and more expensive) U.S. plants. Sony also exploits a similar strategy using flexible but high-cost plants in Japan, and low-cost but specialized plants in Malaysia and China.

As much as possible, a specialized and decentralized approach offers the best way to keep capacity for *high-volume* commodity items with *low forecast risk*. Doing so should produce greater responsiveness and lower transportation costs—but only if doing so maintains adequate economies of scale. In contrast, capacity for

Table 4.4 Tailoring reserves for a given company's context

General Risk Mitigation Strategy	Tailored Strategies
Increase capacity	• Focus on low-cost, decentralized capacity for predictable demand. • Build centralized capacity for unpredictable demand. Increase decentralization as cost of capacity drops.
Get redundant suppliers	• More redundant supply for high-volume products with stable demand, less redundancy for low-volume products with uncertain demand • Centralize redundancy for low-volume products with uncertain demand in a few flexible suppliers.
Increase responsiveness	• Favor cost over responsiveness for commodity products. Favor responsiveness over cost for short- lifecycle products.
Increase inventory	• Decentralize inventory for low-value products with stable demand. Centralize inventory for high-value products with uncertain demand.
Increase flexibility	• Favor cost over flexibility for predictable, high- volume products. Favor flexibility for low-volume unpredictable products. Centralize flexibility in a few locations if it is expensive.
Pool or aggregate demand	• Increase aggregation as unpredictability grows.
Increase capability	• Prefer capability over cost for high-value, high-risk products. Favor cost over capability for low-value commodity products. Centralize high capability in flexible source if possible.

low-volume, short life-cycle products with *high* forecast risk should be made more flexible and centralized to pool demand. That helps explain why auto makers, for example, build specialized plants in each major market for fast-moving products. But plants for high-end, slower-selling cars should be centralized and more flexible.

When capacity is expensive, managers can reduce supply-chain costs by centralizing capacity to pool risk. As costs decline, capacity can be decentralized further to be more responsive to local markets. Consider the personal computer industry. PCs can be assembled to order in two different ways: One is the Dell model in which capacity is centralized. The other is the model widely used in India, in which several companies sell component kits to local assemblers for assembly on demand. Given the low cost of assembly capacity in India, it is economical to decentralize capacity, even though this action reduces pooling and increases the overall size of assembly capacity across the supply-chain. In contrast, given the higher cost of capacity in the United States, centralizing buffer capacity is more effective.

In addition to separating products with different risk characteristics, managers must also consider separating capacity for the low-risk and high-risk aspects of each product. Utility companies use this strategy by employing low-cost, coal-fired power plants to handle predictable base demand, and utilizing responsive but high-cost gas- and oil-fired power plants to handle uncertain peak demand. Similarly, General Electric (GE) ships bulbs by sea on a weekly basis from its plant in China to cover the predictable portion of demand. But the company also maintains an inventory of bulbs in the United States, or flies them in from China, to cover unpredictable demand.

4.5 Conclusion

Successful management of supply chain risks begins with an understanding of the various threats, alone and collectively. By continually stress testing their supply-chains and tailoring reserves, managers can protect and improve the bottom line in the face of many types of supply-chain risks.

Chapter 5
Response as a Mitigation Approach

Abstract As supply chains become more complex, companies find their supply chains more vulnerable to unforeseen disruptions in their supply chains. Without responsive systems in place, such disruptions can have huge impact in terms of cost and recovery time to the company (and its customers as well as suppliers). This chapter presents a *time-based risk management* framework to motivate improved practice and research into systematic and pre-planned response to risk incidents. Our time-based risk management framework focuses on *time* and *processes* as regards the companys response to disruptions when they occur rather than on their cost, likelihood or impact.

5.1 Introduction

Many companies are expanding their supply chains to more external partners in different countries as a way to reduce cost and product development cycle and to explore new markets. For example, Boeing increased the outsourced content from 50% to 70% when it was developing the 787 model, spreading its supplier base across 20 countries. Also, according to an industry study conducted by AMR in 2006, over 42% of the companies manage more than 5 different global supply chains for different products in different markets. As supply chains become more complex, companies find their supply chains more vulnerable to unforeseen disruptions—rare but severe events that disrupt the flow of goods and information in a supply chain. Without a disruption management system put in place, these disruptions can have huge impact (in terms of cost and recovery time) to the company (and its customers).

5.1.1 Examples of Disruptions and Their Impacts

Notable examples of disruptions and their impacts include:

M.S. Sodhi, C.S. Tang, *Managing Supply Chain Risk*, 65
International Series in Operations Research & Management Science 172,
DOI 10.1007/978-1-4614-3238-8_5, © Springer Science+Business Media, LLC 2012

1. Ericsson lost 400 million euros in the quarter following a minor fire at their supplier's semiconductor plant in 2000 (Chapter 4). In addition, due to a design flaw of the Pentium microprocessors, the recall of 5.3 million chips has cost Intel $500 million in 1994.
2. New Orleans did not fully recover even six years after the landfall of Hurricane Katrina in 2005.
3. Based on an analysis of 827 disruption announcements made over a 10-year period, Hendricks and Singhal (2005a) found that companies suffering from supply chain disruptions experienced 33-40% lower stock returns relative to their industry benchmarks over a 3-year time period that starts one year before and ends two years after the disruption announcement date.
4. Over 100 patients have died in 2008 as a result of blood thinning drug Heparin contaminated with unsafe substance (Pyke and Tang (2008)).

Other examples of significant supply chain disruptions include: Mattel's recall of over 18 millions of toys in 2007 (Casey and Pasztor, 2007), Dell's recall of 4 million laptop computer batteries made by Sony in 2006. Land Rover had to lay off 1400 workers after their supplier became insolvent in 2001 as production could not continue without parts. Dole suffered a large revenue decline after their banana plantations were destroyed after Hurricane Mitch hit South America in 1998. And, after 9/11 attacks in 2001, Ford had to close five plants for several days owing to the suspension of all air traffic.[1]

5.1.2 Dealing with Disruptions

Supply-chain disruptions are getting CEOs' attention these days because of their short term effects (negative publicity, low consumer confidence, market share loss, etc.) and long-term effects (stock prices and equity risk). Despite these potentially detrimental effects, not many firms may be willing to invest in initiatives to decrease disruption risk. According to a study conducted by Computer Sciences Corporation in 2004, 60% of the firms reported that their supply chains are vulnerable to disruption (Poirier and Quinn, 2003). Another survey conducted by CFO Research Services concluded that 38% of 247 companies acknowledged that they have too much unmanaged supply chain risk (c.f., Eskew (2004)). While the exact reasons are not known, Rice and Caniato (2003) and Zsidisin et al. (2000) conjecture two key reasons: (1) firms are not familiar with how to manage supply chain risk; and that (2) firms find it difficult to justify investment in risk reduction programs or contingency plans.

To garner support for implementing certain risk reduction programs *without* exact cost/benefit analyses of certain risk reduction programs, effective risk reduction programs must provide strategic value *and* reduce supply chain risks at the same

---

[1] See, for example, Christopher (2004), Martha and Subbakrishna (2002), and Chopra and Sodhi (2004) for more details.

time. Therefore, the biggest challenge is to determine ways to mitigate supply chain risks and increase profits simultaneously so that companies can achieve a higher level of efficiency while reducing risk—this is also our aim with time-based risk management.

One stream of the risk-mitigation literature focuses on preventing rare risk events. For instance, different initiatives developed by Homeland Security (e.g., smart containers, Customs -Trade Partnership against Terrorism) would prevent terrorist attacks at various ports in the United States. However, these prevention initiatives may not be economical for preventing rare disruptions that could occur anywhere in a complex supply chain (Lee and Wolfe, 2003). When a company cannot prevent a risk incident from occurring, it has to figure out ways to respond quickly so as to contain the damage. The focus is really on time and therefore we call it time-based risk management.

5.2 Time-Based Risk Management

Rare events that can cause huge disruption to the supply chain are costly to prevent and companies may be reluctant to invest in prevention as the returns are unclear. However, companies may be more willing to develop ways to respond to risk incidents more effectively *after* their occurrence by containing their impact through "quick response." Our aim in this chapter is to motivate improved practice and research into systematic and pre-planned response to risk incidents. We provide a simple framework to think about response and motivate further research and improved practice through a variety of examples both within and outside supply chain management.

Our proposed framework extends *business continuity* efforts in practice from a local context to a supply chain wide one. Akin to various time-based initiatives such as *time-based competition* (c.f., Blackburn (1990) and Stalk and Hout (1990)), our "time-based risk management" concept focuses on **time** and response processes instead of **cost**, **probabilities** or **impact**. Specifically, we break up a "response" into three time elements—time to *detect* the event across the supply chain ($D1$), time to *design* a response ($D2$), and time to *deploy* the response ($D3$)—that we refer to as the **3-D framework**. By focusing its efforts on ensuring that systems and processes are in place to reduce these three time elements, a company reduces the overall *response* time ($R1$) and thus *recovery* time ($R2$) and total impact. We illustrate this concept through examples from a variety of contexts. Specifically, we shall argue that companies can reduce the impact of supply chain risk incidents by shortening these three elements of time and hence the response time. Increasing responsiveness can help in general and may help increase market competitiveness for the company.

Our time-based risk management framework is intended to help with planning and setting up procedures and protocols *before* a risk event occurs: detection systems and procedures to reduce $D1$, pre-packaged designs to reduce $D2$, and identified communication channels for deployment to reduce $D3$. Just as 80% of the total cost of a product is determined during the product design phase, the activities that take place for designing response can have a significant effect on the overall impact of a disruption.

Our contribution is to highlight a potentially rich area of empirical and modelling research. This should add to the literature that has focused on prevention of delays and disruptions through various means rather than on planning for post-incident recovery. Managerial implications are that time-based risk management dovetails into the company's business continuity efforts and provides a basis for risk reporting for its lenders and investors. With time-based risk management, investment in risk management is low while still increasing competitiveness due to improved responsiveness through more awareness of supply chain processes as well as more communication across the supply chain. This chapter is limited to presenting the concept, although we provide avenues for research and the basis for improved practice.

5.2.1 The 3-D Framework: Detect, Design, and Deploy

Once the supply chain partners have identified and assessed certain types of supply chain risks and developed certain risk mitigation plans based on either the alignment or the agility strategies, it remains to develop adaptive capabilities for quick response to risk incidents; i.e., reduce response lead time comprising of the time to *detect* a risk incident ($D1$), the time to *design* one or more solutions as well as *selecting* one solution in response to the incident ($D2$), and the time to *deploy* the solution ($D3$). Companies can reduce the impact of supply chain risk incidents by shortening these three elements of time and hence the response time ($R1 = D1 + D2 + D3$). After deployment, the time it takes to restore the supply chain operations is the recovery time.

The three "D" components of time can be illustrated by using the failed relief efforts associated with Hurricane Katrina. Despite live TV coverage of Katrina's aftermath in late August of 2005, it took 3 days for FEMA director Michael Brown to learn of the 3000 stranded evacuees at New Orleans' Convention Center. In our terminology, the detection lead time $D1 = 3$ days. According to the Reynolds (2005), coordination between FEMA and local authorities was poor: it took days to sort out who should do what, when and how. For example, it took two days for Louisiana Governor Blanco to decide on the use of school buses to remove the stranded evacuees. In our context, the design time $D2$ is therefore two days. However, as seen on live TV, most school buses were stranded in the flooded parking lots. FEMA requested over 1000 buses to help out but only a dozen or so arrived the day after; hence, the deploy lead time $D3$ was quite long. As a result, New Orleans had not fully recovered even after six years and displaced people continue to live in temporary accommodation so the recovery time $R2$ exceeds six years. The Katrina fiasco suggests that one can reduce the impact—number of deaths, costs, and recovery time $R2$—associated with a disruption by reducing the response lead time $R1 = D1 + D2 + D3$.

There are various ways to shorten the following time elements:

1. *Detection time* (*D*1) can be reduced by developing mechanisms to discover a risk incident quickly when it occurs or even to predict it before it occurs. Companies must also identify ways to share the information with their supply-chain partners. Monitoring and advance warning systems can enable firms to reduce the detection lead time. For instance, many firms have various IT systems for monitoring the material flows (delivery and sales), information flows (demand forecasts, production schedule, inventory level, quality) along the supply chain on a regular basis. For example, Nike has a "virtual radar screen" for monitoring its supply chain operations[2]; Nokia monitors the delivery schedule of suppliers; and Seven-Eleven Japan monitors the production/delivery schedule from their vendors (suppliers) as well as the point of sales data from different stores throughout the day (Lee and Whang 2006). Such monitoring systems typically use various types of control charts to monitor the operations and to issue an alert should any anomaly occur. Advance warning systems, by contrast, are intended to detect an undesirable event before it actually occurs thus reducing detection time to a minimum. For example, smart alert systems enable GE to conduct remote-sensing and diagnostics so it can deploy engineers to service turbines before catastrophic failures occur.
2. *Design time* (*D*2) can be reduced if the company and its partners can develop contingent recovery plans for different types of disruptions *in advance*. Many organizations seek to do so through *business continuity* efforts. For example, Li and Fung has a variety of different contingent supply plans to enable them to switch from one supplier in one country to another supplier in a different country (St. George 1998). Seven-Eleven Japan has developed contingency delivery plans to enable them to switch from one transportation mode (trucks) to another (motorcycles), depending on traffic conditions (Lee and Whang 2006). A company may find it useful to use decision analysis tools such as decision-tree analysis to refine and select a recovery plan. Sarin (2001) presents a decision-analysis framework for evaluating different earthquake-safety plans.
3. *Deployment lead time* (*D*3) can be shortened by improving communication and coordination among supply chain entities within the company or within supply chain partner firms, once a recovery plan has been selected. Van Wassenhove (2006) suggested three forms of coordination: (1) by command (central coordination), (2) by consensus (information sharing), and (3) by default (routine communication). In the event of a major disruption, coordination by command seems to be more appropriate during the design phase of a recovery plan and its deployment. Once the recovery plan is deployed, coordination by consensus would seem appropriate especially when each party has a clear role and responsibility established in advance.

Communication, an important aspect of crisis management, is also a time-based response approach motivated by public perception. Companies need to develop crisis management plans to manage the public perception so as to reduce the long term

---

[2] Hartwigsen, 2005.

impact on the company's reputation. Immediately after a risk incident, the involved supply chain parties need to develop plans to restore customer relationship and communicate their approach for rebuilding public confidence. For example, soon after issuing multiple recalls in 2007, Mattel created simple ways for consumers to return the recalled products using downloadable forms, and free shipping mailers. To restore trust, the company acknowledged problems to stakeholders and apologized to customers publicly in the press. In exchange for returned toys, it offered coupons for customers to buy other Mattel products of the same value. It also announced the three-stage safety check system in September 2007: test every batch of inputs (paint); test every batch of production; and test samples of finished goods from each production run. In addition to external communication, internal communication is important to better manage supply chain risk in the future (c.f., Zsidisin et al., 2005). At Mattel, a new position was created reporting directly to the Mattel's CEO to ensure that potential recall risks were quickly addressed, and, if necessary, elevated within the company for improved internal communication (c.f., Pyke and Tang 2008).

5.3 Response Time and Impact

To understand the importance of shortening the response time $R1$, consider the following examples:

Eradicating med flies in California in 1980:   Despite the initial med fly eradication efforts in the mid-70s, med flies were detected in California again in the early part of 1980. Instead of calling for aerial spray of Malathion in a small area (30 square miles) that is proven to be effective but costly, Governor Brown approved the release of sterile male flies and traps. Unfortunately, these methods were not effective and the area of infestation had expanded more than 20-fold within 1 year—from 30 square miles in June of 1980 to 620 square miles in July of 1981. As Japan and other countries imposed import restrictions, Governor Brown was under political pressure to approve the aerial spray over an area of 1500 square miles starting July 14, 1981. This delayed action came at a significant cost: an expenditure of over $100 million and Governor Brown's political career (Dawson et al., 1998; Denardo, 2002).

Ground shipping after September 11:   Immediately after September 11, Chrysler was the first to request their logistics providers to switch the mode of transportation from air to ground. Speedy deployment of this strategy has enabled Chrysler to get the parts from their supplier such as TRW via ground transportation. Ford was unable to deploy the same strategy because, by the time Ford decided to switch to ground shipping, all ground transportation capacity has been taken up. Facing part delivery problems, Ford had to shut down 5 of the U.S. plants for weeks and reduce its production volume by 13% in the fourth quarter of 2001 (c.f., Hicks, 2002).

Recovering after supply disruption: Both Ericsson and Nokia were facing supply shortage of a critical cellular phone component (radio frequency chips) after their key supplier, Philip's Electronics semiconductor plant in New Mexico,
caught on fire in March of 2000. Nokia recovered quickly by doing the following. First, Nokia immediately sent an executive team to visit Philip's in New
Mexico so as to assess the situation. Second, Nokia reconfigured the design of
their basic phones so that the modified phones can accept slightly different chips
from other Philip's plants; and third, Nokia requested Philip's to produce these
alternative chips immediately at other locations. Consequently, Nokia satisfied
customer demand smoothly and obtained a stronger market position mainly due
to their speedy deployment of their recovery plan. On the contrary, Ericsson was
unable to deploy a similar strategy later because all Philip's production capacity
at other plants has been taken up by Nokia and other existing customers. Facing
with supply delay, Ericsson lost $400 million in sales (Hopkins, 2005).

The above examples suggest that the recovery lead time $R2$ is by and large increasing in the response lead time $R1$. This is mainly because the impact of the event
can escalate exponentially before an effective response is made.

5.3.1 Modeling Disruption Impact over Time

To study the impact of a risk incident over time, we could look into the epidemiology
and the forest fire literature. Specifically, the total impact of a natural disruption
(an epidemic or a forest fire) tends to increase super-linearly or even exponentially
with time initially and then to taper off. Thus, as shown in Fig. 5.1, shortening the
response time $R1$ ($= D1 + D2 + D3$) can reduce the total recovery time $R2$ and
hence the total eventual impact of the disruption.

Epidemic models: The simplest form of epidemic model is the *exponential
model* that can be explained as follows. Let $I(t)$ be the number of people infected
at time t. In this case, the rate of infection can be defined by the differential
equation: $dI(t)/dt = kI(t)$, where the parameter $k > 0$. This differential equation
yields $I(t) = I_0 e^{kt}$, where I_0 is the number of people infected at time 0. Therefore,
the number of people infected grows exponentially overtime. By contrast, the
logistic model is another simple model that stipulates that the infection rate
depends on the number of people infected and the number of people who is susceptible to the infection. The number of people infected $I(t)$ grows exponentially
initially and plateaus later on (Mollison, 2003).

Fire impact models: There are many different types of fire models based on a
system of differential equations for estimating the burned areas over time—see,
for example, Richards (1995) and Janssens (2000) for various fire spread models.
The simplest model is the elliptical fire spread model presented in Arora and
Boer (2005). Essentially, they assume that the burned area takes on the form of
an ellipse with the point of ignition at one of the foci. By assuming that the fire

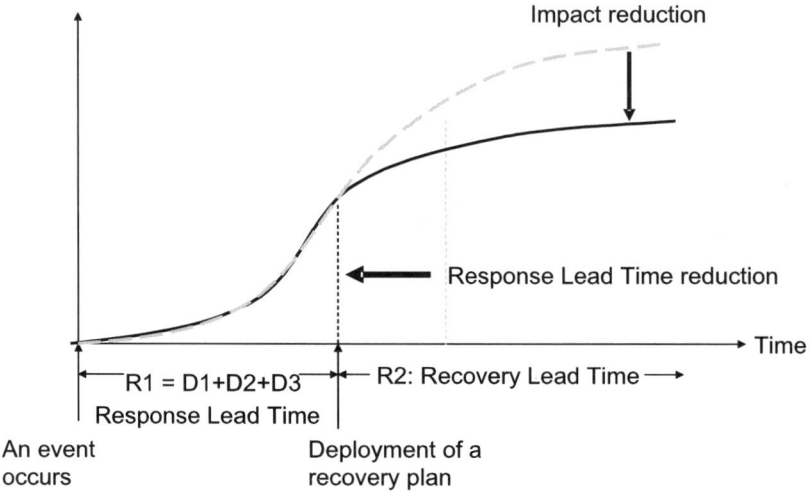

Fig. 5.1 The effect of reducing the response lead time $R1$

spreads linearly over time along a 2-dimensional space, they show that the total area burned (size of the ellipse) grows as a squared function of time elapsed since the start of the fire. Thus, the total area burned grows exponentially over time (but slows down as the area of the remaining forest decreases).

In addition to these two models, there are hazards analysis reports highlighting that the magnitude of the problems associated with many hazards (fire, terrorism, earthquake, etc.) tend to grow super-linearly or exponentially over time (Anderson, 2002).

While we are not aware of any scientific study of modeling impact over time in the context of supply chain risk, it is plausible that a pattern similar to the logistic model will emerge (Fig. 5.1). For instance, a week's supply delay would have caused little damage to Ericsson but several weeks of supply delay resulted in $400 million loss in sales in the first quarter and $2 billion eventually in the first year after the dust had settled.

5.4 Time-Based Risk Management in Practice

We now present five time-based disruption management activities that would enable a company to reduce the response time $R1 = D1 + D2 + D3$ (and consequently the recovery time $R2$ and the total impact):

1. **Work with suppliers and customers to map risks.** Many companies already identify potential disruptions according to its impact and likelihood as part of business continuity efforts. They can further that effort in two ways by tracing

the impact of each disruption along the supply chain from upstream partners and to downstream customers. Doing so requires discussion among supply-chain partners to creates shared awareness of different types of disruptions and their impacts on different parties. This generates support for collaborative efforts for mitigating risks for all parties.

2. **Define roles and responsibilities.** Companies should work with all key supply chain partners to define the roles and responsibilities to improve communication and coordination when responding to a disruption. To coordinate response efforts effectively, Van Wassenhove (2006) suggested three forms of coordination: coordination by command (central coordination), coordination by consensus (information sharing), and coordination by default (routine communication). The coordination mechanism, agreed among the parties *before* a risk event has taken place, can then be used without further discussion for design and for deployment of solutions *after* the risk event has occurred. Coordination by command seems to be appropriate during the design and deployment phases (i.e., during D2 and D3). This is because, during these two phases, an identified group within the firm or comprising partners' representatives as well, needs to take central command for collecting and analyzing information to design a recovery plan, and then for disseminating information regarding the deployment of the selected recovery plan. However, to get to this point of agreed upon procedures, coordination by consensus is more effective in agreeing on detection mechanisms and in designing pre-packed solutions that anticipate disruptions.

3. **Develop monitoring/advance warning systems for detection.** Companies need to develop mechanisms to discover a disruption quickly when it occurs and/or to predict a disruption before it occurs. They must also identify ways to share the information with their supply-chain partners and to get similar information from them. Monitoring and advance warning systems can enable firms to reduce detection time. For instance, many firms have IT systems for monitoring the material flows (delivery and sales), information flows (demand forecasts, production schedule, inventory level, quality) along the supply chain on a regular basis. For example, the monitoring systems at Nike, Nokia and Seven Eleven Japan mentioned earlier use various types of control charts to monitor the operations and to issue an alert should any anomaly occur. Hence, these monitoring systems would reduce the detection time $D1$. Besides monitoring systems, advance warning systems are intended to detect an undesirable event before it actually occurs. For example, Allmendinger and Lombreglia (2005) described how different smart alert systems enabled GE to conduct remote sensing and diagnostics so that it can deploy engineers to service their turbines before failures occur.

4. **Design recovery plans.** Develop contingent recovery plans for different types of disruptions. Establishing contingent recovery plans for different types of disruptions *in advance* would certainly reduce the design time $D2$. Many firms have developed various recovery plans (or contingency plans) in advance. For example, Li and Fung has different contingent supply plans that will enable them to switch from one supplier in one country to another supplier in a different country (St. George (1998)). Also, Seven-Eleven Japan has developed different contin-

gent delivery plans that will allow them to switch from one transportation mode (trucks) to a different transportation mode (motorcycles), depending on the traffic condition (Lee and Whang (2006)).

5. **Develop scenario plans and conduct stress tests.** Companies need to create different scenarios and rehearse different simulation runs/drills based on different scenarios with all key supply chain partners (Chapter 4). Because the deployment time $D3$ accounts for the preparation time to launch the selected recovery plan, scenario planning and stress tests are effective mechanisms for reducing $D3$. For example, having rehearsed response and recovery plans at each P&G site under different scenarios previously, P&G restored the operations of its coffee plant in New Orleans by only mid-October 2005 after the Katrina's landfall in late August 2005. P&G attributed their quick recovery ($R1 + R2 = 2$ months) to its readiness (Cottrill, 2006).

5.5 Risk and Reward Considerations

Besides reducing disruption risks, time-based disruption management can increase a firm's competitiveness as well. As we mentioned earlier, effective risk reduction programs must provide strategic value and reduce supply chain risks at the same time (Tang, 2006a). In other words, we should look for ways to mitigate supply chain risks and increase profits simultaneously so that companies can achieve a higher level of efficiency by reducing risk while increasing reward (Chopra and Sodhi, 2004). Time-based risk management may help achieve this (Fig. 5.2).

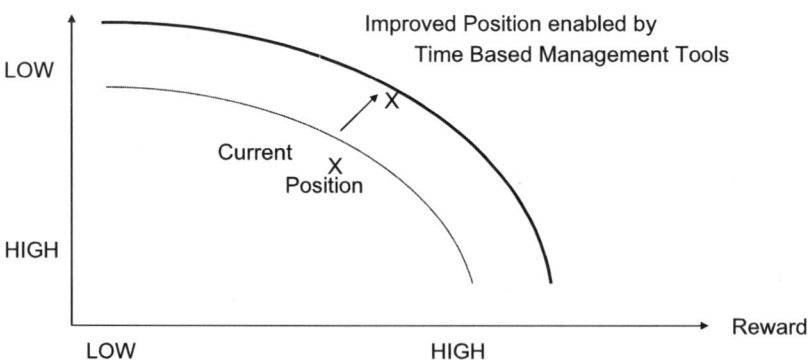

Fig. 5.2 Time-based risk management enables a firm to shift its position to a higher efficiency risk-reward curve (refer to Chapter 4)

We can use Spanish apparel maker Zara's success to illustrate how time-based disruption management can enable firms to increase their competitiveness by responding to dynamic changes quickly (as articulated in the time-based competition literature). Despite the fact that fashion retailers are susceptible to many supply chain risks such as uncertain supply, transportation delays, shrinkage and theft, uncertain demand, Zara continues to be the most profitable European fashion retailer with sales and net incoming growing at an annual rate of over 20%. Ferdows et al. (2004) attributed Zara's success to its "rapid fire fulfilment" strategy that enables Zara to increase its competitiveness while reducing risks (Fig. 5.2). Specifically, Zara's claim to fame is time reduction: Zara is capable of design, manufacture and ship a new line of clothing within 2 weeks, while most traditional fashion retailer will take more than 24 weeks. To reduce various measures of time ($D1$, $D2$, $D3$ and $R2$), Zara performs some of those activities we described in Section 5.4.

As reported in Ghemawat and Nueno (2003), Zara operates as a vertically integrated supply chain by co-locating their designers and the factory in Spain, by managing their own warehouses and distribution centers in Spain, and by running their own stores worldwide. Not only does this integrated supply chain provide Zara supply chain visibility, it enables the company to facilitate close communication and coordination with all supply chain partners including their suppliers. By receiving point of sales data from their own stores on a regular basis, Zara has a well established process for analyzing the sales data to detect sudden changes in demand and/or fashion trends. As such, Zara's detect lead time $D1$ is short. Next, by managing centrally and by working closely with all partners, Zara can analyze the situation and prescribe a response should the market change suddenly. Also, by co-locating the designers and the factory, Zara has the capability to deploy different recovery plans by designing and manufacturing new designed clothes very quickly should the sales of existing designs are below expectations. Hence, the design time $D2$ is short.

Next, the close proximity of suppliers, designers, factories, and distribution centres enable Zara to communicate, coordinate, and deploy the selected recovery plan quickly with all supply chain partners (from suppliers to the stores). Hence, the deploy lead time $D3$ is short. In addition, Zara also implements some of the recovery plans to reduce the deploy time $D3$ and recovery time $R2$. For example, Zara engages in "flexible supply contracts" with "multiple suppliers" so that Zara can adjust the order quantity quickly should a demand disruption occur.

Not only do these mechanisms help Zara to reduce $D1$, $D2$, $D3$ and consequently $R2$, they enable Zara to increase its competitiveness and profit by: (a) generating more accurate demand forecasts in a timely manner; (b) designing, producing, and distributing newly designed products in small batches quickly; and (c) reducing the costs associated with price markdowns (due to over-stocking) and lost sales (due to under-stocking). Relative to its competitors, time-based management has enabled Zara to achieve higher profitable growth and lower supply chain disruption risk simultaneously.

5.6 Conclusion

We have used diverse examples and natural disruption models (epidemic model and fire model) and anecdotal evidence to argue that firms can use time-based management to reduce the response lead time and recovery lead time, which will in turn reduce the impact of a disruption. We suggested five activities that would enable a firm to reduce the response lead time $R1$. Finally, we suggested using Zara's example that we may be able to achieve both risk reduction and extra rewards through increased responsiveness enabled by time-based risk management.

However, we have presented a concept that requires validation in practice and further research. For instance, we presented impact models from the epidemiology literature—we need to develop similar models for supply chains. Likewise, we need empirical research to study how and to what extent companies are extending business continuity efforts to respond to risk events. We believe time-based risk management to be a rich area for modeling and for empirical research. For instance, we can propose two hypotheses for further research: (1) the recovery lead time $R2$ and cost can be significantly higher if the deployment of a recovery plan is delayed; and (2) the execution of a recovery plan can become much more difficult if its deployment is delayed.

Chapter 6
How do Global Manufacturing Companies Mitigate Supply Chain Risks

Abstract In this chapter, we describe risk management at Samsung Electronics for managing different categories of supply chain risks to illustrate how a global manufacturing company manages supply chain risks. Using the concepts presented in earlier chapters for identifying and mitigating supply chain, this chapter takes steps towards empirically understanding *how global manufacturing companies manage supply-chain risk.*

6.1 Introduction

There are many cases of catastrophic losses triggered by some problems inside the company's extended supply chain (c.f., the industry examples described by Sheffi, 2005 a and b). As such, there is a growing awareness that the relationships between supply chain entities are a source of risk. These relationships may also be getting more complex over time owing to due to growing product/service complexity, outsourcing, globalization, and e-business (Harland, Brenchley and Walker 2003).

Have companies incorporated the complex relationships between the different supply chain entities in their formal risk management processes (i.e., identification, assessment, and mitigation)? The extant research literature may not be much of a guide as businesses have grown to view risk much more broadly than they did earlier (Hunt 2000). Empirical work on this topic is in rather short supply (Harland et al., 2003). Of the 90 articles on supply chain risk studied by Paulsson (2004), only eight dealt with supply chains of any complexity, whether empirically or analytically. All the articles in the special issue on risk management (Seshadri and Subrahmanyam, 2005) dealt with the entire supply chain as a single entity or, at most, as a supplier-manufacturer pair.

This chapter presents our initial steps towards a descriptive understanding towards *how global manufacturing companies manage risk*. Risks include those that

M.S. Sodhi, C.S. Tang, *Managing Supply Chain Risk*,   77
International Series in Operations Research & Management Science 172,
DOI 10.1007/978-1-4614-3238-8_6, © Springer Science+Business Media, LLC 2012

stem directly from the structure of their supply chains and from the relationships between the supply chain entities.[1]

To explore how a global manufacturing company manages supply chain risks, consider Samsung Electronics—a vertically integrated global manufacturing company with a complex supply chain in the electronics sector that faces a multitude of risks (Sodhi and Lee 2007). We expand the 2×2 risk-management framework presented in Chapter 2 that deals with risk drivers originating locally or globally and their corresponding consequences being local or global as well. We use this to categorize Samsung Electronics' risk management efforts to build our understanding of how they manage risk.

6.2 Literature Review

Our exploration is based on a risk framework for studying risk management practices in companies. A "theory" in this sense is "a system of constructs and variables" in which constructs are related to each other by propositions, and the variables by hypotheses (Baccarach 1989). Van de Ven (1989) quotes K. Lewin in saying "nothing is so practical as a good theory."

Towards building theory, case study research is well established in the management literature especially for new research areas that require exploration (Eisenhardt 1989). Although usually associated with the strategy literature, case study research has been advocated in the operations management literature (Voss, Tsikriktsis and Frohlich 2002) and the supply chain literature (e.g., Seuring 2005). Essentially, case studies can be used for exploration, theory building, theory testing and theory extension/refinement (Voss et al., 2002). Certainly, case studies can further "engaged scholarship", a "collaborative form of enquiry in which academics and practitioners leverage their different perspectives and competencies to co-produce knowledge about a complex problem. . . " (Ven de Ven and Johnson 2006).

Companies usually manage risks from a financial viewpoint or from an operational viewpoint. The finance literature deals with stochastic processes and mathematical models to make the appropriately hedged decisions, as part of what is called "integrated risk management" or "enterprise risk management"; see for example the textbooks by Doherty (2000) and by Bell and Schleifer (1995). Wang, Barney and Reuer (2003) discuss why risk management is good for companies from an investment and financial market perspective.

From an operational perspective, risk management may involve strategic or tactical decision-making. Sodhi (2003) proposes scenario planning to make strategic decisions pertaining to capacity-, demand-, and environmental regulation-related risks using a polyvinylchloride manufacturer as an example. However, when dealing with a tactical supply chain planning problem with demand uncertainty motivated by a

---

[1] We use the word "structure" rather than "design" because the latter term suggests a conscious effort, whereas in reality supply chains evolve and only a part of the supply chain may be under the company's control anyway.

Japanese consumer electronics company, Sodhi (2005) uses stochastic programming and proposes risk metrics for use at the corporate level. Chopra and Sodhi (2004) present a typology of risks that Sodhi and Lee (2007) apply to Samsung Electronics and some other electronics companies. Some types of risks manifest themselves frequently with small per incident losses but can still present significant accumulated losses. Such risks are usually the subject of the operations management literature on inventory (see for example the textbook by Zipkin 2000). There are other views cutting across the strategic-tactical divide: Johnson (2001) and some others find that risk lies in making matching supply to demand, illustrating this with a study of the toy industry.

We referred at the outset to the paucity of papers on supply chain risk explicitly dealing with the structure of supply chains and the dependencies among supply-chain entities but there are a few articles. Bovet (2005) advocates taking a strategic view in designing the supply chain to mitigate risks but also lists tactical steps companies should take to prevent, control, or transfer losses. Harland et al. (2003) present a framework on how to manage risks in complex supply chains. There are also papers with Monte Carlo simulation models for supply chains to understand inherent risk. For instance, Deleris, Elkins and Paté-Cornell (2004) analyze a supply chain connecting risk modeling from an insurance perspective with supply chain network design. Choi, Dooley and Runtusanatham (2001) take an interesting view of the supply chain as a "complex adaptive system" and therefore explicitly consider supply-chain entities but they do not discuss risk. A particular category of risks, bullwhip effect has been studied quite well taking into account the dependencies of supply-chain entities although this has been done only for simple linear supply chains, e.g., by Lee et al. (1997) and by Sodhi and Tang (2011).

6.3 Motivation

Supply chains are becoming increasing global due to opening of global markets, the search for cheaper suppliers, outsourcing, and the search for skilled labor. Companies are also increasingly at risk to global events like war, terrorism and the major currency fluctuations like the decline of the dollar against major world currencies over the period 2003–2007. Indeed, there are many supply chain related global catastrophes documented for manufacturing companies (Sheffi 2005 a and b).

Recognizing the risks involved, companies are motivated to manage risks. Moreover, insurance is becoming expensive resulting in companies taking on the risk themselves. Insurance companies themselves do not believe they have the right tools to understand risks tied to companies' operations. The Sarbanes-Oxley Act holds the CEO and CFO for financial reporting and audit committees for the responsible for understanding how the company identifies, assesses, and manages risks. Risk management is good for the bottom line and the stock markets rewards companies with good risk management practice (e.g., Wang et al., 2003) just as it punishes companies for poor risk management (Hendricks and Singhal, 2005a).

Global manufacturing companies certainly manage risk centrally at the corporate level as integrated or enterprise risk management. At the same time, they also manage risk at the local level (i.e., at the plant or region level) operationally using inventory and supplier management where the impact is generally local. They use better forecasting tools to reduce demand uncertainty to reduce the risk of supply not matching demand. Despite these efforts, companies could still be vulnerable because there may not be any formal organization or processes to manage such risks that bridge operations at the local level with finance at the corporate level.

More importantly, global manufacturing companies may still be vulnerable to another category of risks that have their origins at the local level but can affect the entire supply chain, resulting in catastrophic losses for the company and for its supply chain partners. Such losses stem from the dynamic dependencies between supply chain entities including those over which the company has no control (e.g., a supplier's supplier having no capacity left because of a surge in demand by one of its other customers). Such risks stem from the complexity created by the network of suppliers, the company's own plants and warehouses, and those of the customers—and possibly supporting technology and inter-company processes. It is possible that the structure is behind extremely unlikely events but with disruptive and sizable impacts when they do happen. Furthermore, many companies may not have formal processes or groups to manage such risks even though some clearly do. In the Ericsson example, competitor Nokia that also sourced from the same plant sensed trouble early on and was able to find alternatives without disruption to its supply chain (Chapter 4).

This has motivated our research into *how global manufacturing companies manage risk* including the processes and the organization with various roles and responsibilities to manage risk. This would set the stage for understanding the extent to which supply-chain network thinking is part of risk management and how global manufacturing companies *should* manage risk in terms of processes, analytical models, and the organization at the corporate, region, and plant levels.

6.4 Methodology

Our research question is *how global manufacturing companies manage risk*. Eventually the answers would be the basis for prescriptive questions about how companies should manage risk but our goal is descriptive in this chapter. Given that our research is in an exploratory stage, we chose to do a case study to understand processes, organization and partnerships. We studied the electronics industry because it has large supply-chain risk exposure, short product life cycles, and has global companies such as Samsung and Sony. Moreover, the industry has had well-publicized failures that had huge impact on market valuation such as delayed launches of the Xbox and more recently the PS3 (especially for the European market). We chose global electronics giant Samsung Electronics along with its wholly owned subsidiary, Samsung Electronics UK, because of Samsung's vertical integration (one of the businesses within

Samsung Electronics is a chips supplier to the other businesses within Samsung Electronics).

Our framework is based on the view that companies face risk drivers that are: local in the sense that they originate in a specific supply-chain entity; or global in the sense they are not specific to any entity. These risks can have consequences (eventual impact) that may also be local or global, thus giving rise to four types of risks—(I) operational risk, (II) localization risk, (III) enterprise risk, and (IV) network risk (Table 6.1)—all of which require to different risk management practices (recall Chapter 2).

Table 6.1 A framework for viewing supply-chain risk and risk management practices in a global manufacturing company with four types of risks corresponding to quadrants I–IV

Risk consequence (Region of possible eventual impact)	**Global** (impact to entire supply chain or to most or all markets)	IV. *Network risk*: Risks originating in a plant or region whose consequences eventually impact the entire supply chain or markets *View*: A connected network of supply chain entities (suppliers' suppliers onwards) and markets (including customers' customers)	III. *Enterprise risk*: Risks to the company as a whole, part of finance *View*: Company as a single entity
	Local (impact to particular supply chain entity or to particular market)	I. *Operational risk*: Local risks stemming from supply, demand, or failure match supply and demand *View*: Specific plant or plants or a specific region or market as a single entity	II. *Localization risk*: Unintended or unavoidable consequences of corporate policies or decisions to specific markets or regions *View*: Corporate center with regions that report to it
		Local (originating at a supply-chain entity or in a particular market)	**Global** (originating supply-chain-wide or at the corporate level)
		Risk Driver (Point of possible occurrence)	

Managing these risk drivers requires a different view of the company: The first requires focusing on a specific plant or plants or a specific geographic region or market as a single entity. The second requires a view of the company as having a corporate center that has regions/markets (possibly run by subsidiaries) reporting to it. The third requires a view of the company as a single entity, which fits in with the role of the corporate center. Finally, the one of most interest to us is the view of the risk management in the context of a network of supply chain entities (suppliers' suppliers onwards) and markets (including customers' customers).

We can understand risk management in these terms so these four categories are a way to collect and analyze information in the case study.

For this initial study, we mainly relied on interviews over the phone of managers in Korea and in the UK conducted by us and/or by a senior manager from Samsung. We describe this in the next section.

6.5 Samsung Electronics

Samsung Electronics is part of the Samsung Group and was ranked 22nd in the Fortune Global 500 list in 2011—it was 46th in the same list in 2006. The company consists of five main business units: Digital Appliance Business, Digital Media Business, LCD Business, Semiconductor Business and Telecommunication Network Business. The Semiconductor Business supplies the other four businesses as well as non-Samsung customers. For 2011, Samsung Electronics had consolidated sales of $138 billion (up from $67.6 billion in 2007) with a net income of $13.669 billion.

Samsung Electronics employed 187,800 people in 2010 across the globe. Not including facilities in Korea, it has 24 manufacturing complexes, 40 distribution bases and 15 branches spread over all continents (except Antarctica). Countries with manufacturing facilities include the US, Malaysia, China, India, and Hungary. Globally, it was ranked 17th in the best 100 global brands in the Interbrand-Business Week list for 2011, with a brand value estimated at more than US$ 23.4 billion, the third highest among all companies in the broad consumer electronics sector after Apple and Hewlett Packard.

I. Operational Risk

Samsung plants and regions face have many risks when it comes to ensuring supply, having demand, and matching supply to demand. The impact of any risk is also plant-or region-specific. The risks are managed at the plant or region level.

Faced with high demand uncertainty pertaining to a large number of products in all of its five businesses, Samsung uses excess capacity, flexible capacity *and* responsive capacity. Having excess capacity can help deal with demand surges (Fig. 6.1). Excess capacity, particularly combined with flexibility as in Samsung's case, can also lower the amount of finished goods inventory required as production can be delayed until the last minute to fulfill firm orders.

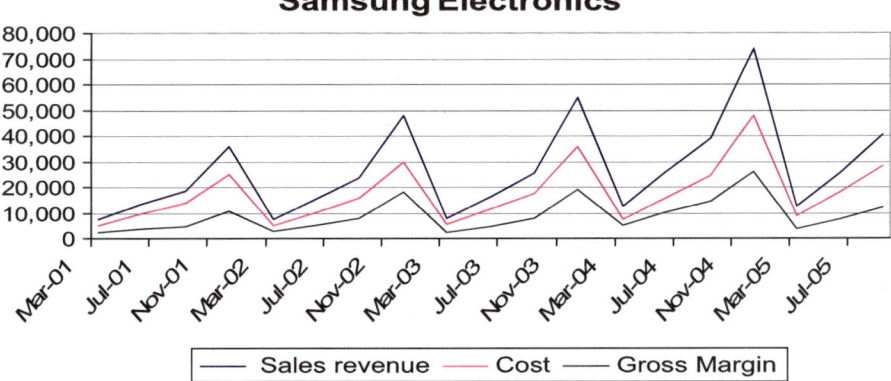

Fig. 6.1 Quarterly numbers for revenues, costs, and gross margin (US$ MM) for Samsung Electronics 2001–2005 illustrating quarter-to-quarter fluctuations. (Source: Datastream)

Samsung uses multiple platforms in its own plants to allow switching from one product to another quickly depending on demand. multi-product platforms for assembly, i.e., producing different products on the same line so there is flexibility to meet demand. The system allows change from assembling one product to another at minimum cost. For example, Samsung's color TV and the computer monitor lines share the same platform to allow quick change from one to the other depending on market demand.

Postponement is yet another form of flexible capacity at manufacturing locations fulfilling orders from different countries with the difference being essentially only in power plugs. Many manuals and packaging labels are multilingual and therefore the same across many countries. This "pools" the risk of demand uncertainty from different regions at the plant level.

To make its supply chain responsive, Samsung co-locates suppliers on its plants with supplier plants are in the same manufacturing complex as the Samsung plant or at least having inventory. Even if the suppliers are not co-located, Samsung and its suppliers share the information from the integrated SAP-based system to figure out the best level of inventory of components and act as if they were one company. Such real or virtual co-location helps mitigate the risk of *supplier delays* that occur when a supplier cannot respond to changes in demand. This is not uncommon in the electronics industry as the same supplier may be supplying to competing companies who are all chasing the same surge in demand for competing products—this is particularly true for Samsung's telecommunication business.

The electronics industry has high value products, high rate of obsolescence, high uncertainty of demand and supply, and large product variety. Therefore, there is risk of having too much or too little inventory. To mitigate inventory risk, Samsung limits core-component *manufacturing* sites to a handful to serve *assembly* sites worldwide. This way the inventory of core-components is "pooled". As regards finished-goods inventory, managers responsible for manufacturing have a clear inventive not to have too much of it. If such inventory is held for 60 days without any action, the SAP system sends an alert to the manager who "owns" the inventory and automatically deducts the value of this inventory. This deduction reduces the "profit" attributed to that manager, thus creating a disincentive for holding such inventory.

Long lead times, seasonal demand, high product variety, and short product life cycles—all attributes of the consumer electronics industry—increase forecast error. To increase the visibility of demand information across the supply-chain in different markets, Samsung is updating an initial implementation of a Rosetta-Net to retailers like Dixon's in the UK to determine appropriate stock levels similar to continuous replenishment programs in the grocery industry. The system is used to monitor the "shelf occupation" ratio (shelf space for Samsung Electronics products to shelf space for all products) in retail stores manually to pick up trends. Samsung also operates the Merchandizing Control System (MCS) to determine the difference between sell-in (to the store) and sell-out (to the end consumer). It uses this information for forecasting and to minimize any information distortion.

To avoid glitches in supply or demand delivery in the UK, Samsung Electronics using a logistics provider, NYK Logistics, to take care of customs clearance, inland haulage, receipt, storage at and dispatch from a distribution center, and trade deliveries. Two members of Samsung's management team are based onsite at the distribution center, 90,000 sq ft of which are dedicated for Samsung's UK operations. NYK also supports Samsung's home delivery system from booking orders, customer liaison via an on-site call center, delivery, and product installation.

Besides risks to material and information flows, Samsung faces risks to its cash flows in the form of receivables risks. Samsung and its subsidiaries sell to distributors in many parts of the world using "hard" currency (US $, euro, or UK pound sterling) to reduce both receivables risk and exchange rate risk. More than that, it uses "information of delivery" technology to reduce receivables risk by reducing errors and by cutting down on the amount of time between delivery and invoicing. This time could be as much as 15 days if it were paper-based. Instead, a driver scans packages upon delivery as proof of receipt. The information is uploaded to the SAP system triggering the invoice.

II. Localization Risk

This is a set of risks facing any global company where the risks originate in the company at corporate headquarters and the resulting decisions (or image) have unintended or unavoidable consequences for a particular region or market. For instance, decisions made by Samsung Electronics across the globe can have implications for supply and demand in the UK.

Cultural differences, whether real or perceived, can be a risk not only to demand but also to supply. Samsung's representatives in a foreign country can misread demand, fail to form business relationships, or be too conservative or too lenient in qualifying distributors and others for credit. Friction between local and expatriate staff may lower operational efficiency. Managers may incorrectly to attribute poor performance to "culture". To mitigate this risk, Samsung operates a huge in-house training center, the Samsung Human Resources Development Center, in Korea. The center offers many multi-national training courses for staff to deal with cultural conflict. Samsung requires its senior expatriate staff to undergo compulsory specialist training to understand and overcome cultural differences. For instance, staff expatriated to the UK have to study British history.

Well-publicized investments in plant and equipment lead to loss of reputation when the company relocates manufacturing out of the country. The decision by Samsung Electronics' decision to move its manufacturing plant in Wynyard, UK to Hungary generated public anger in the North England region. When the Wynyard plant was opened in 1995, it had attracted favorable media coverage and its subsequent closure likely damaged Samsung's reputation in this region (Tighe 2004).

Besides such publicity, being perceived as "foreign" could lead to decreased demand in the UK. To counter this, Samsung Electronics UK supports British interests such as football, tennis, and equestrian shows in the UK. Similar risk mitigation is done globally as well: Samsung Electronics was a major sponsor of the 2004 Olympics in Athens. It also places products in science fiction movies such as "Matrix Reloaded".

There are decisions that are immediately pertinent for only the region in question but are important enough to be made at the corporate level because they apply or could apply to other regions. Two such decisions involve the environment and accounting.

Environmental standards in the UK and the rest of EU (before recent expansion into central Europe) are higher than those of other European countries and the US. Anticipating EU requirements on manufacturers, Samsung has proactively implemented an electronics waste collection system in cooperation with some of its competitors. The company can collect and dispose electronic goods off at the end of the life without cost to local government or consumer. Within the company, there are design projects aimed at reducing or even eliminating electronic and/or hazardous waste including lead, mercury, and cadmium in accordance with EU regulation. Samsung also has started certifying suppliers as "Eco-partners" in Korea. Samsung Europe set up a new environment team as early as 2003. Environment specialists and regulation specialists are specifically recruited for the UK and for other EU countries. These specialists work with government regulators and co-operate with other electronics manufacturers.

Compliance on accounting and in particular on transfer pricing has implications on reputation risk as consumers may not want to buy products from a company with a reputation of not having transparent transactions. Samsung uses PriceWaterhouseCoopers as auditors in the UK and uses a transparent system from SAP for financials.

III. Enterprise Risk

Risks in this category are present for Samsung Electronics as a whole and therefore the company addresses these risks at the corporate level. For instance, mergers-and-acquisitions in the electronics industry, although not very common, still pose the threat that a competitor may acquire a supplier and potentially lock out supplies or increase prices for competitors. To mitigate it, Samsung has long-term relationships with its suppliers. Core suppliers get financial and technological help and even human-resource training. There is information integration as well: Samsung's suppliers use the same intranet network and enterprise resource planning (ERP) as Samsung. As already mentioned, the supplier plants may be physically located in a Samsung Electronics manufacturing complex. As already noted, the deep relationship makes the supply chain more efficient by reducing leadtime and any errors pertaining to orders but it also reduces the risk of a competitor acquiring a key supplier.

Samsung mitigates the threat of excess capacity across the globe by making existing capacity more flexible as noted earlier. It also seeks to minimize excess capacity by serving geographically scattered customers from the same location. In Samsung's case, many core components are (mostly) single sourced from Samsung plants and from core suppliers in Korea for assembly in other parts of the world. This arrangement provides the company to take advantage of economies of scale. A further advantage is decreasing the likelihood of loss of intellectual property as this approach allows it to limit the number of suppliers of core components with whom it shares advanced technology information.

Centralizing capacity this way does have its risks. It increases the risk due to potential labor strife in Korea, although Samsung does not have any experience of labor action. Moreover, some of the neighboring countries pose a long-term challenge for this location.

A company's reputation is global. Samsung has moved up the price curve—for instance, it moved up the reputation curve in the mobile phone handset industry by having phones at the top end of the price range. It now competes with Japanese and European companies on an equal footing on quality and design. For instance, the clamshell and the slider designs for mobile phones were Samsung Electronics innovations.

The electronics industry is constantly facing competition. To reduce the chance of losing its technological edge, Samsung Electronics spends more on research as a percentage of annual revenues than most of its competitors. Samsung Electronics has 42 R&D Centres with around 26,000 employees and their total R&D budget in 2005 was nearly 5.5 percent of 2005 sales.

One consequence is that Samsung had the smallest 1 Gigabyte device—the YEPP YP-ST5Z—offering MP3 player, FM radio, voice recorder, and data storage in 2004. Samsung also announced in September 2004 an industry-first 60 nanometer 8 Gigabit memory device (used in Apple's iPod Nano). However, 2006 fourth quarter results of a decrease in profits of 8% despite an increase of 1% in revenue compared to the same quarter a year ago reflect severe margin erosion for flash memory chips as well mobile phones (Ramstad 2007). Apple's iPod dominates in the US and the UK despite Samsung's technological prowess. These challenges underscore the pressure on Samsung to innovate continually.

Indeed, in December 2006 Samsung announced it has succeeded in developing a 1Gb Mobile DDR DRAM memory chip using 80nm technology, an industry first. Because it consumes 30% less power that existing chips, such a chip would be important for advanced mobile handsets. In January 2007, Samsung launched in Korea and displayed in Las Vegas Consumer Electronics show, the first global positioning system (GPS), Bluetooth Navigator, to connect to a mobile phone using Bluetooth. This enables hand-off functioning while driving (or walking) because the Navigator has a speakerphone along with a host of other functions. At the same time, Samsung announced another industry first: have a single mobile LCD screen display two independent images in the front and back of the screen, obviating the need for two screens. It is also creating designs (mobile phone and flat-panel TVs for instance) that have been copied by other companies. It has also established a design school, SADI.

Besides technology innovation, Samsung continues to revise its strategy in the fast moving consumer electronics sector. For instance, in the MP3 player sector, Samsung has decided to focus on niche products in these markets but compete directly in regions where iPod is not selling much, e.g., Russia and Latin America. For the products where Samsung is doing well, it is continuing the strategy. For instance, it is number one in the industry for overall televisions with 2006 sales of $10.8 billion and is number one for flat-panel in the US. It is now seeking to double the sales of both LCD and plasma TV sets to counter decreasing prices and margins, for LCD TVs using a new brand "Nouveau Bordeaux". In the mobile phone market, having established its brand at the top end of the price range, it has revised its strategy to provide handsets at the low price end market especially with high growth in emerging markets like India and Russia.

Samsung Electronics has joint R&D projects with many technology companies. In 2004 alone, it formed alliances with Sony—Samsung was the first to produce a Blu-Ray DVD player in June 2006—IBM, Maytag, EMC, Sanyo, and Dell while continuing work initiated in previous years with other companies. It also has alliances with many universities and even with former military scientists in Russia to tap into their electronics and physics acumen. Recently, Samsung has formed alliances with Google (e.g., the Galaxy Nexus line of smartphones) and with Yahoo.

Samsung also has to monitor changes in customer tastes globally because it is not clear which technology is on the rise as regards customer preferences and which is on its way out, e.g., DVD vs. Divx (for playing downloadable movies), CD player vs. MP3 player, etc. Samsung Electronics Europe operates the European Customer Care Centre to use customer data to try to catch changes in customer preferences from this call center. The center uses an integrated customer-relationship-management (CRM) tool combining customer information, campaign management, and sales-force automation.

Samsung also operates the Samsung Economic Research Institute (SERI) to provide global analysis of interest to the company. Such analysis informs and gives more confidence to company plans. For instance, a global recession would result in a drop in demand across the globe so different markets would not offset each other so this is a global issue for Samsung and SERI.

Exchange rate fluctuations are always an issue but for a global company like Samsung, related issues are quite complicated. For a market-based subsidiary like Samsung Electronics UK, the sterling-won exchange rate at which profits can be sent back to (or at least reported in) Samsung Electronics is an issue. Moreover, a country-based subsidiary with assembly plants may source parts from nearby countries (with different currencies) and from the company's own core-component plants or suppliers in Korea. It may distribute and sell products in neighboring countries as well. Samsung Electronics uses futures to dampen operational effect of currency fluctuations but eschews complicated derivatives that carry their own risks.

Besides daily exchange rate movements, long-term trends like the decline of the US dollar along with outlook of further declines, Britain not joining the euro, and the multiple currencies of the continually expanding EU with emerging economies of Central and Eastern Europe who will not or cannot join the euro-zone also provide a challenge. Samsung can do business with local distributors in these emerging economies in euros. Still, the long-term decline of the US dollar and the consequent appreciation of the won have hurt revenue growth along with price erosion.

Growing companies like Samsung Electronics face financial risk by tending to focus on market share, "presence", name recognition, and other objective and subjective measures rather than profitability. However, Samsung Electronics has focused on profitability and sold unprofitable businesses. In fact, the move from the UK to Hungary was motivated by a desire to cut costs although in doing so risked considerable loss of reputation.

Moreover, individual companies within conglomerates like Samsung also face the threat that one company going bankrupt owing to currency or credit crises can have a domino effect within the conglomerate as money is moved around within companies because of shareholder interest. Samsung Electronics moved to a more transparent corporate governance system to allow for greater corporate oversight. This has resulted in having less of a conglomerate structure.

IV. Network Risk

These are risks originating in a plant or region whose consequences eventually impact the entire supply chain or all the company's markets. They are particularly challenging to manage not just at the strategic level but also at the operational level. Moreover, managing these risks requires a deep understanding of the entire supply network of supply chain entities (suppliers' suppliers onwards) and markets (including customers' customers)

Political risk in Korea is a big threat because of the presence of core component plants and suppliers are all in Korea and it is difficult and undesirable for Samsung to decrease involvement there. There is always the possibility that strikes can shut down production in a supplier's plant, crippling distribution worldwide. More-over, senior management, distracted by labor action, would not be able to follow up on opportunities in the rest of the world. This is not idle speculation: there have been strikes at Korean electronic companies with enormous consequences in the late 1980s and early 1990s. For example, the 72-day shutdown of LG Electronics' Changwon manufacturing plant in 1989 resulted in a $750 million loss because of the disruption to their domestic and global distribution (Oh 2004).

Samsung has assembly plants and local suppliers in multiple countries, the main motivation being proximity to major markets like the US and EU. To mitigate the risk of labor action, Samsung has profit sharing and stock-ownership schemes and consciously seeks to maintain good relations with employees. As a possible con-sequence, Samsung Electronics is non-union and has never had strikes. Even its suppliers do not have much history of labor strife.

As regards suppliers, Samsung Electronics implements a multiple-sourcing pol-icy for components with non-core components that are 70 percent of all procurement and requires the suppliers to be local. The remaining (core) 30 percent components are single-sourced from the home country in terms of location for greater reliability and economies of scale. However, the supplier is backed up by usually one other supplier and sometimes even by two.

Samsung Electronics uses logistics execution system technology to counter the impact of delays of reacting by increasing visibility into the supply chain. It has a global tracking network system, called the In-Transit Tracking System (ITTS). This system helps track delivering from factory to the ordering party just like the courier service DHL's tracking system anywhere in the globe. Every global internal ordering party can track down their order until it is received. From this system, the company can not only monitor and reduce the risk of delays, but also reduce the risk associated with forecasting, inventory, and procurement. Within the company, the global command center (GCC) in headquarters can monitor the delivery situation globally in real time and deal with any emergency.

Another network risk is related to the information network rather than the supply network. The more a company networks its systems, the greater the threat that a fail-ure anywhere can cause failure everywhere. Samsung Electronics uses such vendors as SAP with experience in dealing with corporate security and reliable backup. In Europe, Samsung Europe operates the Business Recovery Services (BRS) system. It has two centers in different locations that are the data centre and the backup centre separately for the UK. The BRS system can back up the entire network of all of its European subsidiaries' data in real time.

6.6 Conclusions

In our quest to understanding how global manufacturing companies manage risk in terms of the processes and organization, we presented a framework with risk drivers being local or global and their consequences being local of global. We used this framework to understand risk management at Samsung Electronics and its wholly owned subsidiary, Samsung Electronics UK. While we need further work in terms of organizational and process details for this case study as well as for others, there is benefit in using the framework to understand risk management in companies.

Our findings are that successful companies like Samsung Electronics seem to display good coverage of risk management practices to manage a diverse set of risks in all four quadrants of the framework. This indicates strong IT-supported operations at the operational level to manage operational risk (quadrant I), close cooperation between regions and headquarters to manage localization risk (quadrant II), a solid philosophy of risk management as demonstrated in policies towards exchange rates as a way to manage enterprise risk (quadrant III), and strong cooperation, again backed by IT, between local operations and corporate, to manage network risk (quadrant IV).

Therefore, unlike concerns expressed in some of the supply chain literature, successful companies like Samsung Electronics do take risk associated with the supply chain into account as part of supply-chain management. Such companies have risk management procedures at the local (plant or region) level to prevent or otherwise mitigate losses at the local level as well as formal risk management processes for managing risks faced by the company as a whole. Indeed, Samsung's extensive supply-chain thinking in risk management was positively surprising, at least for us.

However, we cannot simply recommend these practices for other companies. For instance, it is not clear if the close relationship between operations and finance-driven headquarters for both strategic and tactical purposes is due to the location of core-component manufacturing being in the proximity to headquarters in Korea. For instance, we can not how this could apply to a company like aerospace giant, Boeing, who moved headquarters from Seattle where it has manufacturing, to Chicago to be closer to headquarters of airlines and other customers.

For this reason and because of the early stage of the research presented here, much is needed for future research. A multi-case study with other companies in the same and other sectors is very much needed to understand the specifics of risk management processes and the organization. Eventually, we should be able to make generic risk management strategies at the level of the four quadrants including quadrant-specific risk metrics.

Specific things to investigate, in particular regarding "network" risks in quadrant IV for our framework can include the following:

1. Managers in global companies using offshore manufacturing and outsourcing, feel more vulnerable to large unexpected losses. Furthermore, the greater the extent of outsourcing and/or offshoring, and the greater the number of supply chain entities (including suppliers' suppliers and customers' customers) spread across the globe, the greater the sense of risk perceived by managers and greater the risk management effort in quadrant IV.

2. Network risks are generally low occurrence frequency with high impact, thus making it difficult for managers to tame such risks. Models for such risks for the purpose of insurance/self-insurance, whether in terms of theoretically-chosen long-tailed loss distributions (as in operational risk, e.g., Holland and Sodhi

2008) or simulation models of supply-chain networks (e.g., Deleris et al., 2004) may be quite useful.

3. In general, companies need to trade off risk and rewards (higher long-term profits). However, doing so assumes that the companies are already on an "efficient frontier" which may not be true for network risks. The implication then is that companies may be able to get both higher rewards and lower risks. For instance, Samsung's ITTS system is not only good for rewards in this sense but also lowers risks.

4. To study "network" or quadrant IV losses, it may be useful to study other networks like those of phone lines and electricity grids rather than hazards of plant failures etc.

5. How do organizations (recall Boeing's case mentioned earlier in contrast to Samsung's case) that have headquarters away from manufacturing handle network risks?

6. Just as there are many metrics for measuring efficiency-related performance in a company, we need to devise appropriate metrics for risk exposure and risk performance for all four quadrants. Some risk metrics can be associated with the financial concept of value-at-risk, for instance, demand-value-at-risk and inventory-value-at-risk (Sodhi 2005). We also need processes spanning financial and operational planning that use these metrics and analytical models used to produce such metrics (Sodhi and Tang, 2011). Devising metrics for network risks could be challenging.

Thus, there is much modeling and empirical work yet to be explored in the area of supply-chain risk management.

Part II
Mitigation Approaches and Applications

Chapter 7
Strategic Approaches for Mitigating Supply Chain Risks

Abstract While many companies recognize serious supply chain risks, they also acknowledge that their actions for managing these risks are not commensurable in part because they find it difficult to justify costly strategies for mitigating potential supply chain disruptions that have occurred rarely in the past and may not occur again in the foreseeable future. Hence, to encourage a firm to adopt supply chain risk mitigation, the supply chain strategies need to be "robust" in the following sense: These strategies should help to efficiently manage the "normal" risks or fluctuations inherent in matching supply and demand and should also help the supply chain become more resilient in the face of "abnormal" risks, i.e., disruptions. This chapter presents various robust supply chain mitigation strategies that many firms have adopted in managing their supply chain risks.

7.1 Introduction

In part I (Chapters 1–6), we presented a general risk management framework (identify, assess, mitigate, and respond) and basic approaches for managing supply chain risks. In this part (Chapters 7–12), we discuss different mitigation approaches that are supply-chain specific (Chapters 7 and 8) and describe how these approaches can be implemented by companies who face different types of supply chain risks (Chapters 9–12).

This chapter presents supply chain risk mitigation strategies that are "robust" in the following two senses: First, these strategies enable a supply chain to manage the "normal risks" or fluctuations inherent in matching supply and demand efficiently. Second, these strategies can help make a supply chain become more resilient in the face of "abnormal risks", i.e., major disruptions. To be sure, a company implementing these strategies will incur costs but the upside may be additional revenues by acquiring and retaining customers apprehensive about supply chain risk, especially after a major disruption.

M.S. Sodhi, C.S. Tang, *Managing Supply Chain Risk*,
International Series in Operations Research & Management Science 172,
DOI 10.1007/978-1-4614-3238-8_7, © Springer Science+Business Media, LLC 2012

Consider the following three well-cited examples in this context, all showing how these companies—Nokia, Li and Fung and Dell—were able to leverage their supply chain strategies to react quickly to disruptions, thus enjoying long-term sales growth as a result:

1. Nokia changed product configurations in the nick of time to meet customer demand during a supply disruption. Both Ericsson and Nokia were facing a shortage of a critical cellular phone component (radio frequency chips) after a key supplier in New Mexico (Philip's semiconductor plant), caught on fire during March of 2000. Ericsson was slow in reacting to this crisis and lost 400 million euros in sales. In contrast, Nokia had the foresight to design their mobile phones based on the modular product design concept and to source their chips from multiple suppliers. After learning about Philip's supply disruption, Nokia responded immediately by reconfiguring the design of their basic phones so that the modified phones could accept slightly different chips from other Philip's plants and from other suppliers. Consequently, Nokia satisfied customer demand smoothly and obtained a stronger market position (Hopkins, 2005).

2. Li and Fung changed its supply plan rapidly to meet customer demand during a currency crisis. When Indonesia Rupiah devalued by more than 50% in 1997[1], many Indonesian suppliers were unable to pay for the imported components or materials and hence, were unable to produce the finished items for their U.S. customers.[2] This event sent a shock wave to many U.S. customers who had outsourced their manufacturing operations to Indonesia. In contrast, The Limited and Warner Bros. continued to receive their shipments of clothes and toys from their Indonesian suppliers without noticing any problem during the currency crisis in Indonesia. They were unaffected because they had outsourced their sourcing and production operations to Li and Fung, the largest trading company in Hong Kong for such durable goods as textiles and toys. Instead of passing the problems back to their U.S. customers, Li and Fung shifted some production to other suppliers in Asia and provided financial assistance such as line of credit, loans, etc., to affected suppliers in Indonesia to ensure that their U.S. customers would receive their orders as planned[3]. With a supply network of 4,000 suppliers throughout Asia in 1997, Li and Fung were able to serve their customers in a cost-effective and time-efficient manner. This capability has enabled Li and Fung to grow nearly 25 times from 5 billion Hong Kong dollars in revenues in

---

[1] Indonesia Rupiah opened the year 1997 at 2363 to the US dollars and closed at 5550 against the dollar. However, in July 1997, Indonesia Rupiah was traded at 10,000 against the US dollars.

[2] The currency crisis affected Indonesia in a very serious manner in 1997. For instance, Indonesia's national car manufacturer, Astra, suspended their production because they were unable to pay for imported components. Also, 60% of Jakarta's public transport system was suspended, because of the soaring price of the spare parts needed to repair the city's buses. Moreover, 40% of the country's 1500 chemical plants were forced to halt production because of the soaring cost of imported raw materials.

[3] Huchzermeier and Cohen (1996) develop a quantitative model to analyze the value of flexible supply base under uncertain exchange rate.

1993 to 124 billion HK\$ (US \$16 billion) in 2010.[4]

3. Dell changed its pricing strategy just in time to satisfy customer during supply shortage. After an earthquake hit Taiwan in 1999, several Taiwanese factories informed Apple and Dell that they were unable to deliver computer components for a few weeks. When Apple faced component shortages for its iBook and G4 computers, it encountered major complaints from customers after it tried to convince customers to accept a slower version of G4 computers. In contrast, Dell's customers continued to receive Dell computers without even noticing any component shortage problem. Instead of alerting their customers to shortages of certain components, Dell offered special price incentives to entice their online customers to buy computers that used components from other countries. The capability to influence customer choice enabled Dell to improve its earnings in 1999 by 41% even during a supply crunch (cf. Martha and Subbakrishna, 2002; Veverka, 1999).

Thus, Nokia's, Li and Fung's and Dell's supply chains are not only resilient to major disruptions but are also efficient in responding to normal fluctuations. These companies have created these supply chains following different robust supply-chain strategies.

7.2 Robust Supply Chain Strategies

Supply chain issues can be generally classified into two major categories: supply related, and demand related. Supply related issues include supplier selection, supplier relationship, supply planning, transportation and logistics, etc., while demand related issues include new product introduction, product line management, demand planning, product pricing and promotion planning, etc. Keeping this in mind, we now describe eleven different robust supply-chain strategies that aim to improve a firm's capability to better manage supply and/or demand under normal circumstances and to enhance a firm's capability to sustain its operations when a major disruption hits. Table 7.1 summarizes the key features of these nine robust supply chain strategies.

1. Postponement

Postponement uses product or process design concepts such as standardization, commonality, modular design, and operations reversal, to delay the point of product differentiation. This strategy enables a firm to first produce a generic product based on the total aggregate forecasted demand across all products in a family, and then

---

[4] Li and Fung, annual report, 2010.

Table 7.1 Robust Supply Chain Strategies

	Robust Supply Chain Strategy	Main Objective	Benefit(s) under normal risk: Improves the company's capability to manage…	Benefit(s) under abnormal risk, i.e., after a major disruption: Enables the company to..
1	Postponement	Increases product flexibility	Supply	Change the configurations of different products quickly
2	Strategic stock	Increases product availability	Supply	Respond to market demand quickly during a major disruption
3	Flexible supply base	Increases supply flexibility	Supply	Shift production among suppliers promptly
4	Make-and-Buy	Increases supply flexibility	Supply	Shift production between in-house production facility and suppliers rapidly
5	Economic supply incentives	Increases product availability	Supply	Adjust order quantities quickly
6	Flexible transportation	Increases flexibility in transportation	Supply	Change the mode of transportation rapidly
7	Revenue management	Increases control of product demand	Demand	Influence the customer product selection dynamically
8	Dynamic assortment planning	Increases control of product demand	Demand	Influence the demands of different products quickly
9	Silent product rollover	Increases control of product exposure to customers	Supply and demand	Manage the demands of different products swiftly
10	Flexible supply contracts	Increase replenishment flexibility	Supply	Shift order quantities across time
11	Flexible manufacturing process	Increase flexibility in producing different products	Demand	Shift production quantities across internal resources (plants or machines)

customize the generic product later on as demand for each specific product becomes known. The postponement strategy has been proven to be a cost-effective mass customization tool to handle regular demand fluctuations under normal circumstances at companies such as Xilinx, Hewlett Packard (HP), and Benetton.[5]

---

[5] Recently, Xilinx, the leading innovator of programmable logic chips, revealed their postponement strategy that enables their customers to use software to fully configure the function of their chips (c.f., Brown et al. (2000)). Next, consider HP. In order to produce 500,000 different config-

In the context of recovering from a supply chain disruption, postponement offers a cost-effective and time-efficient contingency plan that allows a supply chain to re-configure the product quickly in the event of supply disruption. For example, when Philip's informed Nokia that it was not possible to deliver certain components for a certain period after the fire in the Albuquerque plant in March 2000, postpone-ment enabled Nokia to deploy a contingency plan by reconfiguring their generic cell phone quickly. The reconfigured generic phone accepted a component that was slightly different from the one being delivered by the Philip's plant. This product flexibility enabled Nokia to recover from a serious disruption without any signifi-cant problem in delivering the different specific products based on the generic call phone.

2. Strategic Stock

In the "pre-JIT" era, a company would consider carrying additional "just in case" safety stock inventories of certain critical components to ensure that the supply chain can continue to function smoothly when facing a disruption or a delay in supply. However, as product life cycle shortens and as product variety increases, the inven-tory holding and obsolescence costs of these additional safety stock inventories can become exorbitant.

To do this better, instead of simply carrying more safety stock at every location, a firm should consider storing extra inventory only at certain "strategic" locations (warehouse, logistics hubs, distribution centers) where the inventory can be shared by multiple supply chain partners, say retailers or repair centers. For example, Toy-ota and Sears keep certain inventories of cars and appliances at certain locations so that all retailers in the nearby region share these inventories. By doing so, Toy-ota and Sears can achieve a higher customer service level without incurring high inventory cost when dealing with regular demand fluctuations.

When a disruption occurs, these shared inventories at strategic locations will al-low a firm to deploy these strategic stocks quickly to the affected area as well. For example, Center for Disease Control (CDC) keeps large quantities of medicine and medical supplies known as Strategic National Stockpile (SNS) at certain strategic locations in the United States. This strategic stockpile is intended to protect the American public if there is a public health emergency like a terrorist attack, flu out-

---

urations of workstations at HP in an effective manner, HP used postponement by mass producing a generic version of the workstation in a make-to-stock manner. This enabled HP to respond to customer orders quickly by inserting certain product specific components into these generic work-stations (c.f., Feitzinger and Lee (1997)). Finally, consider Benetton. By re-sequencing the dyeing and knitting process at Benetton, Benetton was able to postpone the color specification of the sweater by knitting the undyed sweaters first and then dye the sweaters into different colors after receiving customer orders (c.f., Haskett and Signorelli (1984)). For technical evaluation of differ-ent postponement strategies, the reader is referred to Lee (1996), Lee and Tang (1997), Lee and Tang (1998a) and Swaminathan and Tayur (1999) for details.

break, or earthquake that is severe enough to cause local or regional supplies to run out.[6]

3. Flexible Supply Base

Although sourcing from a single supplier will enable a firm to reduce cost (lower supply management cost, lower unit cost due to quantity discount, etc.), it could create problems for managing inherent demand fluctuations or major disruptions. To mitigate the risk associated with sole sourcing, Billington and Johnson (2002) described how HP used their plants in the state of Washington and in Singapore as their supply base to produce inkjet printers. HP used the Singapore plant for the base volume production and used the Washington plant to produce the excess on top of the base volume, thus handling regular demand fluctuations, i.e., the "normal" risks, smoothly.

Besides enabling a firm to handle regular demand fluctuations, a flexible supply base can help maintain continuity in supply of materials if a a major disruption were to occur. For example, Li and Fung's 4000-supplier network offers Li and Fung great flexibility to shift production among suppliers in different countries quickly when a disruption occurs at a particular country as we described earlier in this chapter.

An extreme form of such flexibility is to help create or join an existing *supply alliance network*. Suppliers (contract manufacturers, airlines cargo companies, trucking companies, logistics providers) can proactively form strategic alliances with other suppliers in different countries. These alliances can serve as a "safety net" for each member, who will receive help from other members if a disruption strikes.

4. Make-and-Buy

When facing potential supply disruptions, a supply chain is more resilient if certain products are produced in-house while other products (or a proportion of the same products) are outsourced to other suppliers. This enables production to be shifted to different locations and/or demand to be shifted to other products depending on whether the outsourced products are from the same family or are entirely different. For instance, HP used to make some of their DeskJet printers at their Singapore factory and outsourced the rest of the products in the DeskJet family to a contract manufacturer in Malaysia (c.f., Lee and Tang (1996)). Apparel-makers Brooks Brothers and Zara produce their fashion items in-house while outsourcing basic items to low-cost suppliers in China or elsewhere (c.f., Ghemawat, 2003).

---

[6] This emergency medical supply is loaded on a wide-body aircraft that can be sent to a disaster area within a 12-hour time window. The reader is referred to http://www.bt.cdc.gov/stockpile/ for more details.

Such make-and-buy strategies not only help match supply and demand efficiently under normal risks, they also offer flexibility to shift production quickly should a supply disruption occur (when the same product or product family is outsourced) or at least allow some revenues continue (when outsourcing entirely different products) during the period of recovery.

However, outsourcing introduces its own risks and we shall discuss ways to mitigate outsourcing risks in Chapter 10.

5. Economic Supply Incentives

In many instances, the buyer does not have the luxury to shift production among different suppliers because of the limited number of suppliers available in the market. To gain the flexibility of shifting production among suppliers, the buyer can offer economic incentives to cultivate additional suppliers. For example, due to the uncertainties of producing a specific flu vaccine formula in any given year, uncertain demand, and price pressure from the U.S. government, many flu vaccine makers exited the market. The number of flu vaccine makers dropped from 25 in the 1970s and 1980s to only 3 in 2003.

In 2003, Chiron entered this market acquiring another company with European plants and became one of US government's two suppliers. However, in October 2004, British regulators (MHRA) suspended operations at Chiron's Liverpool plant after finding bacterial contamination to be excessive. The US regulators (FDA) concurred in December, issuing a warning to Chiron about its inability to meet its commitments. Facing a shortage of 48 million flu shots from Chiron, the U.S. government could initially offer flu shots only to those who belonged to certain high-risk groups (c.f., Brown (2004)). Moreover, being unable to meet its commitments, Chiron allowed itself to be acquired by Switzerland-based Novartis a few months later.[7]

To avoid this kind of fiasco in the future, the U.S. government could consider offering certain economic incentives to entice more suppliers to re-enter the flu vaccine market. For instance, the government could share some financial risks with the suppliers by committing to a certain quantity of flu vaccine in advance at a certain price and to buying back the unsold stocks at the end of the flu season at a lower price.[8] With more potential suppliers, the U.S. government would have the flexibility to change their orders from different suppliers quickly when facing major disruptions.

Even without major disruptions, economic supply incentives can be beneficial. For example, when Intercon Japan became concerned about the "monopoly" mindset of their key supplier, Asahi Metal, they offered economic incentives to entice

---

[7] A. Shanley, 2004. "Chiron's curse", Pharmaceutical Manufacturing, accessed at http://www.pharmamanufacturing.com/articles/2004/187.html on 15th May 2011.

[8] The issues of "risk sharing" and "revenue sharing" has been studied by Narayanan and Raman (2004) in the context of aligning the incentives among supply chain partners so that the entire supply chain can focus on the performance of the entire supply chain.

a new supplier, Nagoya Steel, to develop a new steel process technology for producing different types of cable connectors. To make Nagoya Steel become more competitive, these incentives included a minimum order quantity, technical advice about the new steel process technology, and information about the market demand for this new process technology. By establishing another supplier that used a different process technology, Intercon Japan was able to keep pressure on both suppliers to keep the cost low (c.f., Mishina (1991) and Tang (1999)).[9]

6. Flexible Transportation

In supply chain management, transportation can be the Achilles' heel that could make the supply chain snap. Although, as we described in Chapter 6, Samsung Electronics has never suffered a strike in Korea, it has still faced disruption as a trucking strike disrupted suppliers to its suppliers thus in turn affecting supplies to its plants (Lee and Sodhi, 2007). As such, one should consider adding more flexibility proactively. Here are three approaches for doing so:

1. **Multi-modal transportation.** To prevent the supply chain operations from coming to a halt when disruptions occur in the ocean, in the air, on the road, etc., some companies utilize a flexible logistics strategy that relies on multiple modes of transportation. For example, Seven-Eleven Japan urges its logistics partner to diversify its mode of transportation that includes trucks, motorcycles, bicycles, ships, and helicopters. This flexible logistics strategy has won the hearts of many Japanese when Seven-Eleven Japan used 125 motorcycles and 7 helicopters to make rush deliveries of 64,000 rice balls to earthquake victims in Kobe shortly after the earthquake that destroyed many roads in the late 80's (c.f., Lee (2004)).[10]

---

[9] With a flexible supply base, many firms can enter different supply contracts such as backup supply contracts, quantity flexibility contracts, etc., with their suppliers. For instance, in a backup contract, the buyer is committed to a certain order quantity with the supplier ahead of time. The supplier delivers a pre-specified fraction of this committed quantity before the start of the selling season and reserves the capacity for producing and delivering the remaining units (i.e., the backup units). After observing early demand, the buyer can order up to the backup units by paying the original purchase cost and receive quick delivery. However, the buyer will pay a penalty cost for any of the backup units it does not buy (c.f., Eppen and Iyer (1997a)). In a quantity flexibility contract, the buyer is committed to a certain quantity ahead of time, but the buyer has the flexibility to adjust this quantity upward or downward up to a certain amount at certain specified time frame (c.f., Tsay and Lovejoy (1999)).

[10] Chrysler selected a multi-modal transportation strategy with a third party logistics provider, which would allow Chrysler to switch the mode of transportation from air to ground immediately. This strategy enabled Chrysler to get the parts from their suppliers such as TRW via ground transportation instead of air transportation immediately after September 11. In contrast, Ford did not establish such strategy. As such, Ford was unable to switch the mode of delivery after September 11, due to a surge in demand for ground transportation. Facing part delivery problems, Ford closed 5 of the U.S. plants for weeks and reduced its production volume by 13% in the fourth quarter of 2001 (c.f., Hicks (2002)).

2. **Multi-carrier transportation.** To ensure continuous flow of materials in case of political disruptions (landing rights, labor strikes, etc.), various air cargo companies such as Aeroméxico Cargo, KLM Cargo, Delta Air Logistics, Air France Cargo, CSA Czech Airline Cargo, Korean Air Cargo, etc., have formed an alliance called SkyTeam Cargo that will enable them to switch carriers quickly in the event of political disruptions. Moreover, this alliance enables SkyTeam Cargo to provide low cost global deliveries to 500 destinations in 110 countries.[11]

3. **Multiple routes.** To avoid a complete shutdown, many companies are considering alternative routes to ensure smooth material flows within the U.S. For example, due to long delays at the west-coast ports and heavy traffic jams along various west-coast freeways, some east-coast companies are encouraging shippers to develop alternatives to shipping through the west coast. Specifically, after the west coast ports were shut down for two weeks in 2002, some shippers considered shipping manufacturing goods from Asia to east-coast ports directly via the Panama Canal.

7. Revenue Management via Dynamic Pricing and Promotion

Dynamic pricing is a common mechanism to sell perishable products/services. For instance, when selling limited seats on an airplane with uncertain demand, airlines adjust ticket price dynamically to meet uncertain demand with limited supply. Cook (1998) reported that revenue management via dynamic pricing generated "almost $1 billion of incremental annual revenue" at American Airlines.

Revenue management via dynamic pricing and promotion can also be an effective way to manage demand when the supply of a particular product is disrupted. Specifically, a retailer can use pricing mechanism to entice customers to choose products that are widely available.[12] For example, as mentioned at the outset of this chapter, when Dell was facing supply disruptions from their Taiwanese suppliers after an earthquake in 1999, Dell immediately deployed a contingency plan by offering special "low cost upgrade" options to customers if they choose similar computers with components from other suppliers. This dynamic pricing and promotion strategy enabled Dell to satisfy their customers during a supply crisis (c.f., Martha and Subbakrishna (2002)). For technical analysis of revenue management via dynamic pricing, see Talluri and van Ryzin (2005).

---

[11] Along the same vein, a group of global freight forwarders launched the World Freight Alliance in 2004 that will provide shippers maximum flexibility to switch carriers quickly should a disruption occurs. The reader is referred to Harrington (2004) for details.

[12] In the context of e-commerce, savvy on-line retailers can utilize the profile of each on-line customer such as past click sequence, past purchasing history, etc., to develop a personalized pricing and promotion strategy so as to influence each customer's product choice.

8. Assortment Planning

Similar to the idea of revenue management above, brick-and-mortar retailers have used assortment planning (the set of products on display, the location of each product on the shelves, and the number of "facings" for each product) to influence consumer product choice and customer demand. A study conducted by Chong, Ho and Tang (2001) at five supermarkets in the U.S. revealed that a store manager can manipulate customer's product choice and customer's demand by reconfiguring the set of products on display, the location of each product on the shelves and the number of facings for each product. Their findings suggested that one can use assortment planning to entice customers to purchase products that are widely available when certain products are facing supply disruptions.

9. Silent Product Rollover

Under the silent product rollover strategy, new products are "leaked" slowly into the market without any formal announcement. As such, customers are not fully aware of the unique features of each specific product and they are more likely to choose the products that are available instead of those products that are out of stock or being phased out. For instance, as Swatch produces each watch model only once, Swatch uses the silent product rollover strategy to launch new watches so that their customers view all available Swatch watches as collectibles, not just the recently introduced ones (c.f., Billington et al. (1998) and Moon (2003)). Using the same approach as Swatch, Zara launches their new fashion collection quietly. Since Zara does not usually repeat the production run for the same design of clothes, many Zara's fashion conscious customers purchase the clothes available at their stores right away (c.f., Ghemawat (2003)).[13] Ultimately, all products are essentially "substitutable" at Swatch and Zara. Substitutable products are desirable for handling demand fluctuations under normal circumstances and this is even more true when there is a supply or demand disruption that affects a subset of existing products.

10. Flexible Supply Contracts

Typically, companies are not allowed to adjust their order quantities with their suppliers once they placed the orders. For example, under the partnership agreement with HP, Canon is the sole supplier of the engines for the HP LaserJet printers. To keep supply costs down, HP had to place its orders six months in advance and could

---

[13] Since Swatch and Zara produce each particular design only once, their production process and their product design have to be flexible so that they can switch from producing one product to the next without incurring significant setup time or cost.

not change the order quantity once the order was placed. Due to uncertain demand and inflexible supply contracts, HP was facing major difficulties in responding to dynamic and uncertain demand and HP was keeping a large amount of safety stock (Lee, 2004).. To reduce supply risks, Canon agreed to allow HP to adjust their order quantity upward or downward by no more than a few percent (specified in advance). In this case, the upward/downward adjustment limits capture the flexibility level of this supply contract. This type of supply contract is called a Quantity Flexible (QF) contract and it has been investigated by Tsay and Lovejoy (1999).

11. Flexible Manufacturing Process

To reduce production cost, Texas Instrument (TI) organized their LCD watch manufacturing facility according to an assembly line. This assembly line was efficient for producing high volume of a few models of LCD watches at low cost. Unfortunately, as the LCD watch market matured and as Seiko introduced many different models of LCD watches, TI was unable to use this inflexible assembly line to compete on cost and product variety as the inflexibility of assembly line inhibited TI's ability to increase product variety to compete with Seiko. Consequently, TI had to exit the watch market completely.[14]

To produce multiple products efficiently and effectively, a firm needs to increase process flexibility by installing systems such as Flexible Manufacturing Systems (c.f., Sethi and Sethi, 1990).[15] Consider a situation in which a firm has multiple plants for producing multiple products. As processes become more flexible, different types of products can be manufactured in the same plant. While total process flexibility to produce all products enables a firm to reduce process risks significantly, the cost of implementing such flexibility can be exorbitant and possibly unnecessary. Jordan and Graves (1995) show that the company can mitigate demand risks effectively by establishing only some process flexibility at each plant (e.g., two or three products can be produced in each plant).

An extreme form of flexibility is the deployment of *recovery planning systems* so that a supply chain can recover quickly from a major disruption. For example, even before the September 11 terrorist attacks, Continental Airlines had worked with Caleb Technologies to develop the CrewSolver decision support system to generate globally optimal recovery solutions. The optimal recovery solution enables Continental Airlines to reassign crews quickly to cover open flights and to return them to their original schedules in a cost-effective manner while honoring government regulations, contractual obligations, and customer expectations. The CrewSolver system has helped Continental to save approximately $40 million over disruptions during a five-year period (Yu et al., 2003).

---

[14] The reader is referred to the Harvard Business School case titled "Texas Instruments—Time Products Division" published in 1991 for details.

[15] Besides FMS, cross-training of workers is another effective mechanism to increase process flexibility. The reader is referred to So et al. (2003) for details.

Considerations for Adopting These Strategies

While these eleven robust strategies are beneficial under normal circumstances (without risk or with only "normal" risks of supply and demand) as well as during major disruptions, they also raise the following concerns:

Costs vs. benefits.   Some firms may express concerns regarding the requisite costs associated with these robust strategies, while others may recognize the additional benefits. On a conceptual level, these robust strategies would enhance the competitive position of a firm, especially when other firms' supply chains are more vulnerable to disruptions. However, it is difficult to quantify the value of competitiveness. Theoretically speaking, the costs for implementing these proactive strategies can be viewed as "insurance premiums" that will safeguard the supply chains from major disruptions (c.f., Sheffi (2001)). However, it is difficult to evaluate the return on these insurance premiums especially in the absence of reliable data for probability that a disruption would occur, potential loss due to a disruption, etc.

Strategic fit.   Even though these robust strategies enhance a firm's capability to better manage supply and demand, they may not fit the firm's overall business strategy. Consider two cases. First, suppose a firm has chosen to reduce product variety as a way to rationalize their product lines, then the value of the postponement strategy is diminished. Second, if a retailer has positioned itself as an "every day low price" store, then the dynamic pricing or promotion strategies would seem inconsistent with the retailer's strategic position in the marketplace.

Proactive execution.   A robust strategy is useless unless a firm can execute the strategy proactively. For example, around the time the Los Angeles-port longshoreman contract was due for renewal in 2002, NUMMI (a Toyota-GM joint venture), Ralph Lauren and Tommy Hilfiger developed alternative transportation plans in case there was a strike. As the longshoreman union and the port administration were disputing over the labor contract, Ralph Lauren and Tommy Hilfiger deployed their contingency plans proactively by re-routing their shipments through the east coast. Instead of using an alternative route, NUMMI decided to stock an extra six days of inventory. Unfortunately, by the time NUMMI exhausted this stockpile, truckers were still unable to unload the components from Los Angeles and deliver them to the NUMMI plant. By that time, it was already too late for NUMMI to re-route shipments. Consequently, NUMMI was forced to shut down for several days (c.f., Zsidisin et al., 2004).

Evaluating flexible supply contracts.   As discussed earlier, Canon allowed HP to adjust their order quantity upward or downward by no more than a few percent in a type of supply contract called a Quantity Flexible (QF) contract. However, due to the multi-period nature of their model, an analytical characterization of the value of flexibility is not possible.

The cost of postponement.   HP delays the point of differentiation in manufacturing its DeskJet line (at its in-house production facilities) until the last stage of the process that takes place at the distribution centers instead. As the generic printers produced at the plant are completely customizable to any SKU in the DeskJet family, delaying the point of product differentiation to the DCs until the last stage of the process offers HP the highest level of product flexibility for mitigating demand risks across the different specific SKUs within the DeskJet family. However, the cost of such postponement can be high and raises the question whether it is necessary for HP to delay product differentiation until the last stage of the process which is then carried out in DCs rather than in the plant.

Reducing lead time. A supply chain is more vulnerable to disruption when the lead time is long so to reduce the risk exposure, a company can shorten the lead time rather than (or in addition to) following any of the eleven strategies. For instance, Liz Claiborne launched a campus in China by bringing all stages of the textile supply chain to the campus. Doing so enabled Liz Claiborne to reduce the lead time from concept to retail store from 10-50 weeks to fewer than 60 days (c.f., Sheffi, 2005 a and b). However, doing so requires re-designing the supply chain network.

7.3 Conclusion

This particular chapter has presented robust strategies that have two desirable characteristics from the viewpoint of supply chain management: First, these strategies enable a supply chain to manage the "normal risks" or fluctuations inherent in matching supply and demand efficiently. Second, these strategies can help make a supply chain become more resilient in the face of "abnormal risks," i.e., major disruptions.

A company implementing these strategies will incur costs. Not all of these strategies will be consistent with the company's positioning in the market. It may not be clear how to evaluate the benefits. And certain basic approaches like reducing the lead time may be more useful initially. However, the upside of these robust strategies is additional revenues by acquiring and retaining customers especially after a major disruption.

There are a few research opportunities here. *First*, it would be of interest to develop quantitative models to address the following pertinent questions: How should one measure the effectiveness of a robust strategy? What are the underlying conditions for one robust strategy dominating another robust strategy? What will happen if a firm adopts multiple robust strategies? *Second*, retailers such as Metro and Wal-Mart are pushing for auto-ID technology such as RFID (Radio Frequency Identification Technology) to improve supply chain visibility (c.f., Heinrich (2005)). While the RFID tags can also help a supply chain to reduce shrinkage, misplacement and transaction errors, it would be of interest to see whether RFID technology can be used to develop other form of robust supply chain strategies (c.f., Atah et al., 2005). *Finally*, many firms are reluctant to share private information with their supply chain partners due to the fear of information leakage. To address this information security issue, Atallah et al. (2004) have developed Secure-Supply-Chain-Collaboration protocols to enable supply chain partners to develop collaborative plans without revealing private information to any of the parties. It would be of interest to evaluate the costs and benefits associated with such information security protocols.

Chapter 8
Tactical Approaches for Mitigating Supply Chain Risks: Financial and Operational Hedging

Abstract In addition to supply and demand risks, global supply chains are vulnerable to financial risks arising from uncertain costs of input materials, labor, currency exchange rates, and supplier defaults. This chapter presents both *financial instruments* and *operational mechanisms* for mitigating financial risks, including those tied to input prices and currency exchange, as well as supply and demand risks. These financial instruments are designed to hedge against short-term risks, while the operational mechanisms are intended to manage medium- and long-term risks. We also illustrate how firms can reduce risk significantly by combining financial and operational hedging strategies.

8.1 Introduction

In Chapter 7, we presented eleven robust strategies for mitigating supply and demand risks. However, besides supply and demand risks, global supply chains are vulnerable to financial risks arising from uncertain costs of input materials, labor, currency exchange rates, and supplier defaults. It is important for firms to develop proactive processes to avoid or reduce some of these risks.[1]

Besides the disruptions that affect the supply operations along a global supply chain, there are financial risks associated with the input material costs, labor costs and supplier defaults that firms need to pay close attention. According to recent in-depth interviews with 58 executives from retail and distribution industries conducted by PwC in Australia, Canada, China, France, Germany, India, and United Kingdom, and the United States, 75% of the respondents were concerned about fuel cost and currency exchange (that affects input material cost). However, 75% of these companies track and measure currency exchange risks (c.f., Pricewaterhouse-Cooper (2008)). In a separate survey study of 273 executives conducted by McKin-

---

[1] We would like to thank Volodymyr Babich, Michael Henke, and Joshua Zimmerman for their thoughtful comments on this chapter.

sey (2008), 43% of the respondents were concerned about rising energy prices, 56% concerned about rising wage rates in the global labor market, and 63% concerned about financial volatility. Despite these concerns, 57% of the respondents focused on cost reduction and few mentioned the need to reduce risks associated with rising prices of inputs (energy and labor cost) or other financial matters.

In this chapter, we present various financial instruments and operational mechanisms for mitigating financial risks, including those tied to input prices and currency exchange, as well as supply and demand risks. By and large, these operational mechanisms are intended to manage medium- and long-term risks, while the financial instruments presented in this chapter are designed to hedge against short-term risks. We also illustrate how firms can reduce risk significantly by combining financial and operational hedging strategies.

8.2 Different Types of Financial Risks

Consider four types of risks: (1) risks associated with input material costs, (2) risks associated with labor costs, (3) risks associated with foreign currency exchange, and (4) risks associated with supplier default.

8.2.1 Risks Associated with Input Material Costs

The concern over fuel cost was serious as the crude oil price dramatically increased from $80 to $140 per barrel during the first half of 2008 (Fig. 8.1). (Throughout this chapter, we use the $ sign to denote US dollars.) Due to the significant increase in oil prices, the jet fuel prices have skyrocketed resulting in seven airlines filing for bankruptcy protection in the first five months in 2008. In the first quarter of 2008, Southwest airlines reported a profit of $34 million mainly due to its $302 million in hedges of the oil prices. By late 2007, Southwest was the only large US carrier that had hedged 70% of its fuel at $51 per barrel. In May 2008, due to its hedging strategy, Southwest was getting its jet fuel at a much lower price when its competitors were paying $122 per barrel (c.f., Herbst (2008)). As articulated in Giddy (2009), because hedging strategies designed to reduce risk often receive a great deal of scrutiny by the board of directors, many companies tend to not hedge at all thereby exposing their companies to risk.

Indeed, hedging can backfire. As the world speculated that oil prices would continue to rise, more airlines followed suit to hedge against the skyrocketing jet fuel prices in 2008. As the oil prices dropped drastically in the second half of 2008, many airlines incurred major hedge-related loss in 2009. For example, American Airlines and United Airlines hedged 34% and 29% of their expected 2009 fuel consumption at $99 and $114 per barrel, respectively. Due to the drop in oil prices from $146 barrel in July 2008 to $45 per barrel in March 11, 2009, many airlines were

Fig. 8.1 Crude Oil Price ($ per barrel) from January 1, 2008 to March 28, 2009

facing major hedge-related losses in addition to their operations-related losses. For example, Cathay Pacific airlines posted $1 billion loss in March, 2009—the biggest annual loss in its 63-year history—mainly due to its oil hedge-related loss, and even Southwest airlines reported its third consecutive loss in April 2009—first time in its history (c.f., Esterl (2009)).

Thus, companies need to carefully consider the costs of hedging versus the costs of not hedging. If hedging is called for, companies need to decide how much to hedge.

8.2.2 Risks Associated with Labor Costs

As increasing numbers of multinational firms outsource or offshore their manufacturing (or software development) operations to countries such as China and India, a shortage of labor is created. As multinational firms face significant increases in labor costs (Table 8.1), some companies seek to develop more skilled labor in China or India as an effective strategy. For example, Motorola established its "Motorola University" in Beijing to train their new recruits.

Even though the labor cost is rising significantly in various developing countries, most multinational firms are slow to respond. However, many firms are starting to examine whether to relocate their facilities in the interior of China (where salaries and land are lower) or other low cost countries such as Vietnam and Malaysia. For example, General Motors, Honda, Motorola, and Intel have all shifted some of their manufacturing operations to the interior locations, while Flextronics has recently established a new production facility in India (c.f., Roberts (2006)). A company could also relocate facilities back to the home country. Indeed, GE CEO, Jeff Immelt, went on record in March 2011 to say that GE can operate call centres in the US at a cost that was only 10 percent more than that in India owing to technology (although a

Table 8.1 Average monthly labor cost in selected countries

	2006	2007	2008	2009 (estimated)	2010 (estimated)	Average Increase (%)
China	$ 70.0	$ 80.5	$ 127	$ 146	$ 168	15%
Thailand	$ 69.0	$ 79.4	$ 91.3	$ 104.9	$ 120.7	15%
Vietnam	$ 63.0	$ 71.2	$ 80.4	$ 90.9	$ 102.7	13%
India	$ 68.5	$ 76.7	$ 85.9	$ 96.2	$ 107.8	12%
Brazil	$ 73.8	$ 81.2	$ 89.3	$ 98.3	$ 108.1	10%
Malaysia	$ 70.5	$ 76.1	$ 82.2	$ 88.8	$ 95.9	8%
Mexico	$ 440	$ 451	$ 462	$ 474	$ 486	4%

Source: International Labor Organization

part of the reason is certainly the rapidly rising cost of skilled labor in India that continues to be in short supply).[2] By assuming that labor cost continues to grow at around 17% a year in China (Fig. 8.1) Hal Sirkin of Boston Consulting Group (BCG) claims that manufacturers will be indifferent between locating in America or China for production for consumption in America by 2015.[3]

As companies have multiple facilities located in different countries, they have the flexibility of shifting their production from one country to the other. This flexibility would enable a company to develop operational hedging strategies that would allow them to shift their production from a higher labor cost country to a lower labor cost country.

8.2.3 Risks Associated with Foreign Currency Exchange

Many firms use global supply chains to perform different operations along the supply chain in an effort to exploit the cost advantages of different countries. Consequently, these firms have to process accounts payable (payments for the suppliers) and accounts receivable (payments from the distributors, wholesalers, or retailers) in different currencies. As these firms source more from foreign suppliers and collect revenues from more countries, they are vulnerable to the risks associated with foreign currency exchange owing to fluctuations. As reported in Dornier et al. (1998), currency fluctuations of 20% in one year are not uncommon. For example, within a 3-month period, the exchange rate between Euros and US dollars has changed $7.1\% = (0.75 - 0.7)/0.7 \times 100\%$ from December 28, 2008 to March 28, 2009 (Fig. 8.2).

The need for firms to process accounts payable, accounts receivable, or debt payments in various currencies exposes them to currency exchange risks, especially when there is a lag between the time the contract is signed and the time the pay-

---

[2] J. Bartash, 2011. Immelt defends G.E.'s lower tax bill, Defense Today, accessed at http://todaysdefense.com/?p=751 on May 15, 2011.

[3] Economist, May 12, 2011, http://www.economist.com/node/18682182

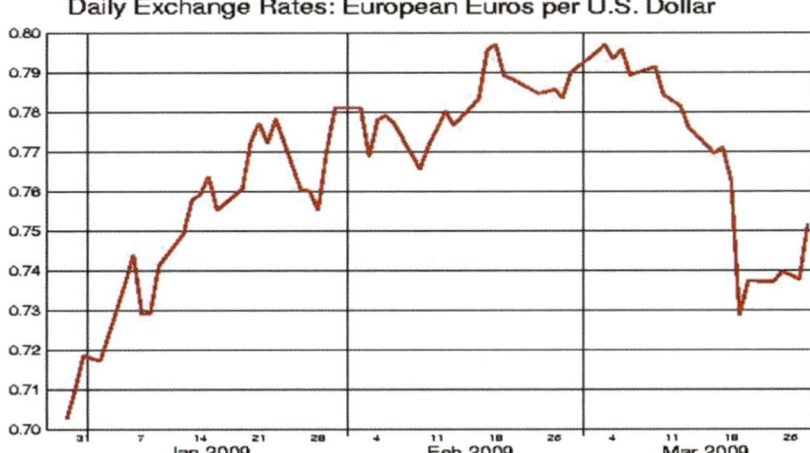

Fig. 8.2 Exchange rate between Euros and US dollars from December 28, 2008 to March 28, 2009. *Source:* Pacific Exchange Rate Service

ment is made. Consider the following *hypothetical* example: A US firm designed an innovative product to be produced by a Japanese supplier (who will charge 100 million Japanese yen) for a Swedish customer (who will pay 8 million Swedish kroner (sek))[4]. Both supply and sales contracts were signed in March 2008, shipment was made in January 2009, and the accounts payable and receivable are due in March 2009. In March 2008, the exchange rates were 99 yen per US dollar and 5.9 sek per US dollar. However, over the course of one year, these rates became 97 yen per US dollar and 8.1 sek per US dollar in March 2009. Had the transactions taken place in March 2008, the profit margin of this product for the US firm would have been US$ (8,000,000 / 5.9) − (100,000,000/99) = 1,355,932 − 1,010,101 = US$345,831. However, as the payments took place in March 2009, the actual profit margin became negative US$ (8,000,000/8.1) − (100,000,000/97) = 987,654 − 1,030,927 = US$(−43,273). Hence, this US firm ended up losing US$ 43,273 due to the devaluation of the Swedish kroner. This example, though hypothetical, highlights the fact that it is important for companies to develop strategies to mitigate this risk.

8.2.4 Risks Associated with Supplier Defaults

Due to the current global economic crisis, consumers in the United States and Europe reduced their spending in late 2008. Consequently, many American and Eu-

---

[4] Clearly, the US firm can negotiate with the Japanese supplier and the Swedish customer to ensure all transactions are conducted in US dollars. By doing so, the US firm transfers the currency exchange risk to their supplier chain partners, which could trigger the supplier to argue for a higher payment and the customer to demand a lower price.

ropean manufacturers either reduced or cancelled their orders with their contract manufacturers (or suppliers). This chain of events has created major problems for China's manufacturing industry. As reported by Ruwitch (2009), over 15,000 factories in the Pearl River Delta were shutting down operations at that time. Most of the factories were labor-intensive, small- or medium-sized producers of metal or plastic products, toys, garments and shoes. Currently, many struggling factories were looking for ways to redevelop products to serve the domestic market. For example, Dongguan's Meng Qiren Fashion, a contract manufacturer that once produced garments for Polo Ralph Lauren, set up an alliance for selling its sweater and woolen garments in department stores (c.f., Quek,2009).

In the United States, the economic downturn of 2007–08 caused many automotive suppliers to declare bankruptcy in the United States. Automotive News (2006) had already warned in 2006 that 38% of US auto parts makers were in fiscal danger and could face bankruptcy by 2008. The financial situation of many US auto suppliers has deteriorated further in 2009. Bailey (2009) reported on a study conducted by A. T. Kearney claiming that over 50% of the US auto suppliers could file for bankruptcy protection by the end of the year in 2009. On February 4, 2009, 400 auto suppliers, including the giant auto suppliers American Axle and Visteon Corp asked for $25.5 billion of federal aid, because of cash flow problems caused by the delayed payments from the US automakers and have asked the government to ensure that payment terms were 10 days rather than the usual 55+ days.

In Europe, when BMW designed its new Z4 convertible, the company decided to source its sun roofs from a reputable German supplier, Edscha. However, Edscha created a dilemma for BMW when it filed for insolvency in February 2009. If BMW were to find an alternative supplier, it would take them at least six months to get qualified and be ready for producing the convertible top. Even so, due to unproven processes, the new supplier might not be able to produce sun roofs that meet BMW's high quality standard. On the other hand, An alternative would be for BMW to provide financial subsidies to Edscha to keep it afloat although it was unclear whether Edscha could survive in the long run even with BMW's support. In March 2009, BMW decided to provide financial support for Edscha to be able to launch the new Z4 according to the original schedule. At the same time, BMW was made keenly aware of initiating the process of developing an alternative supplier just in case (c.f., Milne, 2009).

8.3 Financial Instruments for Managing Input Material Cost and Foreign Exchange Risks

To manage the input material cost and foreign exchange risks, firms usually consider using a combination of three basic financial instruments: *forward contracts*, *futures contracts*, and *options contracts*. The underlying mechanism for each of these three financial instruments is summarized in Table 8.2.

Table 8.2 Comparisons among forward, futures, and options contracts (these apply to selling as well as to buying)

	Instrument		
	Forward contract	**Futures contract**	**Options contract**
Underlying mechanism	The firm agrees to buy an asset (e.g., a commodity or foreign currency) at a *fixed price* at a specific time in the future.	The firm agrees to buy a standardized amount of an asset at a *fixed price* at a standardized delivery time in the future.	The firm pays a premium for the right—but not the obligation—to buy an asset at the *strike price* anytime before the expiration date of the contract.
Underlying risk	Risk avoidance	Risk avoidance	Risk mitigation
Advantages	Customized to meet the needs of the firm (time and quantity).	Can be traded on the commodity (or foreign exchange) market; and therefore less costly.	Can be traded anytime before expiration. Can make an additional profit by exercising the option when the market price is above (or below) the strike price associated with the call (or put) option contract.
Disadvantages	More costly; and inflexible—the contract cannot be changed upon agreement. Cannot be traded on the exchange.	Inflexible (in terms of delivery date and quantity)	Does not eliminate risk entirely. The firm can lose the entire premium associated with the contract.

According to a survey study conducted by Carter and Vickery (1989), 85% of the multi-national companies surveyed conduct some of their transactions in US dollars to avoid currency risk, 20% of the firms reported the use of forward contracts to avoid foreign currency risks, and 35% of the firms reported the use of futures (or options) contracts to manage foreign currency risks. As reported in Headley and Tufano (1994), forward contracts and options are common tools for companies to manage currency exchange risks in the computer industry. For instance, IBM bought over $1 billion worth of forward contracts and $1 billion options and Dell bought $45 million forward contracts and $435 million of options to hedge against foreign currencies. Below we explain each of these three financial instruments.

Forward contracts. A forward contract is an agreement between two parties to buy or sell an asset (e.g., a commodity product or a foreign currency) at a pre-specified price (i.e., forward price) at a specific point of time in the future. In general, if S is the spot (current) price of an asset, and r is the continuously compounded rate, then the forward price N months in the future must satisfy: $F_N = Se^{rN}$. In the context of foreign currency exchange, an N-month forward rate is used in agreements to exchange one currency for another currency N months in the future, and it is calculated based on three factors: spot (current) rate, the interest rate of the base currency, and the interest rate of the quote currency.

Consider a US firm that sells C\$1,000,000 worth of products to a Canadian retailer on March 28, 2009. If the US firm gets the payment immediately, then it should receive US\$808,200 = Spot rate × C\$1,000,000. (Table 8.3 provides the Spot rate and forward rates for Canadian dollars on March 28, 2009). However, due to the terms and conditions stipulated in the contract, the accounts receivable is due 3 months (90 days) later. If this US firm does nothing to hedge against currency risk, then the net payment for the US firm will depend on the actual exchange rate three months from the date of the transaction. If this US firm decides to hedge against this risk by "locking in" the exchange rate, it can do so by entering a forward contract with a bank by paying a premium to exchange C\$1,000,000 for US dollars on June 28, 2009 (i.e., 3 months after March 28, 2009) according to the 3-month forward rate 0.8092. Hence, this US firm is guaranteed to receive US\$809,200 on June 28, 2009. By paying an "insurance" premium, this US firm avoids the currency risk.

Table 8.3 Canadian Dollar Forward Rates (US Dollars per 1 Canadian Dollar) on March 28, 2009

	Rate
Spot	0.8082
1 month forward	0.8083
3 months forward	0.8092
6 months forward	0.8105

Source: Pacific Exchange Rate Service (fx.sauder .ubc.ca). Forward rates for other currencies can be found on www.fxstreet.com.

Forward contracts, by design, are customized to suit an individual firm's need and are therefore easy to create. However, such contracts have two drawbacks. First, once a firm buys a forward contract, it cannot be easily sold in a secondary market. Also, there is an underlying default risk associated with the bank issuing the forward contract.[5] Second, a forward contract is generally more expensive than futures contracts as it is created specifically for a company's specific need at a specific point in time.

Essentially, corporations and investors can buy and sell futures and options products available on the CME group for basic commodity products (wheat, corn, etc.),

---

[5] Of course, one can consider buying credit default swaps to hedge against bank defaults; however, as we witnessed in late 2008, credit default swaps created the global financial crisis. In 2009, the US government is reviewing ways to regulate credit default swaps. We shall discuss credit default swaps in the appendix.

metal products, foreign currencies, energy products (crude oil, ethanol, etc.), and even weather products (US cooling monthly, Hurricane seasonal, etc.).[6]

Futures Contracts. A futures contract is a standardized contract, traded on a futures exchange such as the CME group in the US, to buy or sell a specific commodity or currency of standardized size at a certain date in the future. (For example, wheat is a commodity traded in 5,000 bushels as standard size, and Euro is a currency traded in 125,000 Euros as standard size.)

Consider a bakery that uses wheat (as input materials) to make and sell baked goods in the US. After experiencing a significant increase in commodity prices in 2007 and 2008, this bakery decided to buy futures contracts of wheat to hedge against a price increase of wheat in September 2009. The spot price of wheat is 563 cents per 5,000 bushels on April 3, 2009 (Fig. 8.3). However, the futures price of wheat is 600 cents per 5,000 bushels for the September delivery that takes place on September 15, 2009. (Table 8.4 provides the futures price of wheat for delivery in July, September, December of 2009 and March of 2010.)

If the bakery buys a futures contract of 1,000,000 bushels of wheat on April 3, 2009 for the September delivery, then it must pay $1,200 = 600$ cents $\times$ (1,000,000/5,000) to secure the price of wheat at 600 cents per 5,000 bushels for the September delivery. When September comes, this firm will save some money if the actual wheat price is above 600 cents per 5,000 bushels and will lose some money if the actual wheat price is below 600 cents. Unlike a forward contract, the bakery has the flexibility to sell this futures contract at a later date (e.g. June 2009), if it thinks the price of wheat in September is way below 600 cents.

Table 8.4 Futures price of wheat per 5,000 bushels traded on CME on April 3, 2009

	Jul 2009	Sep 2009	Dec 2009	Mar 2010
Futures Price of Wheat (per bushel)	575 cents	600 cents	622 cents	637 cents

Besides commodity futures, the CME group now trades eight currency futures in the amount of $1 billion on a daily basis. These eight currencies are US dollars, Euros, British Pounds, Japanese Yen, Swiss Francs, Australian Dollars, Canadian dollars and New Zealand dollars. For example, as quoted on CME on April 3, 2009, the spot rate was 1.3488 US dollars per Euro. The futures rates of US dollars per Euro were 1.3485 (Jun 2009), 1.3482 (Sep 2009), 1.3472 (Dec 2009), and 1.3485 (Jun 2010). (The standard size for each trade is 125,000 Euros.)

Futures contracts have two drawbacks: (1) Futures contracts are standardized in terms of time of delivery and the size of contract, which cannot be customized to fit the need of an individual firm. (2) The buyer of a futures contracts is obligated to buy (or take delivery) under the terms of the contract (delivery time and the quantity).

---

[6] The CME group (www.cmegroup.com) was created in 2007 after the Chicago Mercantile Exchange merged with the Chicago Board of Trade. The CME group acquired New York Mercantile Exchange in 2008.

Fig. 8.3 The spot price of wheat per 5,000 bushels (563.4 cents) traded on CME from January 1, 2008 to April 3, 2009

To overcome these two shortcomings, a firm can consider buying an option contract that offers it the right, but not the obligation, to buy or sell a particular asset (or currency) at a later date at an agreed upon price. In return for this option, the buyer of the option contract must pay a premium to the seller. We now explain the option contracts.

Options Contracts. Unlike the forward and futures contracts that enable a firm to lock in the price of a commodity (or a specific rate of a foreign currency), a firm can buy an options contract (or, as referred more typically, an option) by paying a premium to the seller. This option grants the buyer the right—but not the obligation—to buy (or sell, as the case may be) a particular asset at an agreed price (strike price) at a future date. There are two basic types of option contracts: a *call option* that gives the buyer the right to *buy* the underlying asset and a *put option* that gives the buyer the right to *sell* the underlying asset. If the buyer chooses to exercise this right, the seller is obligated to sell (or buy) the asset at the agreed price. If the buyer chooses not to exercise the right, then he can let the contract expire.

Let us consider the bakery example. On April 3, 2009, the spot price of wheat was 563 cents per 5,000 bushels (Fig. 8.3). Instead of buying a futures contract at 600 cents per 5,000 bushels for the September delivery, the bakery would like to consider buying an option contract. For the September delivery, the put and call options for different strike prices are provided in Table 8.5.[7] For example, if the bakery buys a call option of 1,000,000 bushels at the strike price 600 cents per 5,000 bushels. In

---

[7] This example is based on American option that allows a buyer to exercise the option at any time before expiration. This is different from European option that allows the buyer to exercise the option only at a short period of time right before the expiration date.

this case, he has to pay a premium of $1,260 = $0.63 × (1,000,000/5,000) (based on a premium of 63 cents per 5,000 bushels). For a premium of $1,260, he gets the right—but not obligation—to buy wheat at the strike price of 600 cents per 5,000 bushels for up to 1,000,000 bushels anytime before September 15, 2009.

Table 8.5 Premiums for Put and Call Options of wheat per 5,000 bushel (expired on September 15, 2009) quoted on April 3, 2009

Strike Price	580 cents	590 cents	600 cents	610 cents
Put option	59.1	64.4	70.1	75.7
Call option	72.0	67.4	63.0	58.4

If the actual price is below 600 cents per 5,000 bushels between April 3 and September 15, then the bakery would not exercise its right to buy the wheat at the strike price (which is higher). In this case, the bakery will lose the premium. However, if the actual price turns out to be 700 cents per 5,000 bushels before September 15 and the bakery buys the entire 1,000,000 bushels at the strike price (600 cents per 5,000 bushels), then this call option contract yields a profit of US$ (0.7 − 0.6) × (1,000,000/5,000) − premium = $2,000 − 1,260 = $740. (Fig. 8.4 presents the profit and loss associated with this bakery's call option contract at 600 cents strike price.)

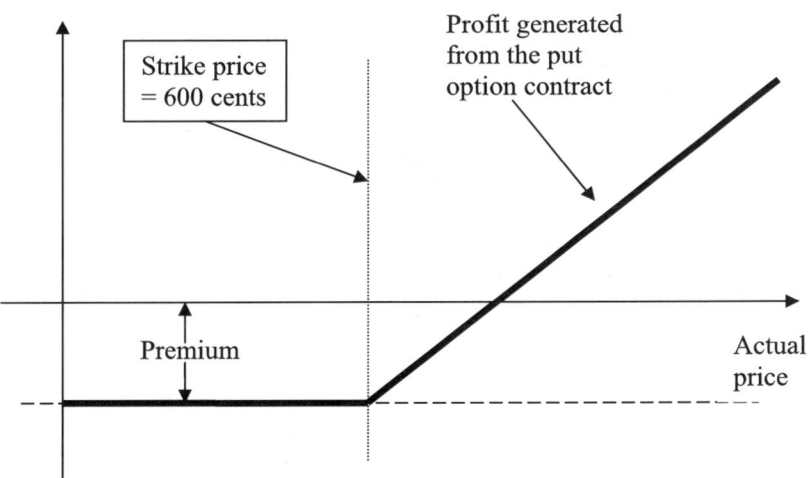

Fig. 8.4 The profit generated by the call option contract with strike price at 600 cents per 5,000 bushels

Now let us consider the case when the bakery buys a put option of 1,000,000 bushels at the strike price of 600 cents per 5,000 bushels. Based on a premium of 70.1 cents per 5,000 bushels (Table 8.5), it pays a premium of $1,402 = $0.701 ×

(1,000,000/5,000) for the right to sell wheat at 600 cents per 5,000 bushels up to 1,000,000 bushels by September 15, 2009. If the actual price is above 600 cents per 5,000 bushels between April 3 and September 15, then the bakery would not exercise its option to sell wheat at the strike price (which is lower). In this case, the bakery will lose the premium. However, if the actual price turns out to be 500 cents per 5,000 bushels before September 15 and the bakery sells the entire 1,000,000 bushels at the strike price (600 cents per 5,000 bushels), then this put option contract yields a profit of US$ $(0.6 - 0.5) \times (1,000,000/5,000) -$ premium $= \$2000 - 1,402 = \$598$. (Fig. 8.5 presents the profit and loss associated with this bakery's put option contract at 600 cents strike price.)

As observed from Figs. 8.4 and 8.5, the call option contract protects the buyer from a price increase while the put option contract protects the buyer from a price decrease. Therefore, it is certainly possible for a firm to buy a combination of call option and put option contracts so as to obtain some protection from price fluctuations. The reader is referred to Hull (2006) for a detailed discussion on ways to determine an optimal combination of option contracts.

8.4 Financial Instruments for Reducing Supplier Default Risks

As more firms outsource their manufacturing operations to developing countries such as China, Vietnam, Indonesia, Malaysia, etc., they face supplier default risks. Due to the fact that the credit market is not well developed in most developing countries, many suppliers (or contract manufacturers) do not have easy access to credit and need working capital to conduct their business transactions. In most cases, these

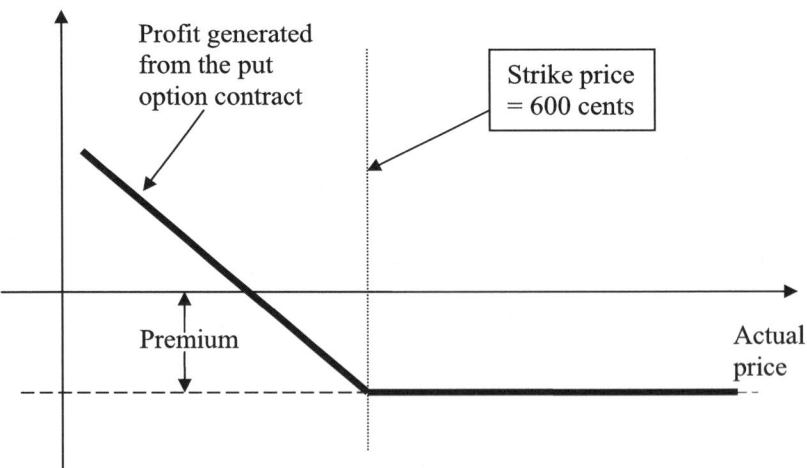

Fig. 8.5 The profit generated by the put option contract with strike price at 600 cents per 5,000 bushels

suppliers are too small to consider using those aforementioned hedging mechanisms to reduce financial risks. Consequently, these suppliers can go bankrupt suddenly, which can disrupt the supply operation for their customers. Consider the case when Indonesia rupiah devalued for over 50% in 1997[8], many Indonesian suppliers were unable to pay for the imported components or materials and hence, unable to produce the finished items for their U.S. customers[9]. For instance, Indonesia's national car manufacturer, Astra, suspended their production because they were unable to pay for imported components. In addition to supply disruption, a supplier may forgo investment in process improvement and sacrifice output quality when facing a credit crunch.

What should a firm do when one of its suppliers is facing a financial crisis? Clearly, if the firm has an established "backup" supplier, it may be able to shift their order quickly to this alternative supplier. We shall discuss this in a later section in the context of operational hedging. However, most firms either do not have backup suppliers or it takes a long time to shift the order from one supplier to another due to the requisite equipment or skills. In this section, we discuss three financial instruments that can be used to reduce the supplier default risks, namely, financial subsidies, factoring, and reverse-factoring. Table 8.6 highlights the differences among these three instruments.

Financial Subsidies. The firm agrees to provide its suppliers with some temporary financial help. Depending on the agreement, the suppliers are obligated to ensure undisrupted supply operations for the firm. In some cases, the suppliers are required to pay back the financial loan at a low interest rate. For example, to ensure smooth supply operations, Li and Fung, the largest trading company for consumer goods (textile and toys) based in Hong Kong, provided financial assistance (line of credit, loans, etc.) to affected suppliers in Indonesia to ensure that their U.S. customers will receive their orders as planned (c.f., Tang (2006a)). When Visteon (Ford's largest supplier) was on the brink of bankruptcy in 2005, Ford subsidized $1.6 billion to Visteon (including a $250 million loan) so that this supplier could stay afloat through major re-structuring (c.f., Babich (2008)). Besides Visteon, the Big Three US auto manufacturers (GM, Ford, and Chrysler) provided financial subsidies to a key supplier Collins & Aikman after it filed for bankruptcy protection in 2005 as a way to ensure undisrupted supply operations.

During the economic downturnv of 2008–09, many suppliers were facing financial difficulties and eventual bankruptcy. As more suppliers are asking for financial help from their customers, many firms need to develop ways to determine effective way to allocate their financial subsidy budget among different suppliers. Based on an empirical study of supplier defaults, Wagner et al. (2009) concluded that the bankruptcy filing suppliers are not independent. As such, when offering financial

---

[8] Indonesia rupiah opened the year 1997 at 2363 to the US dollars and closed at 5550 against the dollar. However, in July 1997, Indonesia rupiah was traded at 10,000 against the US dollars.

[9] The currency crisis affected Indonesia in a very serious manner in 1997. Also, 60% of Jakarta's public transport system was suspended, because of the soaring price of the spare parts needed to repair the city's buses.

Table 8.6 Comparisons among financial subsidies, factoring, and reverse-factoring

Instrument	Financial subsidies	Factoring	Reverse factoring
Underlying mechanism	A firm provides financial aid to its supplier directly.	A supplier sells its accounts receivable to a lender at a discount for quick payment.	A lender establishes a line of credit with a firm in advance in order to provide quick payment to its supplier at a discount.
Underlying risk	Risk avoidance	Risk transfer	Risk transfer
Advantages	The firm may be able to develop a stronger relationship with the supplier.	The firm pays the lender at due date.	The firm pays the lender at due date.
Disadvantages	The firm may incur a loss should the supplier default.	Without credit history of the firm, the lender could face high credit risk. According to accounting rules, accounts payable are rendered as *debts* on the firm's balance sheet.	With an upfront audit of the firm's credit history, the lender can reduce its credit risk. According to accounting rules, accounts payable are rendered as *debts* on the firm's balance sheet.

help to suppliers, a firm should allocate its subsidy to suppliers with independent default risks so as to reduce the risks of losing the entire financial subsidy. Separately, Babich (2008) provides conditions under which it is optimal for the firm to subsidize a supplier up to a certain level by considering a dynamic, stochastic and periodic-review mathematical model.

Factoring. A supplier sells its accounts receivable at a discount (interest plus service fees) to a specialized financial institution (the factor) and receives cash immediately. Factoring is a traditional type of financing intended to provide quick payments to suppliers who do not wish to wait for the direct payments from their customers. In the US, it is common for a supplier to wait for 90 days after delivering their products to its customers. However, the 90-day waiting period is not acceptable for suppliers in developing countries who need working capital (or cash) to conduct their business transactions (such as ordering raw materials) because in many developing countries, such as Russia, smaller suppliers have virtually no access to credit. As such, they either demand short payment terms (e.g., 5 days) or demand a deposit (e.g., a 30% pre-payment) (c.f., Corsten and Saraf, 2008).

While the use of factoring is common in developed countries, it is not very common in emerging markets for the following reasons: First, the factor (lending institution) is reluctant to take on large credit risk especially when the credit history of the supplier is not well established. Second, the factor is reluctant to deal with the legal problems when some of the account receivables are bogus customers especially in countries with an unstable legal system. To overcome this obstacle associated with factoring in Russia, Nestlé Russia and Citibank developed a new scheme that is called "reverse factoring" that can be described as follows. (The reader is referred to Corsten and Saraf, 2008, for details.)

Reverse Factoring. A lender (a bank) buys account receivables from suppliers under an acceptance from a reputable buyer (a multi-national firm) who commits to make his payment to the lender when it is due.

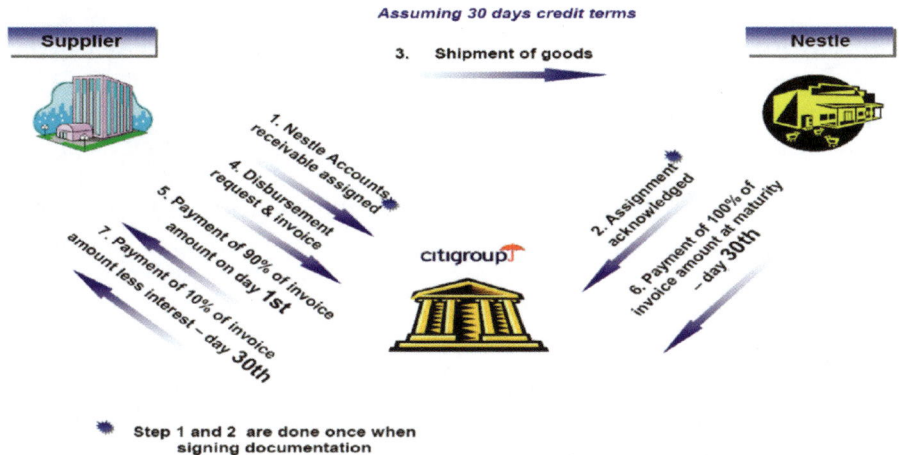

Fig. 8.6 Nestlé and Citibank's Reverse Factoring Scheme. *Source:* Corsten and Saraf (2008).

Let us consider the reverse factoring scheme developed by Nestlé Russia as depicted in Fig. 8.6 (c.f., Corsten and Saraf (2008). In this reverse factoring program, Nestlé Russia (the buyer) provides a list of suppliers who accepted the reverse factoring payment scheme to Citibank (the lender). Each supplier would set up an account with Citibank (Step 1) to be confirmed by Nestlé Russia (Step 2). Once Nestlé Russia received and accepted the shipment from a suppler (Step 3), the supplier will submit a disbursement request along with the invoice to Citi (Step 4). Citi will pay, for example, 90% of the invoice on the first day after submission (Step 5) and the remainder, say, 10% of the invoice less interest to the supplier at the original due date (30 days) after Citi receives 100% payment from Nestlé at the due date (Step 6 and Step 7). The reader is referred to Corsten and Saraf (2008) for details.

Reverse factoring can benefit the lender, the supplier, and the buyer. First, reverse factoring reduces the lender's credit risk because the lender has easy access to the credit history of the buyer. The credit risk is equal to the default risk of the buyer, which is relatively low. Also, the lender can earn a margin based on the interests and the service fees based on the size of the payment.

Reverse factoring enables the suppliers to improve their working capital financing. Specifically, the accredited suppliers can obtain financing from the lender immediately after shipment and invoice submission. Essentially, reverse factoring enables suppliers to transfer their credit risk and borrow on the credit risk of its credit-worthy buyers. Short cash-to-cash cycles are critical for suppliers to function smoothly in the emerging market. As such, reverse factoring is appealing to small to medium size suppliers who do not have sufficient credit history in order to gain access to credit.

Finally, reverse factoring is beneficial to a reputable multi-national firm to establish its presence in an emerging market with a limited supply base. With an easy and quick payment scheme, this firm has a competitive advantage over other firms for getting supplier cooperation. Also, by transferring the payment operations to a single lender, the firm needs to deal with a single lender for all account payables instead of managing different payment terms with different suppliers. By focusing on a single lender instead of thousands of suppliers, this firm can focus on more important issues such as product development and marketing efforts.

We should note that there is some reluctance to adopt reverse factoring in developed countries. Part of the reason may be that accounting rules require that the amount now owed to the lending institution (bank) be treated as debt in the company's balance sheet, discouraging many companies to offer reverse factoring to their suppliers. There are also complications in case the buyer has to return goods to the supplier, owing to say, poor quality.

8.5 Operational Planning for Managing Cost Risks and Supplier Default Risks

We have presented different financial instruments that would enable a manufacturer to reduce the risks associated with costs (input materials, labor, and currency exchange) and supplier default. We now examine different operational strategies that would also reduce these risks. While financial and operational approaches serve similar purposes, they usually require different implementation plans. Specifically, the development of an effective operations strategy requires an in-depth analysis of the supply chain operations and its execution requires collaboration and coordination among supply chain partners. However, unlike a financial approach. an effective operations strategy enables a firm to manage risks for the medium and long term.

Relatively speaking, the aforementioned financial instruments are easier to execute, especially because there are well established institutions (CME and banks) to facilitate the process. However, financial instruments are intended to mitigate short-term risks. Chowdhry and Howe (1999) commented that there is empirical evidence for firms to use financial instruments to hedge short-term risks (but not long-term risks).

This may explain why firms tend to use a combination of various financial instruments and operations strategies to mitigate supply chain risks. We shall discuss the benefit of integrating financial and operational hedging in the next section. Recently, Chod et al. (2009) presented a mathematical model to examine the relationship between financial hedging and operational (flexibility) hedging. They show that, when the risk aversion of a firm is sufficiently low, financial hedging and flexibility are substitutes. However, when the firm is highly risk averse, financial hedging and flexibility are complements.

To effectively manage the risks associated with supply costs and supplier default, a firm usually needs to examine the following issues: (1) supplier selection—who should they source the parts from? (2) supply contract conditions—what is the duration of the contract? (3) capacity planning—how much capacity should they expect the supplier to plan for? (4) production planning—how much should they order from the supplier? (5) allocation planning—in case of changes in cost or supply, how should the firm allocate the production quantity to different markets? The reader is referred to Tang (2006b) for a review of supply chain literature on each of these topics. We now briefly discuss the first two issues and then expand our discussion to the remaining topics.

In the presence of supply chain risks, a firm must first establish selection criteria (cost, quality, reliability, flexibility, technological capability, and financial health) when evaluating suppliers. As reported by PwC (2008) and McKinsey (2008), cost and quality remained the top two supplier selection criteria. However, based on a survey of auto suppliers and automakers in the US, Choi and Hartley (1996) find that financial issues are a primary supply selection factor. Their finding could reflect the financial troubles of the US automotive industry. Babich et al. (2007) presented a mathematical model to examine the impact of supplier default risks on supplier selection. Suppliers competing for business with a single retailer, they become involved in a Stackelberg game[10] in which two competing suppliers act as leaders by setting wholesale prices and the retailer acts as the follower by choosing the order quantity. They show that the equilibrium wholesale prices increase as the correlation to the supplier default risk decreases. Therefore, to ensure the wholesale price is low, the retailer should select suppliers with highly correlated default risks.[11]

Once a firm has selected its supplier(s) to produce certain components (or products), the firm needs to decide whether to establish a long term relationship with the supplier(s). Despite recent trends of supply-base reduction and long-term contracts, non-cooperative supplier relationship and short-term mentality continue to persist especially in the US. Because long term contracts (or relationship) would weaken a manufacturer's bargaining position for price negotiations with the suppliers, there is no incentive for establishing long term contracts unless there is an economic incentive for doing so. Cohen and Agrawal (1999) compare long- and short-term contracts in the presence of risks associated with cost uncertainty. By analyzing a model of one manufacturer and one supplier, they determine the conditions under which the manufacturer should issue a long term contract. Swinney and Netessine (2009) an-

---

[10] A Stackelberg game in economics is one in which a leader firm(s) moves first and follower firms move subsequently. It is named after the German economist Heinrich Freiherr von Stackelberg who described the model in 1934.

[11] This result appears to be counter-intuitive at first glance because, when the suppliers are "passive" (say, fixed wholesale prices), there is a common belief that the retailer can benefit more from selecting a set of suppliers with negatively correlated default risks (c.f., Wagner et al. (2009)). However, when suppliers are active leaders in the Stackelberg game, Babich et al. (2007) show analytically that this common belief does not hold.

alyze a model of one manufacturer and two competing suppliers and show that it is rational for the manufacturer to issue long term contracts in the presence of supplier default risks. This is because long term contracts enable a firm to offer higher prices to the supplier, which reduces the supplier's likelihood of default.

We now discuss how firms can use operational hedging—capacity planning, production planning, and allocation planning—to mitigate the risks associated with uncertain supply costs and supplier defaults. Operational hedging is intended to enable a firm to manage risks associated with uncertain supply costs (due to uncertain input material and labor costs), uncertain demand (due to market uncertainty), and supplier defaults by responding quickly to market dynamics. To develop a responsive supply chain, two fundamental capabilities are critical; namely, the capability for in-season replenishment, and the capability to switch sourcing locations.

8.5.1 Capability for In-Season Replenishment

Without strong incentives in place, most suppliers do not allow in-season replenishments in the fashion industry. As such, most retailers can place a single order before the start of a selling season. Besides inaccurate demand forecasts, the uncertain supply costs and supplier default risks create additional challenges for the buyers (retailers or manufacturers) to decide on the order quantities. As articulated by Fisher et al. (1994), the success of Sports Obermeyer is due to its in-season replenishment capability so that the firm can reduce over-stocking of some products and understocking of others. Clearly, in-season replenishment can reduce the retailer's risks, but it increases the supplier's risk due to long lead times for procuring raw materials and for production and inefficient capacity utilization. As such, suppliers are reluctant to allow in-season replenishment unless there is a strong incentive. In recent years, various researchers have developed models for analyzing different incentive mechanisms for the suppliers to allow in-season replenishments.[12]

Backup agreement with a single supplier. Under this agreement, a buyer "reserves" R units of capacity with a unit cost c from a supplier for the entire selling season. Prior to the start of the season, the buyer orders $X1$ units, where $X1 \leq R$. After observing the market condition (say, demand) during the early part of the season, the buyer can place an backup order (i.e., an in-seasonal replenishment) of units $X2$, where $X1 + X2 \leq R$. For each unit of unused capacity, the buyer has to pay a penalty b. In this case, the buyer's total cost is equal to $c(X1 + X2) + b(R - X1 - X2) = bR + (c - b)(X1 + X2)$. Under this backup agree-

---

[12] Burnestas and Ritchken (2005) investigate the conditions under which the supplier should provide the retailer the right to reorder (return) products at a pre-specified price via call (put) options. By considering the case when the underlying demand curve is downward sloping and by solving a Stackelberg game in which the supplier acts as the leader who sets the premium and the strike price and the retailer acts as the follower who chooses the order quantity and the retail price, they show that this form of option contracts will cause the volatility of the equilibrium retail price to decrease.

ment contracts, the supplier sets the cost parameters c and b, while the buyer selects the reserved capacity R, and the order quantity $X1$ and $X2$. Eppen and Iyer (1997a) reported that backup agreement contracts are used by suppliers such as Anne Klein, DKNY, Liz Claiborne, and buyers such as Catco. Also, they examine the benefits of the backup agreements to the buyer when customer demands are correlated over time.

Pay-to-delay capacity reservation with a single supplier. Under this contract, a buyer "reserves" R units of capacity by paying a fixed cost K prior to the start of the season. Then, the buyer can order X units, where $X \leq R$ anytime during the season by exercising his option at a unit cost c. This pay-to-delay contract allows the supplier to have ample of time to establish the capacity and it allows the buyer to obtain more accurate demand forecasts by postponing the ordering decision. In this case, the buyer's total cost is $K + cX$. Under this contract, the supplier sets K and c, while the buyer decides on the reserve capacity R upfront, and then the order quantity later on. This form of contract has been used in the semiconductor industry by Xilinx. Brown and Lee (1997) analyze a two-period model and establish the conditions under which pay-to-delay is beneficial. Barnes-Schuster (2002) presents a general model that combines the features of backup agreement and pay-to-delay capacity reservation. They show that, in equilibrium, the value of in-season replenishment capabilities depends on the coefficient of variation of the demands and the correlation of the demands over time. In addition, they develop analytical conditions under which backup agreement contracts are beneficial.

Backup supplier(s) with short replenishment lead times. Under this agreement, the buyer establishes arrangements with two different suppliers: a long-term supplier and a "spot" supplier. For the long-term supplier, the buyer can "reserve" R units of capacity by paying cR to the supplier in each period. In return, the long-term supplier agrees to provide up to R units of production in each period. For the spot supplier, the buyer can order any amount at a unit cost b (i.e., place an order on the spot). Serel et al. (2001) analyze a Stackelberg game in which the long-term supplier acts as the leader by setting c, while the buyer acts as the follower by choosing R. (When demand exceeds R, the buyer will order from the spot supplier.) By considering a discrete time model for the case when there is no secondary market for the unused capacity for the long-term supplier, Serel et al. (2001) show that, in equilibrium, the existence of the "spot" supplier will cause the buyer to reduce the reserved capacity R. Along the same vein, Spinler and Huchzermeier (2006) developed a model of two suppliers for non-storable goods (such as electricity). In their model, the buyer can reserve R units from the regular supplier by paying an upfront fixed cost K. In return, the buyer can postpone his order quantity and can order up to R units by paying c per unit in a later period. Then in each period, the buyer first observes the market conditions and the "realized" spot price from the spot supplier. Then the buyer decides on the order quantity from the regular supplier (up to R units) and the order quantity to be purchased from the spot supplier. By analyzing a Stackelberg game in which the regular supplier acts as the leader who sets K and c, and the buyer as the follower who chooses the order quantity from each supplier, Spinler and Huchzermeier (2006) provide closed form expressions for the optimal value of K, c, R, and optimal order quantity in equilibrium.

8.5.2 Flexibility for Switching Sourcing Locations

With a proactive plan to establish a supply network with excess capacity, a buyer (retailer or manufacturer) can shift its production quantity from one supplier located in one geographical region to a different supplier located in a different region in a seamless manner.[13] For example, the success of Li and Fung is due to its capability of shifting its production from one supplier in one country to a different supplier in a different country as we discussed in Chapter 7. With a supply network of over 6,000 suppliers throughout Asia, Li and Fung has the flexibility to ensure smooth production should any single supplier default. Also, because most contracts are conducted in US dollars, Li and Fung can increase its profit margin by exploiting the fluctuating exchange rates of different currencies in Asia. For instance, as Indonesia's rupiah devalued by 50% in 1997, Li and Fung had a record-breaking year of profits by shifting some of the production from other Asian countries to Indonesia (c.f., Tang (2006a)). Many researchers have examined the tradeoff between the cost of excess capacity and the benefit of the flexibility of shifting production from one location to another.

Essentially, excess capacity provides a firm with the flexibility to shift production from one supplier to a different one in another. By using a conceptual framework, Kogut (1985) argues that, by having excess capacity in the supply network, a firm can take advantage of the exchange rates and tax rates by shifting production from one country to another. In a follow-on study, Kogut and Kulatilaka (1994) developed a single-period, two-country model to evaluate the value of the capability to shift production between two suppliers located in two different countries. Also, they showed how this capability would enable a firm to reduce various types of supply risks including supply cost risks and currency exchange risks. (Having the flexibility to shift production among suppliers can also reduce the risk associated with supplier defaults.) Huchzermeier and Cohen (1996) present a stochastic dynamic programming formulation of a multiple-period, multi-country model to examine the benefit of flexible supply in a global supply network. They use numerical examples to illustrate how a flexible supply would enable a firm to reduce the risks associated with uncertain demand and currency exchange rates. Motivated by a problem arising from a multinational firm that has five plants in Latin America producing synthetic fibers, Dasu and Li (1997) present a general multi-period, multi-country model with

---

[13] The notion of excess capacity is one of the enablers that enable Zara to design, manufacture and ship a new line of clothing within 2 weeks. As reported in Ferdows et al. (2004), there is a tradeoff between response time and capacity utilization. In the fashion industry, response time outweighs the cost of excess capacity, which may explain why Zara continues to be the most profitable European fashion retailer with sales and net incoming growing at an annual rate of over 20%.

uncertain demand and exchange rates. They determine the optimal policies for a firm to decide the optimal time to shift production from one supplier to a different one and the optimal production quantity for each.

Kazaz et al. (2005) presented a model motivated by a global electronics manufacturer that sells products in different countries with exchange-rate fluctuations. To mitigate the risk of currency exchange, they examine two types of operational hedging: *production hedging* (i.e., the firm deliberately produces less than the total demand), and *allocation hedging* (i.e., the firm deliberately decides not to serve certain markets). In their model, production hedging can be employed before the exchange rate is realized. Hence, a firm may choose not to fulfill the total demand when the exchange-rate uncertainty is sufficiently high. However, allocation hedging can be employed after the exchange rate is realized. By formulating a two-period, two-country model as a two-stage stochastic program with recourse, they examine the impact of the correlation of the exchange rates on the benefit of production hedging.

In addition to switching sourcing locations, Kouvelis and Gutierrez (1997) examine switching market locations. Specifically, they consider a situation where a firm exploits the difference in timing of the selling season in different geographical markets for "style goods." For example, at the end of the US selling season in March, a U.S. ski-wear manufacturer could sell his remaining winter fashion items in New Zealand for the upcoming winter season there. By considering the uncertain demand and currency exchange rate in two markets (a primary and a secondary market), they determine an optimal production plan (i.e., how much to produce for both market) and allocation plan (i.e., how much leftover inventory to ship from the primary to the second market?) for a centralized system. When the system is decentralized, they develop a non-linear transfer pricing scheme that would enable the firm to achieve channel coordination so that the firm's profit is the same as in the centralized system.

8.6 Integrated Financial and Operational Hedging

So far we have presented common financial instruments and operational mechanisms for reducing the risks associated with costs of input materials, labor, exchange rates, and supply defaults. We now discuss how a firm can benefit from integrating both financial and operational hedging to mitigate risk even further. The notion of integrating financial and operational hedging has received considerable interest in the research community (c.f., Xu and Birge (2004), and Ding et al., 2007). Chowdhry and Howe (1999) present a model in which a multinational firm produces a product in two facilities in two locations (domestic and foreign), and sells it in these two locations. They show that operational hedging can be a powerful tool for mitigating risks when the multinational firm faces uncertain exchange rates and uncertain demand. In terms of operational hedging, they show that it is optimal for the firm to shift its production from the domestic location to a foreign location when the demand in the foreign location is more uncertain than that of the domestic location. In addition, they show that the optimal financial hedging strategy is a combination of foreign currency (call and put) options and forward contracts.

In general, the models that deal with both financial and operational hedging are complex, and the mathematical treatments of these models are beyond the scope of this chapter. For this reason, we shall use simple examples to illustrate the benefit of integrating financial and operational hedging and refer the reader to Ding et al. (2007) and the references therein for more details. Specifically, the following examples are based on Kouvelis (2006).

Example 1. Consider a farmer who grows and sells corn on a yearly basis. Due to the uncertain weather condition, he estimates that there is a 50% chance that his yield is 200,000 bushels (high yield), and a 50% chance that the yield is 100,000 bushels (low yield). According to historical data, the market selling price is $1 per bushel when the yield is high and $2 per bushel when the yield is low.

- **No hedging.** Suppose the farmer let nature takes its course. Then his revenue is equal to $200,000 \times \$1 = \$200,000$ when the yield is high, and equal to $100,000 \times \$2 = \$200,000$ when the yield is low. Hence, his revenue is equal to \$200,000 regardless of the yield.
- **Financial hedging.** Suppose the farmer sells a futures contract in the form of 150,000 bushels at \$1.5 per bushel that is scheduled to deliver after the harvest season. In this case, his revenue generated from this futures contract is $150,000 \times \$1.5 = \$225,000$. Let us consider the following scenarios. First, if the yield is high (i.e., 200,000 bushels), he will get an additional revenue of $50,000 \times \$1 = \$50,000$ that is generated from selling the remaining 50,000 bushels at the market price. Second, if the yield is low (i.e., 100,000 bushels), he needs to buy 50,000 bushels at the market price in order to honor his futures contract. In this case, he will incur a cost of $50,000 \times \$2 = \$100,000$. In summary, by selling a futures contract, this farmer's net revenue equals $\$225,000 + \$50,000 = \$275,000$ when the yield is high, and equals $\$225,000 - \$100,000 = \$125,000$ when the yield is low. By noting that the expected net revenue is equal to $\$200,000 = 0.5 \times \$275,000 + 0.5 \times \$125,000$, which is the same as the case without hedging. Therefore, using futures contracts only increases this farmer's risk in this example.
- **Financial and operational hedging.** Consider the case that the farmer sells a futures contract as before. However, instead of buying and selling in the open market, he develops the following operational policy: store 50,000 bushels of corn in his silo when the yield is high, and use from the silo when the yield is low. This means he does not need to sell his corn when the yield is high or to buy corn when the yield is low. Ignoring the inventory holding cost and spoilage, his revenue is the same as the revenue generated from the futures contract, which is \$225,000.

Although the usage of a financial instrument alone is not beneficial, this example illustrates an important insight: the farmer can reduce his risk by using a combination of financial instruments and operational mechanisms.

Example 2. Consider a multinational firm that can produce and sell its product in the U.S. and Europe under demand and exchange rate uncertainty. Based on an in-depth analysis, this firm believes that one of the two scenarios will occur: (1) there is a 50% chance that the demand in the US is 100,000, demand in Europe is 50,000, and the exchange rate is $1 per euro; and (2) there is a 50% chance that the demand in the US is 50,000, demand in Europe is 100,000, and the exchange rate is $2 per euro. The production cost is $10,000 in the US and is €10,000 in Europe, and the selling price is $20,000 the US and is €20,000 in Europe.

- **Financial hedging**. Suppose the firm produces and sells in each country (i.e., no shipment between US and Europe) so that the profit margin in the US is $10,000 and is €10,000 in Europe. Suppose the firm sells a futures contract in the amount €500 million at $1.5 per euro. Then the firm will get $750 million based on this futures contract.

  If scenario (1) occurs, then this firm will get $1000 million from the US market and use its €500 million profit in Europe to honor its futures contract. The net profit is $1000 million, the profits from US sales.

  If scenario (2) occurs, then this firm will get $500 million from the US market and will generate €1000 million in Europe. After honoring its futures contract, this US firm can sell the remaining €500 million in the open market at $2 per Euro and get $1000 million. Hence, the net profit is equal to $1500 million.

  In summary, the expected profit is equal to $750 million + 0.5 × $1000 million + 0.5 × $1500 million = $2000 million.
- **Operational hedging.** Suppose the firm develops a responsive manufacturing process so that it can produce only in Europe under scenario (1) and produce only in the US under scenario (2).

  If scenario (1) occurs, then the firm will get $1000 million of revenue from the US and $1000 million of revenue from Europe, and will incur a production cost of $1500 million (in Europe given the exchange rate of $1/euro in this scenario). The net profit is equal to $1500 million.

  If scenario (2) occurs, then the firm will get $1000 million of revenue from the US and $4000 million = 100,000 × €20,000 × $2 /€ from Europe, will incur a production cost of $1500 million in the US. The net profit is equal to $3500 million.

  In summary, the expected profit is equal to 0.5 × $1500 million + 0.5 × $3500 million = $2500 million.

This example illustrates an interesting point: a firm can improve returns by using operational hedging albeit with a larger variance in the returns relative to the use of financial hedging.

The above examples illustrate that to mitigate risks arising from input material costs, labor costs, exchange rates, etc., for a firm should develop an effective combination of financial instruments and operational mechanisms. Ding et al. (2007) examined a more general situation than the one in Example 2. Specifically, they consider the case in which the firm has to determine the production capacity in the US and European facilities before the selling season starts. However, by using the excess capacity as discussed earlier, the firm has the flexibility to shift its production from one facility to the other after the demand and exchange rate are realized. They derive the joint optimal capacity and financial option decision, and show that the firm's optimal financial hedging strategy is linked to the firm's optimal operational strategy.

8.7 Conclusions

In this chapter, we presented various financial instruments and operational mechanisms for mitigating financial risks arising from uncertain costs of input materials, labor, currency exchange rates, and supplier defaults. While these operational mechanisms are intended to mitigate medium- and long-term risks, the financial instruments are designed to hedge against short-term risks. Also, we have presented two illustrative examples to show how firms can reduce risk significantly by employing a strategy that combines financial and operational hedging. As articulated in Chowdhry and Howe (1999), it is important for multinational firms to manage supply chain risks by integrating financial and operational hedging mechanisms. Xu and Birge (2004), Ding et al. (2007), Chod et al. (2009), and others have developed a new research stream that examines the inter-relationship between financial and operational hedging. These mathematical models can be used to develop hypotheses for empirical testing in the future.

Appendix: Credit Default Swap

A credit default swap (CDS) is a contract between two counterparties under which a buyer pays an "insurance" premium (in the form of periodic payments) to a seller. In return, the seller agrees to cover any lost interest or principal on bonds or loans issued by a third party (a firm or a country). CDS contracts and insurance are similar in the sense that the buyer pays a premium and, in return, receives a sum of money from the seller if one of the specified events occurs. However, unlike insurance, the seller of CDS need not be a regulated entity[14]; the seller is not required to maintain any reserves of capital to cover the liabilities (i.e., payments to the buyers should a specified event occur); and the buyer does not need to own the underlying bonds or loans.

The original intent of CDS was to allow a buyer to manage the default risk of a bond or a loan issued by a third party. For example, consider a holder of a corporate bond issued by General Motors (GM). Due to the uncertain fate of GM, this bond holder (the buyer) can hedge the default risk of GM by buying a CDS contract from a seller (e.g., CITI bank). CITI bank earns an "insurance" premium from the buyer;

---

[14] After President Clinton signed the bill into Public Law in 2000, CDS became largely exempt from regulation by the Stock Exchange Commission.

however, CITI bank is liable to pay this buyer the full value of the bond should GM default.

Because CDS does not require the buyer to own the underlying bonds or loans, CDS allows investor to speculate on change in an entity's credit quality. For example, consider the case when a hedge fund speculates that American Insurance Group (AIG) will default on its debt. This hedge fund buys $10 million worth of CDS protection for 2 years from CITI bank, with AIG as the reference entity, at 5% premium per annum. In this case, if AIG does not default, then the CDS contract will expire after 2 years and the hedge fund will have ended up paying $1 million worthy of premium and receiving nothing. On the other hand, if AIG defaults after 1 year, then the hedge fund will have paid a premium of $500,000 and received $10 million from CITI bank. In this case, the hedge fund earns $9.5 million, and CITI bank incurs a $9.5 million loss.

The trade volume of CDS has ballooned from $900 billion since its inception in 2000 to $62 trillion in 2008. According to Bajaj (2008), most CDS have been written on countries (Turkey, Italy, Brazil and Russia) and companies (GM, Merrill Lynch, Goldman Sachs, Morgan Stanley, and Countrywide Home Loans). AIG has been selling $440 billion worth of CDS between 2000 and 2008 to buyers who want to hedge the default risks of those mortgage-backed securities. As the housing market collapse and the economy deteriorate, many home owners default on their mortgages. However, AIG did not hedge against the possibility that these mortgage-backed securities might decline in value. In November 2008, the Federal Reserve extended an $85 billion bridge loan to AIG after defaulting $14 billion worth of claims from various CDS buyers (c.f., Philips (2008)). In 2009, US and European regulators are developing ways to regulate CDS contracts.

Chapter 9
Application—Scenario Planning for Mitigating Supply Chain Restructuring Risk at a PVC Manufacturer

Abstract Companies need to manage its long-term capacity-related risk carefully especially after acquisitions and mergers. In this chapter, we use the case of a plastics manufacturing company to show how a company can align its supply chain operations post-acquisition with its strategic direction by combining scenario planning with the use of supply-chain planning computer models to obtain efficient supply-chain configurations. Our approach combines aspects of business-strategy formulation with aspects of tactical supply-chain planning, making each more valuable in planning than either would be by itself.

9.1 Introduction

After presenting different robust strategies and tactical approaches for mitigating supply chain risks in Chapters 7 and 8, we now show how to apply some of these concepts in actual settings in the remainder of Part II (Chapters 9–12). Specifically, we show how a specific company can mitigate its risks by restructuring its post-acquisition supply chain in Chapter 9; mitigate its outsourcing risks by avoiding the outsourcing-related pitfalls in Chapter 10; mitigate new product development risks by managing supply risks proactively in Chapter 11; and mitigate product risks (i.e., product safety) as well as manage the product recall process in Chapter 12. While these applications appear to be different due to different underlying situations, there is a common underlying theme: a company can reduce supply chain risks (i.e., the likelihood of risk incidents to occur and the impact associated with each incident as described in Chapter 3) by being proactive in anticipating different types of risks and by being prepared to deal with such risks including through synergies with its supply chain partners.

Companies face significant long-term capacity-related risk following acquisitions and mergers: they would like to rationalize its capacity so as to realize synergies from the acquisition. However, if they cut too much capacity, they would lose out on growth, and if they cut too little, they would not be able to get the return-

M.S. Sodhi, C.S. Tang, *Managing Supply Chain Risk*,
International Series in Operations Research & Management Science 172,
DOI 10.1007/978-1-4614-3238-8_9, © Springer Science+Business Media, LLC 2012

on-investment on their acquisition. How should they manage this risk and make the right decisions to restructure their supply chains?

In this chapter, we show how a company can align its supply chain operations post-acquisition with its strategic direction by combining scenario planning for outlining different possible futures (that have different demand levels) with the use of supply-chain planning computer models to obtain efficient supply-chain configurations for each of the different scenarios. Our approach combines aspects of business-strategy formulation with aspects of tactical supply-chain planning, making each more valuable in planning than it would be alone. Clearly, companies can do so through early communication between senior managers and supply-chain planners, which shortens strategy-implementation time while letting each group pursue its forte: senior managers formulating strategy to maximize shareholder value; supply-chain planners running optimization models to minimize total supply-chain costs.

9.2 One Company's Story: Restructuring A Post-Acquisition Supply Chain

Consider a strategic supply-chain planning exercise at a polyvinyl chloride (PVC) manufacturer we shall refer to this company as Acme Vinyl Co. Acme's North American revenues came from PVC for building (55%), packaging (15%), consumer goods (10%), the electronic industry (10%), the automotive industry (4%) and from non-PVC products (6%). At the end of the 1990s about 4% of those revenues came from Asia. Acme had been seeing revenue growth for several years, mostly as a result of acquiring other PVC manufacturers. After acquisition, Acme realized there was redundant (or excessive) capacity in different locations within the United States. To rationalize the post-acquisition supply chain, Acme's leaders considered two key options: (a) consolidate manufacturing into one or two new mega-plants; and (b) close some existing plants or lines. To mitigate the risks associated with their ultimate decision, management chose to do a strategic supply-chain-planning exercise to assist decision making.

9.3 The Planning Spectrum

Strategic supply-chain planning is a process that falls in the middle of a decision-making spectrum that has business-strategy formulation at one end and tactical supply-chain planning at the other end (Table 9.1). With a focus on fundamental changes in manufacturing and distribution capacity, it is long-term in scope and impact but can benefit from detailed optimization models and advanced-planning-and-scheduling (APS) technology more associated with medium- and short-term planning. Used in this strategic context, the tools help determine the supply-chain configuration regarding sourcing and which plants or distribution centers to close or keep open.

Table 9.1 Strategic supply-chain planning and the planning spectrum. Companies need the best features of strategy formulation, which involves creative brainstorming, and supply-chain planning, which is more nuts and bolts.

	Strategy Formulation	Strategic Supply-Chain Planning	Tactical Supply-Chain Planning
Scope of decision making	The entire nature of the business; reevaluating the business model	Whether to open or close plants and distribution centers; to modify capacity; to change product offerings; to manufacture in-house or outsource	Determining which plant should produce what product over the coming months on the basis of demand forecast
Decision horizon	Years	Years	Months or weeks
Flexibility to act	Very high	High	Medium
Possible tools	Frameworks such as M.E. Porter's five forces, Boston Consulting Group's matrices or strength-weakness-opportunities-threat (SWOT) analysis. Lower-level analysis may entail spreadsheets, presentations, system-dynamics tools or other simulation tools	Specialized tools are sold by CAPS Logistics, Manugistics (Compass), i2 (Strategist), Insight Inc. (SAILS), SAP (APO) and other companies; general-purpose tools include AMPL and CPLEX from ILOG	Advanced-planning-and-scheduling (APS) tools from Manugistics, i2, SAP (APO) and other vendors

In contrast, tactical supply-chain planning is short- or medium-term in scope and impact, with supply-chain planners using past demand to make forecasts for the near term and adjusting the forecasts per market intelligence or planned promotions. Used in this context, optimization models and APS technology help determine where and when to produce what items and how to best distribute them.

9.4 Planning Processes and Optimization

Although business-strategy formulation also uses tools and frameworks, it requires much more creativity than tactical planning. The optimum route to maximizing shareholder value is rarely obvious. It takes creative thinking and freewheeling negotiations to identify, understand and agree upon possible actions. Computers and spreadsheets support analysis but don't determine strategy.

For tactical supply-chain planning, the decision options and the factors impacting the decisions (production capacity, distribution capacity, related variable costs, demand forecast) are clearly defined. The goal of minimizing total supply-chain costs—for manufacturing, storage and handling, and transportation—is much narrower.

Because tactical planners can identify beforehand possible decisions and factors that impact these decisions—and can build those elements into the software—they can use optimization models that rely on mathematical techniques. The models make recommendations that both minimize costs and help companies meet forecasted demand without exceeding production and distribution capacity. Advanced-planning-and-scheduling technologies come from companies such as SAP (which offers Advanced Planning and Optimization, or APO), i2 (Rhythm), Manugistics and Logility.

Business-strategy planners cannot plug clearly defined factors into software like supply-chain planners, because strategy formulation is in part about trying to identify just what those factors are. But strategic supply-chain planning can benefit from optimization models used appropriately. That is possible because tactical supply-chain models can be extended to include decisions about closing or opening plants and distribution centers for strategic reasons. APS vendors offer appropriate software (Table 9.1). The factors are the same—production capacity, distribution capacity and variable costs. The goal of minimizing cost is extended to include the fixed costs of keeping plants and distribution centers open, the onetime cost of opening new ones and the onetime cost of closing existing ones. These more strategic computer models can then guide decision making.

Optimization models for tactical supply-chain planning and models for strategic supply-chain planning differ only slightly in their design, but markedly in their use. Even if supply-chain planners tweak the production and distribution schedule obtained from tactical models to expedite an order to a key customer, they largely retain the model's recommendation. In contrast, managers use a strategic supply-chain model's recommendation to determine the opportunity cost of decisions that *differ* from the recommendation.

At General Motors, for instance, EDS consultants found that 90% of the time, GM managers used the models' recommendations to *benchmark* actions they considered likely candidates (Breitman and Lucas, 1987). The candidate decisions were based on qualitative criteria not included in the models. Creating benchmarks that way is unique to optimization models using linear programming, a method using advanced mathematics to capture all the constraints, such as capacity, and to find the best possible recommendations that would minimize the total operating cost or some other stated objective. That is because such models guarantee the lowest possible supply-chain cost. Spreadsheet calculations or simulation models cannot provide benchmarking capability.[1]

---

[1] Optimization models can be built into spreadsheets using the solver feature within Excel, for instance, or using add-on software, but these models are too small to take into account the scope and

9.5 Use of Optimization for Strategic Supply-Chain Planning

A variety of industries have successfully implemented optimization-based tools, including one called Strategic Analysis of Integrated Logistics Systems (SAILS) (Geoffrion and Powers, 1995). Baxter International Inc. used SAILS software to evaluate consolidation approaches following its 1985 acquisition of American Hospital Supply Corp. SAILS also helped Pet Inc. assess supply-chain synergies from two potential acquisitions. In another case, a personal-computer manufacturer made successful strategic use of an optimization model for its global manufacturing and distribution network (Cohen and Lee, 1988). GM uses a tool called Production Location Analysis NETwork System (PLANETS) to determine what products to produce—and how, when and where to make them (Breitman and Lucas, 1987). Digital Equipment Corp. (DEC) probably lengthened its life by using supply-chain models to decrease costs (Arntzen et al., 1995).

9.6 The Need for a Combined Process

Clumsy integration following mergers or acquisitions points to the dangers of relying on only one type of decision making. M&A business-strategy formulation rarely entails models to optimize the supply chain before and after a merger, even though most operating costs reside in the supply chain. In one study, researchers surveyed 700 high-value international mergers and acquisitions made between 1996 and 1998 and found that failure to integrate supply chains was the main reason four out of five deals "failed to enhance shareholder value."[2]

Using an optimization model without a good strategy is similarly lacking. DEC's strategic supply-chain-planning system of the early 1990s improved manufacturing, distribution and service, but without a robust business strategy, the company ultimately succumbed to acquisition by Compaq.[3]

As supply chains are becoming increasingly global, managers are making more strategic decisions about supply-chain design and reengineering. Thus supply-chain

---

detail needed for a global supply-chain model. The cost of APS software, including implementation, for such optimization can be in excess of $1 million.

[2] See A. Macdonald and D. Beavis, "Seize The Moment—Radical Supply Chain Integration as a Means of Increasing Shareholder Value and Enabling Acquisitions To Deliver on Their Promises," Manufacturing Engineer, August 2001, http://www.paconsulting.com/news/media_by/200112190.html.; C. Farrell and R. Melcher, "The Lofty Price of Getting Hitched," Business Week, Dec. 8, 1997, http://www.businessweek.com/1997/49/b3556083.htm.; and D. Henry, "Mergers: Why Most Big Deals Don't Pay Off," Business Week, Oct. 14, 2002, 72–78.

[3] J. Muller, "Compaq Will Buy Digital in A Record $9.6b Deal," Boston Globe, Jan. 27, 1998; and "Compaq To Acquire Digital for $9.6 billion," Compaq press release, Jan. 26, 1998, New York, http://www.compaq.com/newsroom/pr/1998/pr260198c.html.

management has evolved from a function garnering little attention or honor to a highly visible and respected one.[4]

But the successful use of technology from SAP, *i2*, Manugistics and other advanced-planning-and-scheduling software vendors to improve *tactical* supply-chain planning has created unwarranted expectations that, with similar software, senior leaders could delegate *strategic* supply-chain planning to supply-chain planners. But executives' aloofness from strategic supply-chain-planning tools creates the same problems as ignorance of tactical advanced-planning-and-scheduling tools. DEC relied too much on supply-chain planners. And after Nike's $400 million APS technology implementation, $100 million in inventory got misdirected, ultimately triggering a $2.5 billion loss in market capitalization.[5]

The reason for senior managers' aloofness is lack of knowledge, even suspicion. Managers' distrust of optimization models at Shell, for example, kept the company from bringing such models into its scenario-planning processes in the early 1980s (de Geus, 1997: p.69). The extensive detail, rigid structure and managers' lack of knowledge of optimization models aren't conducive to the freewheeling discussion that strategy development needs. Moreover, with management education focusing less on analytical strategic planning and operations research, people getting their MBAs since the mid-1980s and rising to senior levels have increasingly shown less interest in the technical side of management (Mintzberg, 1994).

That's why they need input from supply-chain planners familiar with optimization models. Only through a seamless process that encompasses those who are developing strategy and those who are using detailed supply-chain models can companies align their business strategy and their supply chain.

9.7 Scenarios and Scenario Planning

In scenario planning—a flexible process for formulating business strategy—senior managers build internally consistent alternative views of possible future outcomes, including some that are "unthinkable," as Herman Kahn suggested in his 1950s examination of Cold War scenarios.[6] Typically, only two or four business scenarios are developed, to keep the focus on important factors in relation to the long-term future.

---

[4] D.J. Frayer and R.M. Monczka, "Enhanced Strategic Competitiveness Through Global Supply-Chain Management," Annual Conference Proceedings of the Council of Logistics Management (Oak Brook, Illinois: Council of Logistics Management, October 1997): 433–441.

[5] J. Greenbaum, "SCM Is Dead, Long Live SCM," July 16, 2002, http://itmanagement .earthweb.com/columns/entad/article.php/1407831.

[6] See M. Porter, "Competitive Advantage" (The Free Press: New York, 1985); B. Melzer, "The Uncertainty Principle," CIO Insight, June 1, 2001, http://www.cioinsight.com/print_article/ 0,3668,a=7142,00.asp; and H. Kahn, "Thinking About the Unthinkable" (New York: Horizon Press, 1962).

Building scenarios is more art than science. The creative input needed from managers differs too much from one company to another for experts to offer more than guidelines (Larsen, 2000; Schwarz, 1991; Schoemaker, 1993 and 1995). Leaders creating business strategy use what-if situations based on the plausible interplay of factors expected to have long-term effects. The only constraint is *plausibility*.

Larsen (2000) has identified three phases of scenario planning (Table 9.2): *Scenario building* involves identifying issues, driving forces, and factors that produce uncertainty, then coming up with rough scenarios and next fleshing them out. In *scenario planning*, managers evaluate possible decisions, policies and strategies to see what effect they'd have in each scenario, modifying and reevaluating them as necessary. *Scanning the business environment* involves checking early indicators of change in the environment to see which scenario or combination of scenarios is actually unfolding, thus giving time to managers to revise and refine the decisions made earlier in phase two.

Consider Acme's scenario planning. Acme managers saw three decisions as likely candidates: first, rationalizing existing production capacity and shutting down plants or parts of plants, possibly expanding other plants; second, concentrating production at one or two new megaplants; and third, rationalizing the company's distribution network, closing some distribution centers and opening others.

They identified four main drivers affecting the company's prospects: macroeconomic forces that determine U.S. and Canadian gross-national-product growth and the growth of demand in most sectors; efforts of Western European governments and the European Union to phase out PVC, partly in response to Greenpeace activism; oil-price fluctuation and its impact on prices of raw material (and the company's margins); and the cycle of prices for PVC goods.

The managers also predicted business trends: continued U.S. construction growth; slower industrial growth in the United States, Japan and Western Europe; rapid growth from 2000 through 2005 in Asia (not including Japan); a gradual shift from PVC to packaging polymers by major Japanese and Western European producers of household goods, chemicals and construction materials; and a move away from PVC compounds in the auto industry in Western Europe.

After their analysis, the managers developed two *business* scenarios: "Official Future" and "Sunrise-Sunset." The Official Future reflected senior managers' belief that, at least for four or five years, the company's business would grow at the same rate as the growth in U.S. and Canadian gross national product (about 2% per year). Some sectors, such as electronics and consumer goods, would grow faster than others. Asian demand (excluding Japan), a small portion of the total North American production, would grow as Asian countries' gross national product grew and as Acme achieved greater market penetration. The trends would continue for 20 years, and U.S. government policies would remain favorable to the PVC industry regardless of which party was in power. The electronics and consumer-goods businesses and Asian demand growth would eventually slow down.

Sunrise-Sunset anticipated that events in Western Europe and Japan would lead to a sunset in the U.S. and Canadian PVC industry. In this scenario, state governments concerned with dioxin emissions from burning PVC would file lawsuits

Table 9.2 Scenario-planning steps outlined by different proponents

	E. Larsen	P. Schwarz	P.J.H. Schoemaker
Phase One	*Scenario building* – Identify issues, key driving forces and important factors for uncertainty – Sketch out the basic scenarios – Flesh the scenarios out	-Identify focal issue or decision – List the key factors influencing the success or failure of the decision – List the driving forces in the macro-environment that influence those factors – Identify the two or three factors or trends that are most important and most uncertain – Sketch each scenario using the logical relationships among factors, and view each scenario in a graph that has the previously identified drivers as axes – Flesh out the scenarios	-Define the scope of strategy formulation – Identify the major stakeholders (who can be affected and who can influence) – Identify current trends – Identify key uncertainties – Construct extreme scenarios – Assess scenarios' internal consistency and plausibility – Create representative scenarios—internally consistent scenarios covering a wide range of outcomes adjusted to reflect stakeholder behavior – Identify research needs for each scenario
Phase Two	*Scenario planning* – Evaluate policies and strategies against each of the scenarios to see how they hold up – Modify the strategies – Iterate and reevaluate as many times as necessary	– In each scenario, determine the implication of decisions being considered	-Develop and apply quantitative models (for forecasting and optimizing decisions for the uncertain future; these activities will likely lead to scenario refinement) – Describe decisions for different scenarios, combining managerial judgment with results from previous step
Phase Three	*Scanning the business environment* – Identify leading indicators and charge people with monitoring the changes in the environment for early detection of which scenario or combination of scenarios is unfolding	– Select leading indicators and signposts to detect which scenario (or combination of scenarios) is unfolding	

against manufacturers and garbage-incineration companies. Non-PVC-polymer production would become much more important. Meanwhile, the sun would rise over PVC exports to Asia, where an overwhelming need for buildings, water and sewer lines, and the like would increase demand over 20 years despite environmental issues. The total market would expand and Acme would see its market penetration increase, especially in India, China, Thailand and Indonesia, where Acme might need to build or acquire plants.

9.8 Middle Ground

The ideal process adds a step between business-scenario creation and final decisions: using supply-chain planners' optimization models (Table 9.3). For each of the business scenarios, supply-chain planners can create multiple detailed *model scenarios* to run with their supply-chain optimization models. The result: a supply-chain configuration that minimizes the total fixed and variable, long-term supply-chain costs for the particular business scenario.

How do these *model scenarios* of supply-chain planning differ from the more familiar *business scenarios* of scenario planning? Model scenarios feature an array of demand forecasts and tactical production configurations (extra shifts, planned maintenance and the like). The underlying factors are built into optimization models for use with advanced-planning-and-scheduling software. Creating model scenarios is straightforward and involves little creativity or discussion. It is similar to companies' taking a *macro* business scenario and creating *micro* scenarios for business units or regions (Mandel and Wilson, 1993). Only convenience and the time it takes to solve a model scenario constrains how many a company can have for each business scenario.

Linking each business (macro)scenario to multiple model (micro)scenarios and uniting the creative work of strategy formulation with the analytical, optimization-model approach results in a powerful synergy with "right-brained" strategizing joined to "left-brained" planning (Mintzberg, 1976).

Phase One: Scenario Building. In Phase One, the strategy team identifies the candidate decisions and the uncertainties, and then outlines business scenarios. The supply-chain team develops or modifies its supply-chain model with input from the strategy team regarding possible decisions. The two teams then validate the model. The strategy team garners useful information about the supply chain and updates its notions about how long-term supply-chain configurations affect total supply-chain cost. Next, the strategy team fleshes out the business scenarios using this information and data on demand forecasts, plant locations, distribution centers and so on. Finally, the supply-chain team develops multiple model scenarios for each business scenario to run through its software.

Acme's strategy team developed the Official Future and Sunrise-Sunset scenarios, then met with the supply-chain team. The teams agreed that the supply-chain

Table 9.3 Strategic supply-chain planning using scenario planning. The senior team, which uses brainstorming and other hard-to-measure approaches to formulate strategies that increase shareholder value, combines its strengths with those of supply-chain planners, who rely on optimization software to save supply-chain costs.

	Strategy Team (senior managers)	Supply-Chain Planning Team (planners)
Phase One	*Scenario building* – Identify issues, key driving forces and important factors for uncertainty – Sketch out business scenarios – Flesh out the business scenarios	– Develop or modify the optimization model for the supply chain (including any target acquisitions and locations for new plants or distribution centers) – Validate the model with strategy team
Phase Two	*Scenario planning* – Evaluate possible decisions (policies and strategies) against each of the scenarios to see how they hold up in each case, using the scenario-specific recommendations of the supply-chain team as a benchmark – Modify the candidate decisions – Iterate and reevaluate as many times as necessary	– For each business scenario, create model scenarios and run the optimization model with each model scenario – Determine the best decision for the business scenario, taking into account the recommendations of the optimization model for each model scenario – Present this business-specific choice to senior management along with the team's rationale for recommending it – If the strategy team has provided its list of candidate decisions, compare each of them against the scenario-specific recommendations as well
Phase Three	*Scanning the business environment* – Identify leading indicators and charge people with monitoring the changes in the business environment for early detection of which scenario or combination of scenarios is unfolding.	– Refine the model scenarios to reflect the reality of the unfolding scenario and determine the best optimization-model-recommended decision – Present recommendations to the strategy team

group should make recommendations for both scenarios to reconfigure the supply chain to minimize the cost of manufacturing, distribution, closings of old plants and establishment of new ones. They identified possible locations where megaplants could be built and which of the existing plants could be altered or closed—and identified 14 product families that aggregated several hundred products. Then the supply-chain team, focusing on demand for North American production, developed model scenarios.

Phase Two: Scenario Planning. In Phase Two, the supply-chain team runs the model scenarios and makes scenario-specific recommendations. Then the strategy team modifies its pool of possible decisions until it can make one that it then shares with other managers.

Acme's supply-chain team used software from an APS vendor and ran model scenarios that the strategy team requested. The latter group weighed the resulting recommendations with factors relating to long-term profitability of the business. Acme decided to spin off part of the vinyl business as a joint venture with a supplier. It also chose to merge with a non-PVC polymer company.

Phase Three: Scanning the Business Environment. In Phase Three, the strategy team identifies leading indicators (in the case of an Acme, it might be housing starts) that enable early detection of which scenario or combination of scenarios is actually unfolding. When the strategy team has determined which scenario is occurring, it informs the supply-chain team. With the new information, that group revises the scenarios and runs the model again to fine-tune its recommendations. The strategy team then revises its decisions and shares them widely so they can be acted on.

9.9 Conclusion

Using scenario planning and supply chain planning together, Acme preserved shareholder interests following the exercise and the merger, its stock price staying even despite the plunging of other chemical stocks over the same period. Having executives who formulate business strategy do scenario-planning was critical to the final decisions, which would have been different if the company had used supply-chain optimization modeling alone.

But even without optimization-based planning, scenario planning would have elicited the same decision regarding the merger. The main benefit of optimization lies in refining the decisions that emerge from scenario planning, including rationalizing the supply chain.

Whenever a company considers a merger, it must evaluate its own assets accurately (in addition to its target's) so that its shareholders get the best deal. Acme's main assets were plants and long-term customer contracts, and supply-chain modeling and optimization helped senior managers understand which in a merger could be spun off to increase Acme's value.

Like any other management tool, however, both scenario planning and optimization modeling can have their downside. Some managers get seduced by colorful future possibilities in scenario planning when they should be focusing on the particular trends, underlying factors and uncertainties that are relevant to decisions that pertain to their context. Others stay aloof, relying on consultants and then ignoring the scenarios when making decisions. Similarly, development of a strategic supply-chain-planning optimization model sometimes serves no practical use, becoming unwieldy or taking too long to complete because of multiple motives.

Still, the joint use of scenario planning and optimization models is the better road to shareholder value than using the approaches in isolation. Strategic supply-chain planning, combining elements of both strategy and tactical supply-chain planning, is complex. The process we proposed in this chapter can be used to tame this complexity as long as the two teams understand their roles and work together.

Chapter 10
Application—Mitigating Outsourcing-Related Risks

Abstract While outsourcing part of the companys manufacturing can appear attractive owing to reduced assets and headcount, many firms continue to be disappointed with their outsourcing experience. In this chapter, we identify the underlying causes of outsourcing failures and discuss ways to avoid these underlying causes. Specifically, we use real examples to describe eight different outsourcing-related risks and ways to avoid or mitigate them so as to improve the chances of outsourcing success.

10.1 Introduction

In Chapter 7, we mentioned outsourcing as part of a "make-and-buy" robust strategy. In this chapter, we describe some risks of outsourcing, and present ways to avoid eight common pitfalls so that firms can increase the chance of realizing the true value of outsourcing.

Outsourcing is an increasingly important part of supply chain management, and therefore provides an increasingly important context for supply chain risk. Since the 1990s, more firms have been improving the way they select and manage suppliers, and suppliers have become more sophisticated in managing customer relationship. Despite these improvements, disappointing results and failed relationships between firms and their suppliers persisted. As outsourcing is quite common for most firms, it is critical to develop successful outsourcing strategies.

To compete for market share and to improve financial performance such as return on assets quickly, manufacturers often outsource labor- and capital- intensive operations so that they can transform themselves as product designers (e.g., Nike), supply chain integrators (e.g., Boeing's 787), or business solution providers (e.g., IBM). With improved information technologies, tax benefits, and lower labor cost, outsourcing enables firms to exploit suppliers' capabilities and skills. (For IT projects or call centers, firms can take advantage of the time difference between US and Asia to operate around the clock at low cost.) By 2006, the manufacturing sector accounted for less than 13% of the jobs in the United States. Consequently, there is

M.S. Sodhi, C.S. Tang, *Managing Supply Chain Risk*,   147
International Series in Operations Research & Management Science 172,
DOI 10.1007/978-1-4614-3238-8_10, © Springer Science+Business Media, LLC 2012

no surprise that 80% of the toys sold in the US were made in China, and 97% of the HP notebook computers were manufactured by contract manufacturers such as Taiwan-based Quanta and Compal. Another telling statistic is that worldwide spending on business process outsourcing (BPO) projects, whether onshore or offshore, has exceeded $302 billion in 2004, according to a study conducted by the Gartner group.

Manufacturers are under tremendous competitive pressure to develop new products frequently and quickly. To reduce product development time and cost, many manufacturer outsource the product development tasks to external suppliers. Outsourced efforts for new product development include Apple's decision to outsource its IC chip development for iPods to Pinexe Systems in India and its operating system development to Pixo Inc. in California. Besides electronics companies, Boeing outsourced 70% of the development tasks of the Boeing 787 to external suppliers. We shall use the case of Boeing 787 development in Chapter 11 to illustrate different supply chain risks arising from new product development and discuss various strategies to mitigate these risks. In the pharmaceuticals industry, there is an emerging market of a $30 billion contract drug-development with annual growth rates of 20% (c.f., Perry, 2001).

While outsourcing can be beneficial, many firms are disappointed with their outsourcing experience. First, in 2000, Dun & Bradstreet reported that 20 to 25 percent of all outsourcing relationships (manufacturing, finance, information technology, etc.) fail within two years and 50% fail within five years. Then, in 2005, Deloitte Consulting surveyed 25 world-class organizations and found that a quarter of these companies brought business functions back in-house after realizing that they could do the work themselves better and cheaper; 44% of the companies reported that outsourcing did not save any money; and half of the companies found 'hidden' costs to be unexpectedly high. Similar responses were reported in a recent survey study conducted by Bain & Company. Bhalla, Sodhi and Son (2008) looked at Fortune Global 500 companies and did not find any evidence that IT offshoring, including outsourcing, was linked to any of the different performance metrics over five year. These discouraging results raise the following questions:

1. What are the underlying causes of outsourcing failures (or at least lack of success)? Are there ways to avoid these underlying causes?
2. Why was it the case that some companies did not realize that they could perform tasks under consideration for outsourcing better and cheaper in-house?
3. What are the hidden costs associated with outsourcing? Is there a way to get a better handle on these costs to avoid unpleasant surprises down the road?

The answers to the above questions are caused by risks that can be avoided or mitigated. We describe eight such outsourcing-related risks and ways to avoid or mitigate them so as to improve the chances of outsourcing success.

10.2 Risk 1: Underestimating the Strategic Value of Procurement

By and large, the procurement function is undervalued in most companies because it is commonly viewed as an operational level function. However, in many instances, the procurement decisions (make or buy, outsourcing, off-shoring, supplier selection, contract negotiation, supplier relationship) and many strategic level decisions are inter-twined. Without recognizing the strategic importance of the procurement function, the procurement department would focus on transaction cost and not so much on strategic factors when recommending a sourcing arrangement (c.f., Butter and Linse, 2008). In general, the transaction cost of outsourcing that include search cost, unit cost, transportation cost, quality assurance, technology transfer cost, and restructuring cost are relative objective and easy to evaluate. However, without a clear communication and coordination, the procurement division may not fully incorporate strategic factors when making procurement recommendations. These strategic factors include the impact of outsourcing on internal operations, reputation and brand value, legal issues, political and economical issues when dealing with foreign suppliers. Without a strong coordination among a cross-functional group that includes the procurement department, it is difficult for a company to realize the full potential of outsourcing.

Consider Chrysler's turnaround in the mid-90s. In the 80s, Chrysler viewed suppliers as parts producers. As they discovered the potential value of suppliers, Chrysler reduced the supply base by forming long-term partnerships with certain strategic suppliers who perform design, engineering and manufacturing functions. To capture the value of this special partnership with these strategic suppliers, Chrysler changed its procurement function from managing standard transactions to developing and managing collaborative sourcing relationships with key suppliers.

In 1993, Chrysler launched a program called SCORE (Supplier COst Reduction Effort) that involved early supplier involvement during the design phase, solicit suggestions from suppliers, sharing cost savings with suppliers, etc. The collaborative sourcing relationships helped Chrysler to reduce the new product development time from 234 weeks to 160 weeks, and reduce the new product develop cost by 30%. More importantly, this strategic partnership has generated 875 ideas from suppliers worth \$170 million in annual savings. By the end of 1995, Chrysler implemented 5300 ideas that have generated more than \$1.7 billion in annual savings. To capture the value of strategic sourcing, Chrysler created and developed a different kind of procurement function that was aligned with the company's strategic direction (c.f., Dyer (1996) and Tang (1999)).

Mitigating these risks. Executives at global companies need to recognize that the role of procurement professionals goes beyond basic purchasing transactions. These professionals need to identify and select suppliers, undertake contract negotiation, and monitor supplier performance. In today's market, procurement function intertwines with strategic decision-making at the executive level. Therefore, to capture strategic value of outsourcing, firms need to develop and train procurement professionals with the requisite skills so that they can assume the role as the liaison between the executives of the firm and the suppliers' representative. According to a recent survey study conducted by Capgemini in 2008 reported that more firms are creating Chief Procurement Officers (CPO) positions to oversee the strategic procurement issues of the firm. Because CPOs do not belong to any business units, they can conduct objective evaluation of different sourcing options by incorporating different strategic issues. As firms developed these CPOs, they can get them involved earlier so that they can help the executives to make informed decisions. By improving the communication and coordination among different parties internally and externally, these CPOs can create and capture value of outsourcing through the procurement function (c.f., Abery and Stark (2008)).

10.3 Risk 2: Underestimating the Strategic Value of Outsourcing

Lowering the operating cost is usually the primary reason for companies to outsource certain functions. With this mindset, manufacturers can obtain short-term benefits and the limited value of outsourcing. However, the same mindset could hinder the capability of the manufacturers to identify, create and capture the strategic value of outsourcing.

In a market that is changing rapidly, Gottfredson et al. (2005) has argued that it is no longer a company's ownership of capabilities that matters but rather its ability to manage the critical capabilities. In some cases, owning the capabilities to perform certain important functions can impede a company to transform itself. For example, 7-Eleven (formerly known as the Southland Corporation) operated successfully as a vertically integrated convenient chain store since 1961. However, as more gas stations added more mini-marts, 7-Eleven was unable to change its mode of operations to compete on quality (product freshness and product assortments) and price. Consequently, it suffered from profit loss and declining market share in the 80s and filed for bankruptcy in 1990. The turning point began when Ito-Yokado Group of Japan bought 70% of its common stock and transformed the business model of 7-Eleven in 1991.

Under the new business models, 7-Eleven focused on merchandising—product assortment planning and pricing—by tracking customer purchasing behavior, and outsourced all non-core functions (HR, finance, IT management, data processing, distribution, food processing, delivery). More importantly, to induce full cooperation from its vendors, 7-Eleven shared its financial and productivity gains with strategic suppliers. For example, 7-Eleven formed a joint venture partnership with a prepared-foods distributor Sween for in-store delivery of sandwiches and other fresh prepared food items. Because of its skills and scale, Sween can make and deliver freshly prepared items multiple times per day under this partnership in contrast to the once-a-day delivery under 7-Eleven's in-house operation. Under this partnership, 7-Eleven experienced higher revenue (due to fresher products) and lower cost (fewer leftovers). Based on this new business model by outsourcing almost everything, 7-Eleven has created and captured the value of outsourcing and has enjoyed sustainable profitable growth ever since (c.f., Gottfredson et al. (2005), Lee and Whang (2006)).

Outsourcing has enabled other firms to transform themselves. For example, after suffering from a 3-year loss of $16 billion from 1991 to 1993, IBM CEO Lou Gerstner and a management team developed a turnaround plan that entailed outsourcing manufacturing of personal computers, divesting under-performing business units (which included the divesture of IBM notebook computer to Lenovo in 2004), and transforming IBM from a computer company to an integrated business solution provider. As a result, IBM generated a net income of $7.7 billion in 1999 and resumed the leadership in the business computing industry (c.f., Austin and Nolan (2000)).

Besides outsourcing various manufacturing and service operations, companies are beginning to outsource their research and development efforts to suppliers, partners and third-party design/engineering houses. The reason for this trend is because large companies that are strong in commercialization of products for mass markets are often weak in developing innovative products for new markets. As articulated by Kandybin et al. (2004), Coca-Cola would never be able to discover a new market for the Red Bull energy drink by conducting focus group interviews and surveys. (The concept of taurine based energy drink was discovered by an Austrian Dietrich Mateschitz during his international travel, not through focus group interviews or blind tasting experiments.) Similarly, Pfizer's Listerine PocketPaks—the portable strips of breath freshener—did not originate in-house. In truth, it was based on a confection technology developed in Japan. In 2008, Procter & Gamble attributed its double-digit profit growth to its external innovation network through its "Connect & Develop" program. Under this program, P&G has adopted innovative ideas and solutions provided by external partners to develop innovative products such as printable Pringle's chips, Mr. Clean Magic Eraser, and Crest Spin Brush. The idea of "open innovation" is catching on, and more web-based idea exchanges such as InnoCentive.com and NineSigma.com are enabling companies to solicit innovative ideas and solutions from an external network (c.f., Chesbrough (2003)).

Mitigating these risks. While inventing new technologies and developing new products in-house have worked well for decades at powerhouses such as AT&T Bell Labs, GE, IBM, and DuPont, many firms are now getting innovative ideas and solution from external parties. Kandybin et al. (2004) reported a study conducted by Delphi Pharma in 2002, which suggests that firms can increase their return on investment (ROI) by outsourcing innovations in the pharmaceutical industry. In addition, as articulated by Chesborough (2003), the shift toward an open model was in part inspired by information technology and social networks as we witness from the success of Linux and Wikipedia. Besides P&G, global giants such as Eli Lilly and Avery-Dennison are soliciting ideas and solutions using open exchange network InnoCentive.com. Generating new ideas for new products is ripe for outsourcing; however, one key challenge is for the firm to develop mechanism to solicit, capture, and evaluate different ideas generated from external communities.

10.4 Risk 3: Poor Understanding of Internal Capability and Cost Structure

In many instances, companies outsource certain operations (manufacturing, technical support, after sales services) to external suppliers because outsourcing is more cost effective than keeping the operation in-house. However, even when the cost comparison is accurate, the cost benefit of outsourcing can diminish over time for two key reasons: (1) The manufacturer did not take into account that the cost of outsourcing can increase over time; for example, due to demand exceeding supply, the cost of recruiting and retaining skilled labors in China and India is on the rise. (2) The manufacturer overestimated the benefit of outsourcing. This occurs when the manufacturer measures the quality and cost of the current performance of their internal operations without taking the future performance after some improvements and the changing business environment into consideration.

Consider the outsourcing of IT projects at Sainsbury's supermarket in the United Kingdom. In 2000, Sainsbury's signed a 10-year outsourcing contract with Accenture with the intent to save an estimated £35m per year from its £200m annual IT budget. Later on, after a series of mishaps during the early phase of the project, Sainsbury's decided to revaluate its internal capability and discovered that they could save £40m per year if IT projects were conducted in-house. In 2005, Sainsbury's terminated the contract with Accenture by bringing 470 IT staff from Accenture back in-house along with all of its IT assets. Although there was a one-off cost of around £65m for cancelling the contract with Accenture, Sainsbury's commented that they could recover the development cost faster than expected (c.f., McCue (2006)).

After acquiring Bank One Corporation for $58 billion in 2004, JP Morgan Chase decided to terminate its $5 billion IT project contract with IBM that the two companies had signed in 2002. As a result of a series of mergers and acquisitions, JP Morgan Chase believed that it was important for them to conduct the IT project in-house so as to gain better control in terms of scope and cost. To ensure a smooth transition from outsourced to in-house development, JP Morgan Chase brought 4,000 IBM employees in-house.

Mitigating these risks. To avoid this kind of abrupt change of heart about outsourcing, the company needs to do due diligence by looking beyond the common cost factors such as production, labor, transportation, training, etc. Specifically, the firm needs to consider such factors as currency exchange, geopolitical factors, import/export quotas, internal labor issues, compatibility between the manufacturer and the supplier in terms of value and culture. By noting that keeping the operations in-house is always an option, a firm should consider outsourcing certain operation only when the firm can obtain significant improvements in operational measures or capture strategic value beyond in-house capabilities. To make an informed decision whether to outsource, the firm should be in a position to benchmark its internal operations against external suppliers.

10.5 Risk 4: Outsourcing the Wrong Thing

Even when a manufacturer uses the following "Make or Buy" framework (Figure 10.1) to determine if certain operations should be outsourced, it could still end up outsourcing the wrong thing especially because "strategic value" can be misjudged easily. This is because the strategic value of a function depends on how a firm competes in the market place. For example, in the automotive industry, most car makers would consider stamping, welding, and injection molding as low value-added processes that could be outsourced to external suppliers. However, Toyota recognized the strategic value of these basic functions that affect the consumer's perceived quality due to the visible "fit and finish" of its cars. For this reason, Toyota is one of few car manufacturers who chose to retain ownership of these seemingly simple processes in-house (Doig et al., 2001).

It is interesting to compare Nokia's and Ericsson's outsourcing strategies of the telecommunication equipment such as radio base stations and network systems products. In 2000, Nokia adopted modular design of its telecom equipment, outsourced the manufacturing operations of certain non-critical modules, and kept the production of certain "strategic" modules in-house. Because of this sourcing arrangement, Nokia managed to develop close relationship with its suppliers so that they can coordinate their plan for new product development. On the contrary, Ericsson viewed itself as a "knowledge" company and decided to outsource the entire production of its telecom equipment. Without close relationship with its suppliers, Ericsson was slow in developing new telecom equipment. Consequently, Ericsson's sales of telecom equipment declined and the cost of obsolescence increased with the company experiencing financial loss for 11 consecutive quarters (c.f., Bengtsson and Berggren (2004)).

Because the strategic value of apparel manufacturing appears to be low and because there are many capable contract manufacturers in Asia, the above "Make or Buy" framework suggests that apparel companies should outsource their manufacturing operations—indeed, this is the norm for many apparel makers competing on cost. However, if an apparel maker decides to compete on speed, then outsourcing the manufacturing function can become a disadvantage. Consider the success story of Zara that is known for its "fast fashion." Zara is one of the few clothing companies that use its in-house manufacturing operations and, as a result, can design, manufacture and ship a new line of clothing to its stores within two weeks. By recognizing the manufacturing operations is a critical part of the supply chain operations, Zara's in-house manufacturing operations enables Zara to respond to market dynamics quickly and to become the most profitable European fashion retailer with sales and net incoming growing at an annual rate of over 20% (c.f., Ferdows et al., 2004). Successful companies such as Zara, Procter & Gamble, and Intel keep their manufacturing operations in-house mainly because they understand the strategic value of their in-house operations.

Mitigating this risk. To avoid misjudging the strategic value of certain operations, a firm needs to establish a clear strategic direction before making the "make or

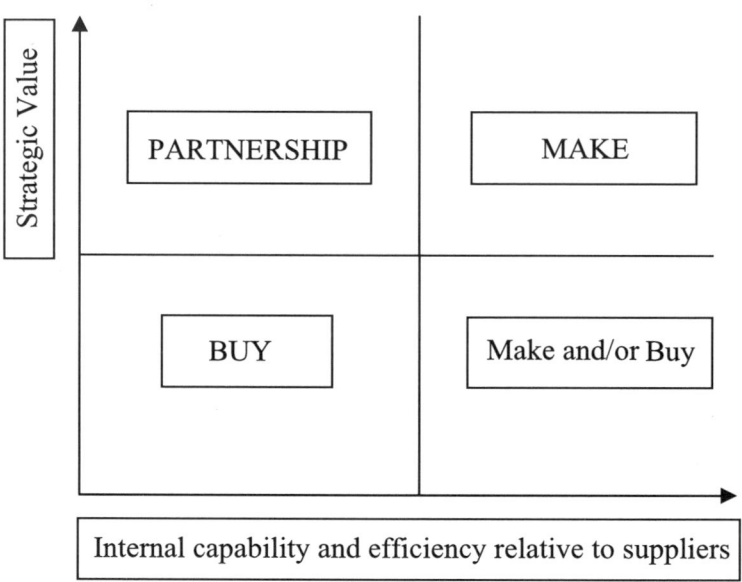

Fig. 10.1 The make or buy framework

buy" decision. For example, many companies view after-sales services ranging from product repair to call center support as low value-added activities. Also, because the requests for this kind of services are sporadic, it is definitely more efficient to out-source to external suppliers who can generate the scale by "pooling" the requests from multiple companies. However, Best Buy, the largest retailer of consumer elec-tronics in the U.S., recognized the strategic value of after sales service operations. This is because, from the customers' perspective, after sales support is very criti-cal mainly because they need these services when they are facing some problems and need help. Hence, a firm can improve customer relationship management, cus-tomer satisfaction and customer loyalty by keeping this operation in-house. Best Buy found a new subsidiary called Geek Squad in 1994 that specialized in com-puter and consumer electronics technical support such as in-home installation and repair of home theatre and computer network systems. Although Geek Squad made up 4% of Best Buy's revenue in 2007, it helped to boost Best Buy's profit because of its operating margin (10-20%) is much higher than that of retailing (5%). By rec-ognizing the strategic value of after sales service and keeping it in-house, Best Buy improved customer satisfaction and increased profit.

10.6 Risk 5: Underestimating the Importance of Incentive Alignment

When establishing an outsourcing arrangement with a supplier, manufacturers often focus on maximizing their own benefits such as cost, quality, and delivery, without thinking about how their action would affect the behavior of their suppliers. There is an inherent conflict of interest that is present in any outsourcing relationship: the buyer wants to squeeze the supplier as much as possible, and the supplier wants to make as much of a profit as possible.

Without a careful process to manage these conflicts carefully and sharing risks, there could be long-term problems down the road for one or even both parties. For example, Cisco outsourced the manufacturing function to Solectron—the largest contract manufacturer in the 90s. As the demand of Cisco's product was growing rapidly in the 90s, Solectron stockpiled work-in-process and finished goods inventories to meet Cisco's growing demand. Also, to boost profit margins, Solectron purchased large volumes of components from suppliers at lower prices than Cisco had negotiated. Unfortunately, as the demand for Cisco's product dropped in 2000 and as the ownership of those work-in-process and finished goods inventories was not clearly defined in the contract, Cisco and Solectron ended up writing off \$2.5 billion and \$1.5 billion of inventories, respectively. Partly because of this fiasco, Solectron was acquired by Flextronics in 2007 (c.f., Narayanan and Raman (2004)).

The notion of self-interest can hinder R&D effort as well. In the late 90s, GM and Ford spun off their parts manufacturing operations to Delphi and Visteon. However, as GM and Ford decided to compete on price, they continued to pressure the suppliers to reduce their cost. To stay in business, there is very little incentive for these suppliers to make long term R&D investments.

Besides R&D, cost cutting pressure can cause some suppliers to cut corners by lowering their internal product safety standards (Roth et al., 2007). For example, in 2007, Mattel had to recall over 10 million toys tainted with lead paint produced by a Chinese contract manufacturer Lee Der. Upon investigation, there were two underlying causes for these unsafe toys. First, the owner of Lee Der purchased cheaper paint from a supplier owned by his close friend. This sourcing practice actually violated the sourcing process stipulated by Mattel that required contract manufacturers to purchase paint from a list of certified suppliers. Second, Lee Der did not perform mandated tests on the incoming paint and the finished toys. Clearly, the contract manufacturer has violated the agreement; however, the risk could have been mitigated had Mattel conducted the safety tests in-house or through a third-party (c.f., Pyke and Tang (2008)).

Mitigating this risk. To align incentives of the firm and its suppliers, Narayanan and Raman (2004) suggested three ideas: win-win contracts, information sharing, and trust building. When each party behaves in ways that maximizes its own interests, developing the right incentive to align the interests of both parties is critical to ensure outsourcing success. For example, after the Santa Monica freeway in Los Angeles was damaged after the Northridge earthquake in 1994, LA residents wanted the government to repair the freeway quickly to get their daily life back to normal; the city government wanted to repair the freeway quickly without overpaying the contractor; and the contractor wanted to maximize its profit. To align the interest of all parties, the City of Los Angeles (the project manager) offered the following time-based contract for open bidding. Specifically, the contractor would receive an extra reward of $200,000 per day if the project is completed before 6 months (180 days); however, the contractor would pay a penalty of $200,000 per day if the project were completed after 6 months (c.f., Kwon et al., 2010). The time-based contract was awarded to Clint Meyers (the contractor), and it provided the right incentive for Clint Meyers to complete the repair 74 days ahead of schedule. In this particular case, the time-based contract aligned the interests of all three parties and the outcome was appreciated by all three parties.

10.7 Risk 6: Outsourcing Operations without Considering the Supply Chain

In many instances, companies examine the efficiency of different operations in isolation. Without a clear understanding of the value of an operation in the context of the entire supply chain, companies can outsource the wrong operation.

For example, in early 1990, IBM San Jose, California (sold to Hitachi in 2002) was an integrated manufacturing facility that produced hard disks for IBM mainframe computers. The manufacturing operations consist of three basic steps: fabrication of disk head (a process similar to the fabrication of integrated circuits), assembly operation (an assembly process of the head and various electro-mechanical parts), and final inspection and repair (c.f., Demeester and Tang (1996)). To compete on price, IBM evaluated the strategic value and the operating cost of each step of the process. Due to the proprietary technology of its fabrication process and inspection process, it became clear that IBM could reduce labor cost by outsourcing the relatively simple but tedious assembly operations to Mexico.

As IBM outsourced its assembly operation (an intermediate step of the manufacturing process), they discovered the cost of outsourcing was much higher than expected for two key reasons: (1) IBM needed to manage the outbound logistics of the heads and the mechanical parts to Mexico, and the inbound logistics of the assembled disks for in-house inspection and repair. These two additional logistics operations was costly, not to mention the additional cost of training and technology transfer. (2) The total lead time resulting from the outsourced assembly operation was much longer than expected because of the roundtrip transportation time, the time to clear customs, and the longer lead time of the assembly operations in Mexico. As the lead time increased, the work-in-process and finished goods inventories increased as well. As computer technology changed rapidly, the obsolescence cost of those inventories increased significantly. In hindsight, the underlying cause of this outsourcing failure was due to the fact that IBM San Jose did not take the supply chain operations into consideration when deciding to outsource the assembly operation to Mexico.

Mitigating this risk. The manufacturer should first map out the entire supply chain operations so as to obtain a clearer understanding about the impact of different outsourcing options on different supply chain partners. Then the manufacturer should form a cross-functional team to evaluate different options: (1) keep production inhouse, (2) outsource specific components, (3) outsource specific modules, or (4) outsource the entire production operation. Once an outsourcing option is decided, the manufacturer should communicate and coordinate its outsourcing plan with the supply chain partners to ensure efficient outsourced operations.

10.8 Risk 7: Outsourcing without the Right Contracts

When the intent of outsourcing is operational improvement, the manufacturer can specify various operational performance measures in the supply contracts. However, if the intent is to gain strategic advantage, then the relationship between the manufacturer and the supplier is no longer simple and requires careful planning. This is mainly because the performance measures are difficult to define, measure, and verify.

For example, as Laura Ashley, a global clothing and furnishings retailer founded in the UK, expanded its product offerings and its stores globally, the company realized it was not equipped to handle the global distribution logistics operations efficiently and effectively. In 1992, Laura Ashley decided to outsource its entire worldwide distribution (inbound and outbound logistics, warehouse operations, and inventory management) to FedEx. The strategic alliance between Laura Ashley and FedEx was a 10-year partnership that was relatively open-ended and based on trust. The objective was to be able to supply 99 percent of Laura Ashley's merchandise to customers anywhere in the world within 48 hours. By 1993, this strategic partnership helped Laura Ashley to reduce out of stock of desirable items and reduce leftover inventory (c.f., Anthony and Loveman (1996)). Despite this success, the alliance came to an end in 1994 when Laura Ashley's new CEO raised suspicions about frequent air shipment of its products instead of ground shipment. Essentially, without a contract with pre-specified expectations, it is very difficult to maintain a stable supplier relationship especially when there is a sudden change in the top management.

Mitigating this risk. When outsourcing certain innovative processes such as design, it is difficult to establish well defined expectations, responsibilities and activities on both sides up front. In this case, both sides should not focus on the outcome of the process, but measures should be based on the process itself. By defining the process steps up front, both parties can define the scope of each step, the expected outcome of each step, the review process, and the process to resolve any conflicts or mismatched expectations. More importantly, doing so would allow both parties to set expectations for contract negotiation and exit options. In terms of contract, Auguste et al. (2002) argued that the cost plus contracts do not provide incentive for the suppliers to reduce cost; however, suppliers do get to keep the rewards from process innovation under fixed-price contracts.

10.9 Risk 8: Not Being Prepared for Supply Disruptions

Many companies underestimate the risks of outsourcing. Many companies have developed mechanisms to handle regular supply risks that happen on a regular basis. For example, to mitigate the risk associated with cost fluctuations due to currency exchange rates or raw materials, many firms usually established various hedging mechanisms to reduce the risk exposure. Also, to mitigate the risk associated with rising labor and operating costs in major cities in China and India, some companies are considering diversifying their sourcing locations to other countries such as Vietnam and Thailand. Moreover, to reduce the risk associated with potential delay in delivery of parts from the suppliers, many firms have established some build-in safety stocks to handle delays.

While companies are doing reasonably well in managing supply risks that occurred regularly, they are vulnerable to supply disruptions that occurred rarely but with major consequences. Despite the detrimental effect of major supply disruptions, most companies invested little time or resources in managing supply chain risks even though they conducted supply chain risk assessment exercises. Two surveys confirm this perplexing dichotomy. First, according to a study conducted by Computer Sciences Corporation in 2003, 43% of 142 companies, ranging from consumer goods to healthcare, reported that their supply chains are vulnerable to disruptions, and 55% of these companies have no documented contingency plans (c.f., Poireir and Quinn (2003)). Next, according to another survey conducted by CFO Research Services, 38% of 247 companies acknowledged that they have too much unmanaged supply chain risk (c.f., Eskew (2004)).

For example, recall the Ericsson and the Apple vs. Dell examples we mentioned in Chapter 7, Ericsson was facing supply shortage of a critical cellular phone component (radio frequency chips) after its key supplier, Philip's Electronics semiconductor plant in New Mexico, caught on fire in March of 2000. After Ericsson failed to recovery the shortage quickly, Ericsson lost 400 million Euros in sales in 2000. Along the same vein, Apple faced component shortages for its iBook and G4 computers after an earthquake hit the suppliers' plants in Taiwan in 1999. Instead of finding ways to recover, Apple tried to convince their customers to accept a slower version of G4 computers and suffered a major decline in sales. Besides the disruptions in the delivery of parts, there are other types of outsourcing risks including intellectual property risks, geopolitical risks, and behavioural risks (Sodhi and Tang, 2009).

Mitigating this risk. To mitigate the consequence of a potential supply disruption, consider the following options: safety stock can reduce potential shortage, backup suppliers can ensure supply availability, and responsive pricing can entice consumers to switch their demand from products in short supply to products with ample supply. To develop a process for managing supply risks, we advocate the following steps stemming from enterprise risk management: (1) identifying risk, (2) assessing risk, (3) mitigating risk, and (4) responding to risk (Chapter 1). The first three entail activities that take place *before* the occurrence of an incident that generates negative (and mostly) unanticipated consequences while the last one applies to actions taken *during* and *after* the occurrence of an incident—the focus here is on time, which is also called *time-based risk management* (Chapter 5) about ways to respond to risk events proactively.

10.10 Conclusion

Given the increasing importance of outsourcing in today's supply chains, we described eight risks pertaining to outsourcing: (1) underestimating the strategic value of procurement; (2) underestimating the strategic value of outsourcing; (3) poor understanding of internal capability and cost structure; (4) outsourcing the wrong thing; (5) underestimating the importance of incentive alignment; (6) outsourcing operations without considering considering the supply chain operations; (7) outsourcing without the right contracts; and (8) not being prepared for supply disruptions. We also presented ways to mitigate the risks associated with these pitfalls.

Chapter 11
Application: Mitigating New Product Development Risks—The Case of the Boeing 787 Dreamliner

Abstract Besides those outsourcing-related risks presented in Chapter 10, there are various supply chain risks arising from new product development. In this chapter, we illustrate these new product development risks by using the case of the Boeing 787 Dreamliner to present how Boeing managed these risks . More importantly, we highlight key lessons for manufacturers to consider when designing supply chains for new products.

11.1 Introduction

In Chapter 10, we presented eight outsourcing-related risks as well as different ways to mitigate these risks. In this chapter, we examine various supply chain risks arising from the specific context of new product development in the case of the Boeing 787 Dreamliner. Specifically, we shall (a) analyze Boeing's rationale for developing the 787's supply chain that was a break from past practice, (b) describe Boeing's challenges for managing this supply chain, and (c) highlight key lessons for manufacturers to consider when designing supply chains for new products.[1]

Boeing decided to develop its latest aircraft—the 787 Dreamliner—to stimulate revenue growth and raise its profile in the commercial aircraft industry. With the 787 Dreamliner, Boeing developed not only a revolutionary aircraft, but also an unconventional supply chain to reduce development cost and time drastically. However, despite significant management effort and capital investment, Boeing experienced delays in its schedule for the maiden flight and for delivery to customers.

Since the US government deregulated air travel in 1977, more airlines entered the market causing fierce price competition. As airfares continued to decline, the total number of US passengers rose from approximately 240 million in 1977 to 679 million in 2008. Since then, traffic has grown significantly in other parts of the world. At the same time, US commercial aircraft manufacturers faced major

---

[1] Joshua Zimmerman contributed significantly to this chapter.

M.S. Sodhi, C.S. Tang, *Managing Supply Chain Risk*,
International Series in Operations Research & Management Science 172,
DOI 10.1007/978-1-4614-3238-8_11, © Springer Science+Business Media, LLC 2012

competition from Europe. After losing market share in the late 1990s to Airbus (owned by EADS, based in Europe), Boeing was under pressure to decide between two basic competitive strategies: reduce the production costs (and the selling prices) of existing types of aircraft or developing a new aircraft that could create more value for its customers.

In 2003, Boeing decided to focus on the latter approach of creating value to its customers (airlines) by developing an innovative aircraft: the 787 Dreamliner. (We use the terms "787 Dreamliner", "787", and Dreamliner" interchangeably.) First, Boeing's value creation strategy for the passengers was to improve their travel experience through redesign of the aircraft and offer significant improvements in comfort as a means for differentiating from competitors, mainly Airbus. For instance, relative to other aircrafts, over 50 percent of the primary structure of the 787 aircraft (including the fuselage and wing) is made of composite materials (c.f., Hawk (2005)). As opposed to the traditional material (aluminum) used in airplane manufacturing, the composite material allows for increased humidity and pressure to be maintained in the passenger cabin offering substantial improvement to the flying experience. Also, light-weight composite materials enable the Dreamliner to take long-haul flights, enabling airlines to offer non-stop flights between distant pairs of cities without layovers, as preferred by most international travelers (c.f., Hucko (2007)). Table 11.1 and Fig. 11.1 compare the 787 aircraft with other popular aircrafts from Boeing itself and the other new airplane for long-haul from Airbus, the A380.

Table 11.1 Comparison of select Boeing (737, 747, 787) and Airbus aircraft (A380-800)

Airline Family	Max Range (nautical miles)	Max Capacity (passengers)	Empty Weight (lbs)	Cruising Speed (mph)	Strategy
737-800	3,000	189	91,000	514	Direct flights to multiple cities
747-8	8,000	467	410,000	570	Hub to hub
787-9	8,500	330	254,000	561	Direct flights to multiple cities
A380-800	8,200	525	610,000	561	Hub to hub

Second, Boeing's value-creation strategy for its customers (the airlines) was to improve efficiency by using midsize airplanes to provide big-jet-type flying ranges while flying at approximately the same speed (Mach 0.85). The resulting efficiency would allow airlines to offer economical non-stop flights to and from more and smaller cities. In addition, with a capacity between 210 and 330 passengers and a range of up to 8,500 nautical miles, the 787 Dreamliner is expected to use 20% less fuel than existing airplanes of similar size. The cost per seat-mile is therefore-expected to be 10% lower than for any other aircraft. Also, unlike the traditional

Up, Up and Away

Weighing in at 280 metric tons and with a wingspan as wide as a
football field is long, the Airbus A380 is the world's largest
passenger jet, designed to carry about 850 passengers between
hub airports. In contrast, Boeing says its smaller, fuel-efficient
787 Dreamliner will allow for direct flights between more
cities even at great distances.

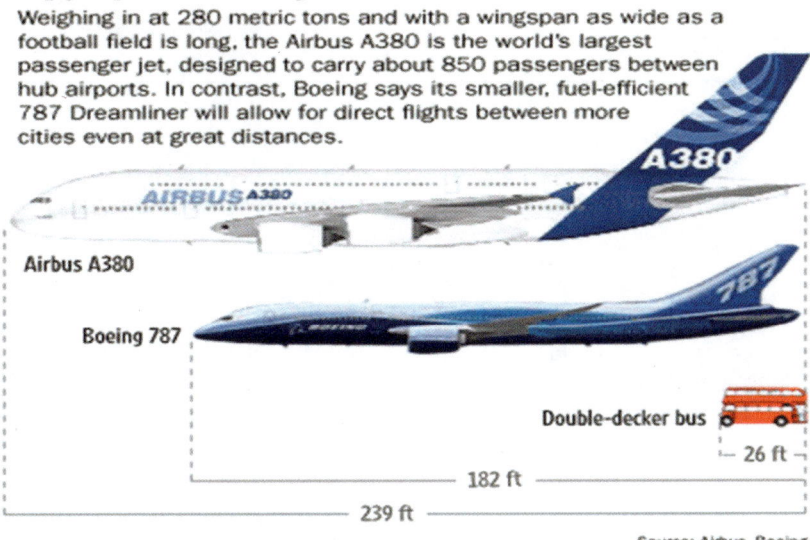

Source: Airbus, Boeing

Fig. 11.1 Dreamliner and A380 size comparison

aluminum fuselages that tend to rust and to experience metal fatigue, 787's fuselages are based on the composite materials, which reduce airlines' maintenance
and replacement costs (c.f., Murray (2007)). Table 11.2 provides a summary of the
Dreamliner's benefits for both the airlines and their passengers.

Due to the unique value that the 787 provides to the airlines and their passengers,
the number of orders exceeded expectations. The Dreamliner is the fastest-selling
plane in aviation history with carriers attracted to its new largely composite design
and innovative next-generation jet engines that will allow the wide-bodied plane to
fly further on less fuel. The Dreamliner program has been considered a model en-
deavor combining novel technology and production strategies. As of November 16,
2008, Boeing (www.boeing.com) had received orders from 7 airlines that accounted
for 895 Dreamliners. The overwhelming response from the airline industry to Boeing's 787 forced Airbus to quickly redesign its competitive wide-bodied jet, the
A350, to make it even wider, which was later re-released as the A350XWB for "extra wide body"[2]. Boeing is currently the second largest global aircraft manufacturer
(behind Airbus) in terms of revenue and deliveries (though having received more orders than Airbus), the second-largest aerospace and defense contractor in the world
(behind Lockheed Martin), and the single largest U.S. exporter (with nearly $29 bil-
lion of exports in 2009). Sales for the fiscal year ending in March 2011 amounted to
$64 billion with net income of $4.46 billion.

Besides airlines, the stock market also responded favorably when Boeing
launched its "game-changing" 787 Dreamliner program in 2003. Between 2003

---

[2] c.f., Wallace (2006 a and b).

Table 11.2 Comparison of select Boeing (737, 747, 787) and Airbus aircraft (A380-800)

Feature	Values to airlines (immediate customers)	Value to passengers (end customers)
Composite material	– Fuel efficiency (lighter material lowers operating cost) – Corrosion resistance (lower maintenance cost) – Stronger components that require fewer fasteners (lower manufacturing cost)	– Faster cruising speed, which enables city-pair non-stop flights – Higher humidity in the air is allowed, which increases comfort level
Modular design that allows for two types of engines (General Electric GEnx and Rolls-Royce Trent 1000)	– Flexibility to respond to future circumstances (market demand) at a reduced cost – Simplicity in design allows for rapid engine changeover	– Cost savings with cheaper and faster engine changeover may be passed on to passengers
Large and light sensitive windows	– Lower operating costs due to less need for interior lighting	– "Smart glass" window panels work like transition lens—controlling the amount of light automatically—decreasing glare and increasing comfort and convenience
Redesigned chevron engine nozzle (serrated edges)	– Reduction in community noise levels	– Reduction in interior cabin decibel level
Easy preventive maintenance	– Boeing provides service so planes are in operation for longer periods of time	– Fewer delays due to mechanical problems

and 2007, Boeing's stock price increased from around $30 to slightly over $100 (Fig. 11.2). Since then however, Boeing has had to announce a series of delays beginning in late 2007 to which the market has reacted negatively (Fig. 11.2). The negative market response is expected as publicity of Boeing's supply chain issues become increasingly evident. As shown in Fig. 11.2, Airbus shared a similar fate after announcing a series of delays for the delivery of its A380 aircraft in early 2006 (c.f., Raman et al., 2008). Despite significant capital investment and management effort, Boeing has continued to face delays its schedule for plane delivery to customers as of this writing although the maiden flight of the Dreamliner finally took place in December 2009 and Boeing began pilot training for the first customer, All Nippon Airways (ANA) in April 2011. This motivates us to examine the underlying causes of Boeing's challenges in managing 787's delivery schedule.

We first examine Boeing's rationale for the 787's unconventional supply chain and then present our analysis of the underlying risks associated with its supply chain. In subsequent sections of this chapter, we describe Boeing's risk mitigation strategies to expedite its development and production process. Then we draw some

Fig. 11.2 Historical stock prices of Boeing and Airbus compared to the S&P500

lessons for other manufacturers to consider when designing their supply chains for new product development.

11.2 The 787 Dreamliner's Unconventional Supply Chain

To reduce development time for the 787 from the typical six to four years and the development cost from $10 billion to $6 billion, Boeing decided to develop and produce the Dreamliner by using a new and unconventional supply chain in the aircraft manufacturing industry. The 787's supply chain was envisioned to keep manufacturing and assembly costs low, while spreading the financial risks of development to Boeing's suppliers. Unlike the 737's supply chain that requires Boeing to play the traditional role of a key manufacturer who assembles different parts and subsystems produced by thousands of suppliers (Fig. 11.3), the 787's supply chain is based on a tiered structure that would allow Boeing to foster strategic partnerships with approximately 50 tier-1 strategic partners. These strategic partners serve as "integrators" who assemble different parts and subsystems produced by tier-2 suppliers (Fig. 11.4). The 787 supply chain depicted in Fig. 11.4 resembles Toyota's supply chain that has enabled Toyota to develop new cars with shorter development cycle time and lower development cost than its competitors (c.f., Tang (1999)). Table 11.3 highlights the key differences between the supply chain for the older 737 and that of the new 787. For instance, under the 787's supply chain structure, tier-1 strategic partners are responsible for delivering complete sections of the aircraft to Boeing to allow allow Boeing to assemble these complete sections in three days at its plant located in Everett, Washington (Fig. 11.5).

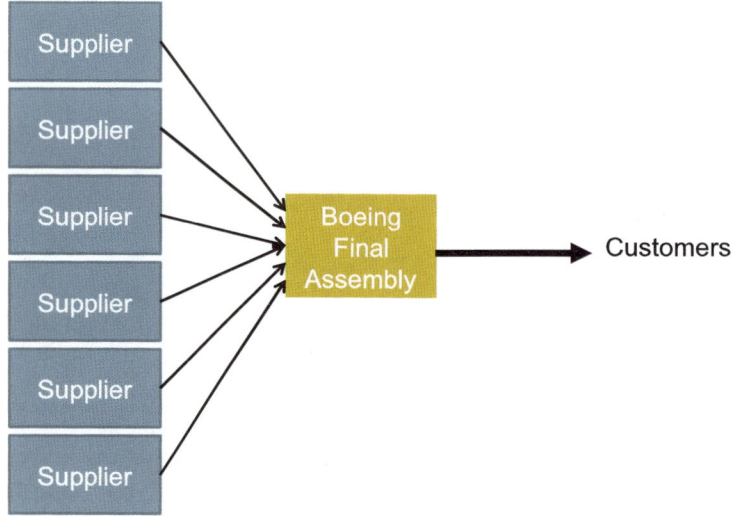

Fig. 11.3 Traditional supply chain for airplane manufacturing

We now explain the rationale behind the 787's supply chain as highlighted in Table 11.3.

Outsource More

By outsourcing 70% of the development and production activities under the 787 program, Boeing sought to shorten the development time and development cost by leveraging suppliers' ability of developing different parts in parallel. As Boeing

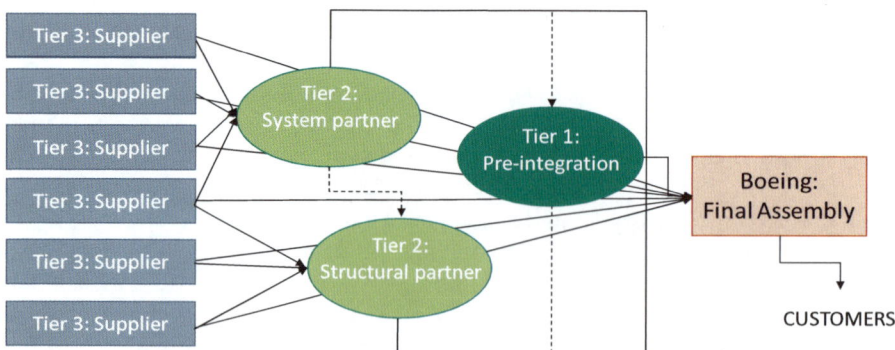

Fig. 11.4 Redesigned supply chain for the Dreamliner program. (Adapted from Steve Georgevitch, Supply Chain Management, Boeing Integrated Defense Systems.)

Table 11.3 Dreamliner features with benefits for airlines and passengers

Component	737 program	787 program
Sourcing strategy	Outsource 35-50%	**Outsourced 70%**
Supplier relationship	Traditional supplier relationship (purely contract based)	**Strategic partners with tier-1 suppliers**
Supplier responsibilities	Developed and produced *parts* for Boeing	**Developed and produced entire *sections* for Boeing**
Number of suppliers	Thousands suppliers supplying directly	**Approximately 50 tier-1 strategic partners**
Supply contracts	Fixed price contracts with delay penalty	**Risk sharing contracts**
Assembly operations	30 days for Boeing to perform final assembly	**3 day assembly of complete sections**

Source: Seattle Post-Intelligencer, Seattle

outsourced more, communication and coordination between Boeing and its suppliers became critical for managing the progress of the 787 development program. To facilitate the coordination and collaboration among suppliers and Boeing, Boeing implemented the Exostar Supply Chain Systems powered by E2Open software that is intended to gain supply chain visibility, improve control and integration of critical business processes, and reduce development time and cost. For example, when delivery problems arise with tier-2 suppliers, Exostar would alert the affected tier-1 partner. Then, if the situation is not resolved appropriately, Exostar would alert Boe-

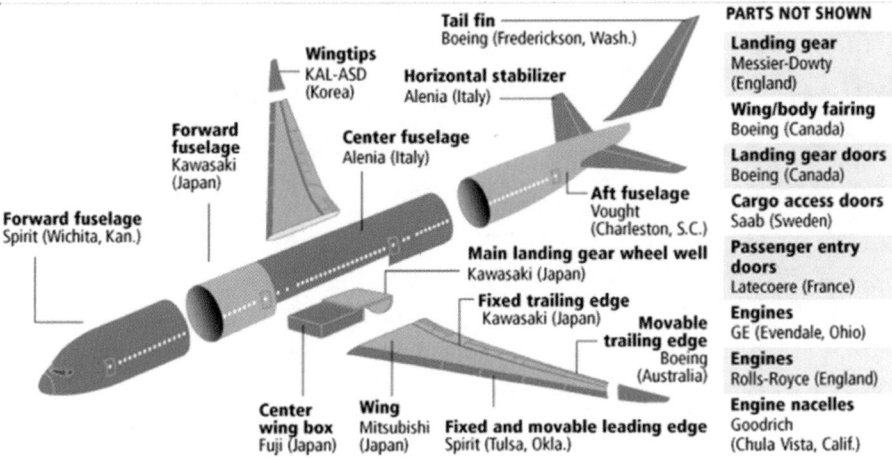

Fig. 11.5 Dreamliner subassembly plan (Source: Seattle Post-Intelligencer, Seattle)

ing so that Boeing has the option of intervening directly to address the issue (c.f., Manufacturing Business Technology, 2007).

Reduce Direct Supply Base, Delegate More, and Focus More

To reduce development time and cost for the Dreamliner, Boeing fostered strategic partnerships with approximately 50 tier-1 suppliers to design and build entire sections of the plane and ship them to Boeing. By reducing its direct supply base from thousands for the 737 to only 50 or so for the new 787, Boeing could focus more of its attention and resources on working with tier-1 suppliers (pre-integration stages) rather than with raw material procurement and early component subassembly. (Dyer and Ouchi, 1993, discuss the benefits of reducing the direct supply base as used by Toyota to develop new cars faster and cheaper than GM in the automotive industry.) The rationale behind this shift is to empower its strategic suppliers to develop and produce different sections in parallel so as to reduce the development time. Also, by shifting more assembly operations to its strategic partners located in different countries, there are potential savings in development costs as well (Fig. 11.6).

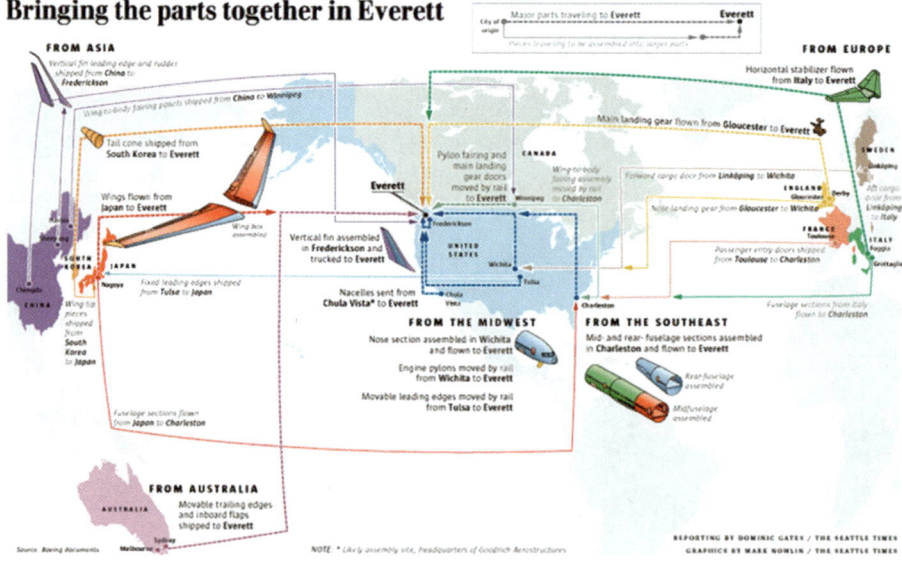

Fig. 11.6 Boeing's global supply partners (Source: The Seattle Times)

Reduce Financial Risks

Under the 787 program, Boeing instituted a new "risk sharing" contract under which no strategic supplier receives the payment for the development cost until Boeing delivers its first 787 to its launch customer, ANA airlines. This contract payment term was intended to provide incentives for strategic partners to collaborate and coordinate their development effort. While this contract imposes certain financial risks for Boeing's strategic suppliers if delivery deadlines are missed, they are incentivized by being allowed to own their intellectual property, which can then be licensed to other companies in the future. Another incentive for the strategic partners to accept this payment term is because it allows them to increase their revenues (and potential profits) by taking up the development and production of the entire section of the plane instead of a small part of the plane.

Reduce Assembly Time Without Incurring Additional Costs

Decentralizing the manufacturing process would allow Boeing to outsource non-critical processes. The intention is to reduce the capital investment for the 787 development program. Also, under the 787 supply chain, Boeing would need only three days to assemble complete sections of the Dreamliner at its plant. Relative to the 737 supply chain, this drastic reduction in cycle time would in turn increase Boeing production capacity without incurring additional investments.

11.3 The Dreamliner's Supply Chain Risks

While the 787 supply chain (Fig. 11.4) has great potential for reducing development time and cost, there are various underlying supply chain risks. As we described in Chapter 2, there are many types of supply chain risks ranging from technology to process risks, from demand to supply risks, and from IT system to labor risks. In this section, we present some of the risks and actual events that caused major delays in the Dreamliner's development program (Table 11.4).

Technology Risks

The 787 Dreamliner involves the use of various unproven technologies. Boeing encountered the following technical problems that led to a series of delays.

- *Composite Fuselage Safety Issues:* The Dreamliner contains 50% composite material (carbon fiber reinforced plastic), 15% aluminum and 12% titanium. The

Table 11.4 Dreamliner features with benefits for airlines and passengers

Risk Factor	Potential risk caused by the 787 supply chain	Risk consequence: what happened at Boeing?
Technology	Infeasibility of material in flight tests, which is untested on this scale	Invisibility of development issues with tier-1 suppliers' partners resulting in major delays
Supply	Tier-1 suppliers outsource development tasks to tier-2 partners, which may not have technical know-how	Lack of knowledge about supplier selection by tier-1 partner, delay in development and manufacturing work
Process	Over reliance on tier-1 partners to coordinate their development tasks with their suppliers further down the supply chain	Need for increased coordination of supplier's activities required "traveled work" by Boeing personnel
Management	Inexperienced management team without supply chain expertise	Management failure, need for reorganization at highest levels
Labor	Union dissatisfaction with Boeing's decision to outsource more	Union strike causing work stoppage
Demand (Customer)	Publicity of problems may cause problems with airline and passenger perceptions of Boeing	Delivery delays may cause financial penalties and cancellation of orders

Source: Seattle Post-Intelligencer, Seattle

composite material has never been used on this scale and there was a perceived risk creating an airplane with this mixture of materials is not feasible. Also, lightning strikes are a safety concern for wings made out of this composite material because the lightning bolt could potentially travel through wing-skin fasteners (c.f., Wallace (2006)).

- *Engine Interchangeability Issues:* One of the key benefits of the 787's modular design concept was to allow airlines to use engines from two different makers (Rolls-Royce and GE) interchangeably. However, due to recent technical difficulties and parts incompatibility, it takes 15 days to change engines from one model to the other instead of the originally intended 24 hours (c.f., Leeham (2005)).
- *Weight Issues:* As of May 2009, Norris (2009) reported that the first 787s have 8%, or 2.2 metric tons of excess weight over that in the original design. Analysts are predicting that the heavier planes will have a 15% reduction in range. This reduction in range negates some of the advantages that the Dreamliner was supposed to provide.
- *Computer Network Security Issues:* The current configuration of electronics on the Dreamliner puts passenger electronic entertainment on the same computer

network as the flight control system. This raises a security concern for terrorist attacks (c.f., Zetter and Kim (2008)).

Supply Risks

Boeing is relying on its tier-1 global strategic partners to develop and build entire sections of the Dreamliner based on unproven technology. Any break in the supply chain can cause significant delays of the overall production. In early September 2007, Boeing announced a delay in the planned first flight of the Dreamliner citing ongoing challenges including parts shortages and remaining software and systems integration activities. Although the Exostar computer system is an efficient tool to manage the information flow within the supply chain, it is only useful if accurate and timely information is provided by different suppliers. First, after realizing insufficient in-house system integration capability, one of the tier-1 suppliers, Vought, hired Advanced Integration Technology (AIT) as a tier-2 supplier to serve as a system integrator without informing Boeing. AIT is supposed to coordinate with other tier-2 and tier-3 suppliers for Vought (c.f., Tang (2007)). Second, due to cultural differences, some tier-2 or tier-3 suppliers did not always enter accurate and timely information into the Exostar system. Consequently, various tier-1 suppliers and Boeing were not aware of the technical problems (such as parts shortages) or delay problems in a timely fashion. These two problems undermined the benefits sought from Exostar.

Process Risks

The design of the supply chain for the 787 is a likely cause of the delays because delivery depends on the synchronized Just-In-Time deliveries of all major sections from Boeing's tier-1 strategic partners: if the delivery of any section is late, the delivery schedule of the entire aircraft is delayed as a consequence. Also, under the risk-sharing contract, none of the strategic partners will get paid until the first completed plane is certified for commercial flight. As strategic partners recognize they may be penalized if they were to complete their tasks before other suppliers, the risk-sharing contract payment may actually tempt these strategic partners to work slower, which undermines the original intent of the risk-sharing contract (c.f., Kwon et al., 2010).

Management Risks

Given the unconventional supply chain for the Dreamliner, it was essential for Boeing to assemble a leadership team that included members with proven supply chain management experience of having prevented and anticipated certain risks as well as having developed contingency plans to mitigate the impact of different types of risks. However, Boeing's original leadership team for the 787 program did not include members with expertise on supply chain risk management. Without the requisite skills to manage an unconventional supply chain, Boeing was undertaking a huge risk in uncharted territory.

Labor Risks

As Boeing increased its outsourcing effort, Boeing workers became concerned about their job security. Their concerns resulted in a strike by more than 25,000 Boeing employees in December 2008. The effects of the worker strike were also felt by Boeing's strategic partners. For example, Spirit Aerosystems, a manufacturer of fuselage parts for the 787, could not deliver parts as a direct result of the strike.

Demand Risks

As Boeing announced a series of delays, some customers lost their confidence in Boeing's aircraft development capability and began canceling orders for the Dreamliner or migrating toward leasing contracts instead of purchasing the airplane outright. As of May 2009, 58 orders for the Dreamliner had been canceled by four different airlines and by August 2010 more orders had been cancelled than booked for that year.

11.4 What Boeing Did to Mitigate Risk

We now present Boeing's reactive response for reducing the negative impact of these problems and for avoiding further delays (Table 11.5).

Mitigating Technology Risks through Redesign

To improve the safety of its composite fuselage, Boeing is redesigning its fuselage by using additional material to strengthen the wing structure; however, this addi-

Table 11.5 Boeing's reactive risk mitigation strategies

Risk Factor	Reactive Risk mitigation strategy
Technology	Modify design
Supply	Purchase company at the bottleneck stage (Vought Aircraft Industries)
Process	Boeing personnel perform "traveled work" to solve issues with underperforming partners
Management	Reorganization of top management—replaced program manager with supply chain expert
Labor	Concessions to labor unions—increased pay and decreased outsourcing
Demand (Customer)	Boeing to pay penalties for delivery delays, public relations campaign to reassure customers

tional material would increase the aircraft's overall weight. Boeing management has continued to assure its customers that it would work diligently to reduce the weight of the final version of the plane. Regarding the time it takes to change engines from one model to the other, Boeing is redesigning its installation process with the hope of reducing changeover time. Finally, to ensure that the computer network is secure, a proper design is required that allows for the separation of the navigation computer systems from the passenger electronic entertainment system.

Mitigating Supply Risks

Attempting to regain control of the delayed program, Boeing purchased one of its tier-1 strategic partners—a unit of Vought Aircraft Industries—that was known to be the weakest link of Boeing's 787 supply chain (c.f., Ray (2008)). This acquisition would provide Boeing direct control of this unit as well as over the tier-2 suppliers for the fuselage development. As a result of continued production delays, some of Boeing's suppliers were facing massive profit losses, which in turn put completion of the entire Dreamliner program at risk. To mitigate this risk, Boeing paid its tier-1 strategic partner, Spirit Aerosystems, approximately $125 million in 2008 to ensure that this partner continued its vital operations (c.f., Reuters (2008)).

Mitigating Process Risks

With suppliers being unable to meet production deadlines, Boeing decided to send key personnel to supplier sites across the globe to understand and address production issues in person. However, reducing supply risks this way created its own risks as personnel were pulled from responsibilities on-site at Boeing to address supply and manufacturing issues at the sites of their outsourced partners. The strategy of relying on suppliers for subassembly proved to be too risky for Boeing in certain circumstances and resulted in Boeing having to perform the work themselves. This "traveled work" to sites such as Finmeccanica SpA's Alenia unit and Carlyle Group's Vought Aircraft Industries Inc. has been a substantial headache for Boeing. Ultimately, Boeing had to redesign the entire aircraft subassembly process (c.f., Gunsalus (2008)).

Mitigating Management Risks

To restore customers' confidence about Boeing's aircraft development capability and to reduce the possibility of further delays, Boeing recognized the need to bring in someone with a proven record of supply chain management expertise. In response, the original 787 program director, Mike Bair (with marketing expertise), was replaced by Patrick Shanahan with expertise in supply chain management. In his new role, Shanahan is now responsible for coordination of all activities for Boeing's major plane families which includes the Dreamliner. Moreover, Boeing changed its top leadership by replacing its Interim CEO, James Bell, with Jim McNerney in 2008.

Mitigating Labor Risks

To bring about an end to the strike after two months of shutdown, Boeing made concessions that would give workers a 15 percent wage boost over four years. On the key issue of job security, which had been the major impediment to reaching an agreement, Boeing agreed to limit the amount of work that outside vendors can perform. Therefore, Boeing's concept of outsourcing a significant amount of work to global partners was put to risk and production costs may eventually rise. In response to the wage increases and limits in outsourcing promised by Boeing, the machinists union conceded to withdraw charges filed with the Department of Labor regarding allegations of unfair bargaining practices at Boeing (c.f., Gates (2008)).

Mitigating Demand (Customer) Risks

As customers began to cancel their 787 orders and as the company's capability of developing the 787 was in question, Boeing has developed the following mitigation strategies. First, as a way to compensate its customers' potential loss due to the late deliveries of their orders, Boeing is supplying replacement aircrafts (new 737 or

747) to various concerned airlines such as Virgin Atlantic (c.f., Lunsford (2007) and Crown (2008))). Second, to restore Boeing's public image, Boeing has improved its communication by sharing its progress updates on its webpage. Finally, Boeing is conducting a publicity campaign to promote the superior technology of the planes and the overall value that the airplane will offer to airlines and passengers (c.f., Crown (2008)).

11.5 What Might Boeing Have Done Differently to Mitigate Risk

With the benefit of hindsight, we can point out certain risk mitigation strategies that Boeing could have used at the outset of the program to better manage potential risks proactively (Table 11.6). These may be useful lessons for other manufacturers seeking to redesign their supply chains for new products.

Table 11.6 Alternative strategies for mitigating program risks

Risk Factor	Proactive actions	Risk affect
Supply chain visibility	Use IT to ensure transparency of entire supply chain	Avoided or reduced
Strategic partner selection and relationship	Proper vetting of all strategic partners to determine their capability of completing tasks	Reduced
Process	Develop better risk sharing opportunities and incentives for strategic partners	Reduced
Management	Establish proper working team with expertise in supply chain logistics	Avoided
Labor	Outreach and communication with union heads to discuss sourcing strategies	Avoided
Demand (Customer)	Treat customers as partners and better communicate the potential for missing delivery deadlines	Avoided or reduced

Source: Seattle Post-Intelligencer, Seattle

Improve Supply Chain Visibility

As described earlier in this chapter, Boeing's supply risk was caused in part by the lack of supply chain visibility. Without accurate and timely information about the supply chain structure and the development progress at each supplier's site, the value of Exostar in terms of providing status updates was compromised significantly. To improve information accuracy, Boeing should have insisted on having all strategic partners and suppliers provide all information imbedded in the supply chain relationships instead of relying on alerts generated from the program only when they were directly affected. Also, Boeing should have provided incentives for all suppliers to use Exostar to communicate accurate information in a timely manner.

Improve Strategic Supplier Section Process and Relationships

Had more effort been spent on evaluating each supplier's technical capability and supply chain management expertise for developing and manufacturing a particular section of the Dreamliner, Boeing could have selected more capable tier-1 strategic suppliers. Doing so would have enabled Boeing to avoid or reduce potential delays caused by tier-1 suppliers with inadequate experience. Also, Boeing should have insisted on participating in the tier-1 partner's vetting process of tier-2 (or tier-3) suppliers. The additional effort of properly vetting key suppliers would certainly enhance communication and coordination and reduce the risks of potential delays, which would in turn reduce the development time and cost (c.f., Lunsford (2007)).

Modify the Risk-Sharing Contract

As we noted already, under Boeing's risk-sharing contract, individual tier-1 strategic partners would not get paid until they had all completed their sections and the first plane was actually certified. While this payment term was intended to reduce Boeing's financial risk, it did not provide proper incentives for tier-1 suppliers to complete their tasks early. When some strategic partners did not develop their respective sections according to the schedule, the entire development schedule was pushed back and Boeing was hit with millions of dollars in penalties that it had to pay out to its customers (c.f., West (2007)). To properly align the incentives among all strategic partners, Boeing should have structured the contracts with reward (penalty) for on-time (late) delivery (c.f., Kwon et al., 2009).

Proactive Management Team

Having the right people for the job at the outset of the program would have helped Boeing better avoid and anticipate the risks associated with its novel supply chain structure. Also, identifying the sources of potential problems and having the right person (or team) in place would mitigate many of the risks, and allow Boeing to respond more quickly and effectively when problems were realized. Had Boeing appointed appropriate persons (including someone like Patrick Shananhan with proven supply chain management expertise) to serve on the original leadership team, Boeing could have avoided and anticipated various types of supply chain risks. Boeing could have reached out to experts in other industries who have managed complex supply chains similar to that of 787 supply chain. For example, Toyota has successfully managed its tiered supply chain that resembles the 787 supply chain. Also, such a leadership team would have the expertise and authority to respond to delays more effectively.

Proactive Labor Relationship Management

Dissatisfaction among Boeing's machinists was caused by Boeing's strategy to increase its outsourced operations to external suppliers. Had the union's general disapproval of Boeing's outsourcing strategy been taken into account, Boeing may not have decided to outsource 70% of its tasks. Even if this outsourcing strategy was justified financially, Boeing could have managed its labor relationship proactively by discussing the strategy, by offering job assurances, and by obtaining buy-in from unions. This proactive labor relationship management would have created a more mutually beneficial partnership, which could have avoided the labor strikes.

Proactive Customer Relationship Management

Recognizing the risks associated with innovative product development, proactive customer relationship management is critical to help customers set proper expectations when placing their orders. Better communication with customers throughout the development process enables a company to manage customers' perceptions throughout the entire product development process. Had Boeing set proper expectations about the delivery schedules of its 787 Dreamliner, the airlines may have managed their aircraft replacement schedule differently, say by ordering more 737s and 747s and delaying or ordering fewer 787s. Without an aggressive delivery schedule to its customers, Boeing could have reduced the penalty caused by the delayed delivery schedule. Through continuous engagement and open communication about the challenges and Boeing's contingency plans, Boeing could have better managed its customers' perception. By earning the trust from its customers, Boeing could have prevented the erosion of Boeing's reputation.

11.6 Conclusion

Boeing's Dreamliner program involves dramatic shifts in supply chain strategy from traditional methods used in the aerospace industry. In addition, the airplane itself required novel manufacturing techniques and used technological advances never before tested on a project of this scale. Such dramatic shifts from convention done in parallel involve significant potential for encountering risks throughout the process. Boeing's issues with meeting delivery deadlines were a direct result of its decision to change many elements of such a complex system simultaneously without having the proper management team in place. Further, this team did not proactively assess the risks that were later realized and did not develop strategies for effectively mitigating them. While it may be impossible to identify all potential risks and create contingency plans for all eventualities before a project begins, Boeing could have done many things differently. It is instructive for managers in any industry to view the issues that Boeing faced to learn from mistakes that were made before engaging in similar supply chain restructuring for a new product. Some of these lessons are:

1. *Assemble a leadership team with requisite supply-chain expertise.* On the surface, it appears that Boeing's fundamental problem was caused by its attempts to take on too many drastic changes simultaneously: unproven technology, unconventional supply chains, unproven supplier's capability to take on new roles and responsibilities, and unproven IT coordination systems. However, the underlying reason for Boeing to take on so many drastic changes was due to the fact that the 787 leadership team underestimated the risks associated with all these changes. Had Boeing constituted a multi-disciplinary team with expertise to identify and evaluate various supply chain risks, Boeing could have avoided and anticipated potential risks, and developed proactive mitigation strategies and contingency plans to reduce the impact of various supply chain disruptions.
2. *Obtain internal support proactively.* Partnerships between management and labor are essential for smooth operations for companies to implement any new initiatives including new product development programs. While their interests are often misaligned, better communication of business strategies with union workers can be step towards avoiding costly worker strikes. Also, aligning the incentives for both parties proactively can reduce potential internal disruptions down the road.
3. *Improve supply chain visibility to facilitate coordination and collaboration.* A company must cultivate strong commitment among its suppliers for accurate and timely information. Overly relying on IT communication is highly risky when managing a new project. To mitigate the risks caused by partners further upstream or downstream, companies should strive for complete visibility of the entire supply chain. Having this capability would enable a company to take corrective ac-

tion quickly, which would certainly reduce the negative impact of a disruption along the supply chain (recall Chapter 5 about responding to risk events).

4. *Proactively manage customer expectation and perceptions.* Due to the inherent risks associated with new product development, it is critical for a company to help its customers set proper expectations proactively, providing a range of potential outcomes regarding product specification and delivery schedule reflecting the uncertainty. Setting proper expectations reduces potential customer dissatisfaction down the road. During the development phase, it is advisable for the company to maintain open and honest communication with customers regarding the actual progress, technical challenges, and corrective measures. Such efforts could gain customers trust, which would improve customer loyalty in the long run.

Chapter 12
Application: Managing Product Recalls—the Case of Mattel, Inc.

Abstract In addition to supply chain risks that occur before sales, companies need to think about the risks that may occur *after* sales. Specifically, product recalls due to safety considerations for consumers can put companies at risk. This chapter deals with product safety and recall and examines how companies manage their product recall process and implement ways to mitigate product risks in the future. We use the case of Mattel, Inc. and its toy recalls in 2007 to illustrate handling of the product recall process.

12.1 Introduction

So far, we have discussed ways to mitigate various types of supply chain risks ranging from supply to demand risks (Chapters 7 and 9), and from financial to outsourcing risks (Chapters 8 and 10). This chapter deals with product risks related to safety and examines how companies manage their product recall process and implement ways to mitigate product risks for the future. To illustrate, we use the case of Mattel, Inc. and its toy recalls in 2007.

Western consumers are becoming increasingly concerned about product/food safety as the total number of product recalls in the U.S. broke a new record in 2007. The period 2003-07 witnessed major product recalls: Merck recalled its pain killer Vioxx due to health risks in 2004; Dell recalled its notebook computer batteries due to fire hazards in 2006; Menu Foods recalled various tainted pet foods with harmful effects on cats and dogs in 2007; and Mattel recalled millions of toys in 2007 because of either noncompliant levels of lead in paint or other product safety issues. Although other toy companies also announced recalls, the public, the media and the U.S. government reacted most strongly and publicly to Mattel.

Mattel was forced to carry out a series of toy recalls in 2007. Since all of the recalls in 2007 involved products or materials made in China, it is understandable that governments as well as the public in western countries got concerned about the safety of other products made in China as well. In the midst of these recalls

M.S. Sodhi, C.S. Tang, *Managing Supply Chain Risk*,
International Series in Operations Research & Management Science 172,
DOI 10.1007/978-1-4614-3238-8_12, © Springer Science+Business Media, LLC 2012

announcements, Mattel needed to *execute* the product recall plan and deploy a strategy quickly to ensure safety and restore consumer confidence, especially about toys made in China.

In this chapter, we analyze and evaluate the way Mattel managed its recalls and planned to strengthen product safety for the future. Also, we highlight some key challenges and opportunities for companies to consider in the future.

Mattel's recalls created major challenges for the company: First, in September 2007, the CEO of Mattel (Mr. Robert Eckert) had to provide testimony to Congress explaining the underlying causes of the recalls and what he was doing to ensure toy safety.

Second, since Mattel's first lead-related recall of toys in August 2007, its stock price fell by 13% (from $23 in early August 2007 to $20 at the end of March 2008), although the Dow Jones U.S. Index also fell by 7% during the same period. Although, the direct impact of product recalls on stock price is unclear, product recalls entail a number of costs. These costs include investigation costs, communication costs, logistics costs for collecting recalled products and for distributing replacements, disposal costs of returned products, personal injury claims associated with the recalled products, lost profits due to lost sales during the recall period, and potentially lost profits due to deteriorating consumer confidence and associated declining sales.

Third, because Mattel recalled its products in August and beyond, it was a big challenge for Mattel to ensure the safety of its toys before the holiday shopping season started in November. This was especially the case as its long supply chain was slow to respond to changes.

Why did Mattel's recalls get so much attention? After all, there were 400 product recalls in 2007 in the U.S. where the US Consumer Product Safety Commission (CPSC) announces product recalls virtually every single day (http://www.cpsc.gov/cpscpub/prerel/prerel.html). There are three possible explanations:

- **Frequency and Magnitude.** Mattel stood out because it announced several recalls involving an unusually large number of products. Specifically, within a few months Mattel issued multiple recalls that involved a total of 20 million toys of different types. These successive recalls caused the public to doubt whether Mattel was making sufficient efforts to ensure product safety. Moreover, Mattel is by far the largest toy maker in the world—twice the size of its nearest competitor—so for many consumers, it is almost synonymous with the industry.
- **Children.** Because children are among the most vulnerable of consumers, the recalls gave rise to heightened concerns and lead to more searching inquiries about the root causes. Indeed, the recalls concerned lead paint and small parts that could come loose and be ingested by children.
- **Other incidents involving Chinese manufacturers.** More than 50% of products subject to recalls in 2007 involved products made, in whole or in part, in China. This is not surprising in itself given the proportion of all products in the U.S. that were made in China in 2007, but it did catch the public's attention. Besides toys

and pet foods, Disney recalled children's pajamas in April 2007 due to a burn hazard, and Panama reported hundreds of deaths in May 2007 caused by cough syrup mixed with a poisonous counterfeit ingredient imported from China. These trends caused public wariness and even outright resentment of products made in China. Because many of Mattel's toys are made in China, it is understandable that Mattel became one of the targets through which the public and the government raised their general concerns regarding the safety of products made in China. Moreover, Mattel's initial public efforts for the first toy recall were to rest all the blame on China and on Chinese manufacturing. However, CEO Eckert later admitted Mattel's own failings and apologized to the Chinese government.[1]

In this chapter, we first examine the underlying causes of the product recalls at Mattel and then analyze the way Mattel managed its product recalls since July 2007. We conclude with challenges and opportunities associated with product recall management.

12.2 Underlying Causes of Mattel's Product Recalls

Product recalls are usually caused by one or more of the following factors: design flaw; production defect; new scientific findings about dangers from products or materials that were earlier assumed safe; accidental contamination; product tampering; unforeseen misuse of products by consumers; and failure to meet safety standards (Berman, 1999). According to our analysis, there were three underlying causes for Mattel's product recalls in 2007.

1. **New data about dangers of magnets.** When designing such toys as Polly Pocket and Doggie Day Care, toy designers at Mattel recommended using a certain type of glue to attach tiny, powerful magnets to different parts of the toys. Despite the use of this glue, these magnets could come loose in play. Although, loose and small but powerful magnets were earlier regarded by industry and regulators alike as a quality issue rather than as a safety issue, it subsequently became clear that, two or more loose magnets were swallowed in a precise sequence could perforate or block a child's intestines. Accordingly, in 2006, ASTM International, a voluntary standard setting organization, began working on a draft design standard for retaining magnets in toys. The draft was completed in December 2006 and published in May 2007. Mattel actively participated in the development of the new standard and, in January 2007, months before it was published, implemented the new standard prospectively. In its August 2007 recall, Mattel retroactively applied the new design standard to 18.2 million toys with magnets attached

---

[1] On Sep, 21 and 22, 2007, this was reported by major newspapers and news programmes. See for instance, http://www.npr.org/templates/story/story.php?storyId=14599222 (National Public Radio, Sep. 21, 2007) and http://articles.latimes.com/2007/sep/22/business/fi-mattel22 (The Los Angeles Times, Sep. 22, 2007).

to plastic that the company had previously produced, recalling them so that participating consumers could receive replacement toys that complied with the new design standard.

2. **Supply chain failure to meet safety standard.** Mattel toys with noncompliant levels of lead in paint were produced by various Chinese contract manufacturers. One of these, Lee Der, was owned by Mr. Zhang, who committed suicide in the midst of the investigation in August 2007. It turned out that the noncompliant paint was produced by a company owned by a close friend of Mr. Zhang. Mattel's investigation uncovered that certain vendors or their sub-contractors violated the company's well-established rules. In some cases, they appear to have been careless. In others, they appear to have deliberately avoided doing what they knew they were required to do. Some vendors failed to identify subcontractors or facility locations, even though they were mandated to do so.

3. **Testing.** Mattel required vendors to purchase paint from a list of certified suppliers or test the paint they used to ensure compliance with the established standards. Moreover, the company audited the certified paint suppliers to ensure compliance with lead level standards and periodically audited vendors to ensure that they were complying with paint requirements. It also conducted lead level safety tests on samples drawn from the initial production run of every product and had protocols for further recertification testing for lead on finished product. Despite these procedures and protocols, in reality, vendors failed to provide certified paint to their subcontractors to use on Mattel products and one vendor was found not to have performed the mandated test on paint.

12.3 Mattel's Product Recall Management Process

First consider a framework that will allow us to evaluate the effectiveness of Mattel's action plan for managing recent recalls. To mitigate the risks associated with product recalls, companies need to develop a sustainable plan that reduces the likelihood and the impact of a product recall. To develop such a plan, we suggest the application of the 3R—Readiness, Responsiveness, and Recovery—framework (Pyke and Tang, 2008 and Fig. 12.1).

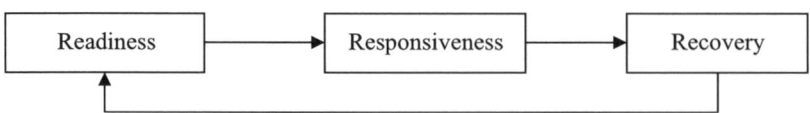

Fig. 12.1 The 3R Framework for Mitigating Product Recall Risks

Essentially, Readiness, Responsiveness and Recovery refer to various actionable items to be taken before, during, and after a product recall, respectively (Table 12.1).

Notice that some items are intended to reduce the likelihood of having a recall, while others are designed to reduce the impact of an imminent recall.

Table 12.1 Actionable items for mitigating the risks associated with product recalls

	Readiness (before)	Responsiveness (during)	Recovery (after)
Policy	Develop product recall policies, and define roles and responsibilities of different business units.	Assemble recall response team and determine recall action items quickly.	Design resolution plan (e.g., product exchange, refund, etc.).
Production / Logistics	Establish effective inspection plans along the supply chain to ensure product safety, and develop plans for managing product returns.	Suspend production and distribution of noncompliant products until root causes are eliminated, and launch information systems for managing product returns.	Track product returns record, and develop improvement plans for producing safer products and for managing product returns in the future.
Product Development	Incorporate product safety and traceability ideas such as Poka-yoke (fool-proof) and RFID technology in product development.	Identify root causes and correct them quickly.	Design new plans to prevent these root causes from recurring.
Communication	Develop product recall communication plans.	Quickly acknowledge and apologize for problems to stakeholders (including the government), announce recall, and report recall progress.	Communicate a specific plan to make products safer and report its progress to stakeholders to rebuild confidence.

The 3R framework is similar to the "butterfly" description of supply chain risks as depicted in Chapter 2 in that it also deals with the issues arising before, during and after the occurrence of a risk event. Also, the underlying philosophy of the 3R framework is based on the continuous improvement concept, which emphasizes that effective product recall management requires continuous effort through the experience of different risk events.

We can use the 3R framework to summarize Mattel's recall management process implemented in 2007 (Table 12.2). It can be seen that Mattel has taken systematic steps to handle these recalls effectively. This conclusion is consistent with the comments made by Quelch (2007), praising Mattel's effective communication process

Table 12.2 Mattel's product recall management plan in 2007

	Readiness (before)	Responsiveness (during)	Resilience (after)
Policy	Before the 2007 recalls, Mattel had established product recall policies, including roles and responsibilities of different business units, and the appointment of Mr. Jim Walter as the senior vice-president of global product integrity.	The recall response team quickly developed a recall plan. In September 2007, the company appointed Mr. Geoff Massingberd to head its corporate responsibility unit that was created after the lead-paint-related recall.	Mattel provided free shipment for returning the recalled toys. In exchange for returning the toys, Mattel offered coupons for customers to buy other Mattel products with the same value.
Product Development	Mattel had established a product development process to ensure safety before the recalls. Catherine H. Pilarz, senior director of product safety at Mattel/Fisher-Price joined the board of ASTM International in 2006.	When the issue of separated magnets emerged in 2006 around a competitor's product, Mattel participated with consumer groups, industry and regulators as part of a voluntary standard setting organization, ASTM International, to develop a new design standard to prevent magnets from separating. At the same time, the company coated magnets attached to plastic with glue slurry in an effort to address magnet separation issues. Later, in November 2006, Mattel recalled some magnet toys for which it received reports of either an injury or a relatively high volume of reports of separated magnets even if no injury had been reported.	In January 2007, Mattel unilaterally adopted the final draft ASTM International design standard, which was not published until May 2007. Today, the ASTM International standard, which requires that magnets in plastic toys be locked in place with a plastic rim or encasement, is the only standard that has been adopted in an effort to eliminate the possibility of having the magnets come loose. In the August 2007, Mattel retroactively applied the new design standard to most of its plastic toys with magnets that were produced prior to January 2007.

Table 12.2 (Continued)

	Readiness (before)	Responsiveness (during)	Resilience (after)
Production / Logistics	Prior to the 2007 recalls, Mattel has established safety standards, testing procedures, and requirements with contract manufacturers to ensure compliance with lead level standards.	Upon recognizing the fact that some long-term contract manufacturers failed to follow Mattel's established safety procedures and requirements with respect to paint, Mattel suspended production of and stopped shipment of products that it believed may have been noncompliant with requirements concerning lead in paint. It also took samples of all products that originated from Asia and tested them before they could reach store shelves in the U.S.. To facilitate the recall, Mattel created channels that provided a simple way for consumers to return products via downloadable forms, free shipping mailers, etc..	In September 2007, Mattel announced a 3-stage safety check system to prevent the manufacture of toys with noncompliant levels of lead in paint (Eckert 2007). The program has three parts: (1) testing every batch of inputs (paint); (2) new requirements, including segregated storage of paint for Mattel products and in process testing, as well as stepped up auditing and inspection of vendors and any subcontractors; and (3) finished goods testing, including from every production run.
Communication	Mattel had established product recall communication plans.	Mattel acknowledged problems to stakeholders and apologized to customers publicly in the press. The company also has a dedicated website (http://service.mattel.com/us/recall.asp) that provided information and photos depicting the recalled toys, as well as clear instructions regarding ways to return the recalled products. The company proactively publicized the recall on websites and other media while its executives reached out to the media.	To restore consumer confidence in Mattel's toys, CEO Eckert of Mattel announced the 3-stage safety check system in September 2007. In addition, a new position reporting directly to the company's CEO was created to ensure that potential recall risks were quickly addressed, and, if necessary, elevated within the company.

for managing the 2007 recalls despite the early ham-handed efforts to lay the blame on Chinese manufacturers.

12.3.1 Challenges Pertaining to Product Recalls

While Mattel had to work hard to ensure toy safety, there are challenges associated with product recalls that Mattel and other manufacturers will need to consider in the near future:

1. **Corporate responsibility.** Under the law, manufacturers are responsible for ensuring product safety. For example, in January 2008, when Baxter recalled its blood-thinning drug Heparin, which was made by a Chinese contract manufacturer, Changzhou SPL, the FDA stated explicitly that safeguarding the legality, safety, and quality of raw materials imported for use in pharmaceuticals is the responsibility of the importing country. However, as companies outsource manufacturing operations to other countries, it is also common for them to outsource inspection and shipping operations to third parties so as to reduce cost and time. To ensure product safety for consumers, manufacturers should re-examine their outsourcing strategies.

2. **Scrutiny of suppliers.** A minority of suppliers may choose to "cut corners." For example, to reduce cost, time, and potential scrutiny by the Chinese government, Baxter's contract manufacturer Changzhou SPL was registered as a chemicals manufacturer, not as a pharmaceuticals company. Nor did Changzhou SPL register its drug Heparin as pharmaceutical materials with the Chinese Food and Drug agency. Without appropriate registration, Changzhou SPL managed to avoid the required inspections. Thus, dealing with a global supply base, can be a big challenge for manufacturers to police their suppliers.

3. **Recall decisions.** When making a recall decision, a company needs to examine the scale and the scope of the underlying problems and identify the root causes. However, many companies may not have accurate and timely information that is needed to make the following recall related decisions: Should the company issue a recall? If yes, what kind of recall? Should the company recall locally or globally? Should it recall selected items or an entire brand? If there is a recall, when should the company announce the recall? Consider for instance the Ford-Firestone tire recall of August 2000 in this context: it was not clear to Firestone whether the problem was with its tires or with a particular Ford vehicle model.

4. **Disposal of recalled products.** After receiving recalled products, a company needs to develop a disposal plan for the returned goods. For example, products containing less than 5 parts of lead per million can be disposed of at regular landfills; however, for products with higher lead concentrations, the Environmental Protection Agency requires extra steps to be taken in order to make sure the lead will not contaminate the environment. Mattel is fully compliant with environmental requirements governing the disposal toys subject to the lead-related recalls and actively looks for opportunities for recycling.

5. **Return rate.** Despite a company's extensive efforts, it can be a challenge to get consumers to return recalled products. For the majority of recalls of consumer products, consumer return rates are only about 1 or 2 percent. Products, including toys, that are relatively low in price and have relatively short product lifecycles typically have even lower rates of consumer participation. Because of the company's proactive approach and media attention, consumer return rates were in excess of two percent despite the characteristics of the recalled products. Some good examples of other successful recalls with relatively high return rates can be found in Smith et al. (1996).

6. **Resale of recalled products.** Despite a company's efforts to suspend the sales of the affected products during and after a recall, there is currently no federal law or regulation against reselling recalled products in the US. Legislation that would prohibit the practice is pending in Congress. As reported in the media, recalled products continue to be sold on auction websites such as eBay or in small shops, despite efforts by companies, including eBay and Mattel, to prevent it. When dealing with sales channels world-wide, manufacturers like Mattel would find it a challenge to police the sales of recalled products in various aftermarket channels such as eBay or small stores in developing countries.

7. **Recovery of costs associated with additional inspections and audits.** To ensure product safety after a product recall, it is common for companies to institute additional measures, including inspections and audits. These measures would improve product safety, but they contradict the TQM philosophy that calls for "inspection at the source" so as to reduce duplication of efforts. These additional safety steps cost time and money. To recover this additional cost, each company needs to decide whether to absorb all of the cost or pass some of it on to consumers. These steps also raise questions about outsourcing itself because the outsourcing company does not have visibility of the manufacturing of the outsourced products.

8. **Third-party inspections.** Inspection to verify procedures and materials may itself be outsourced. As such, in many instances, inspections are conducted by a third party company, such as SGS, Underwriter Laboratories and Bureau Veritas, which specialize in product testing. Besides being knowledgeable about the safety standards of different geographical regions, these testing services have expertise in methods for conducting inspections that have been suggested by independent agencies like the American National Standards Institute. However, it is not always easy for third-parties to inspect products and factories overseas where cultural differences, legal and contractual obstacles, and politics (e.g., visa denials) can frustrate these efforts.

9. **Track-and-trace.** RFID technology enables companies to track each unit or product that is sold and link it back to each step of the supply chain operation. These steps include manufacturing, packaging, distribution, and logistics. With this information on hand, companies can quickly identify when and where problems occur. In addition, RFID technology enables retailers to trace the location (sales floor, store room, warehouse, etc.) of recalled products quickly. RFID technology can also enable a company to retrieve recalled products in the sales

channel. However, due to privacy issues, it is still a serious challenge for companies to track the identity of each consumer who purchased recalled products in the United States. Some companies use the warranty registration as a mechanism to track customers, but the number of consumers that use this registration is quite low.

10. **New certification requirements.** Currently, the CE mark on various products sold in Europe certifying compliance with EU standards is verified by manufacturers in the EU. However, the EU consumer protection commission is considering imposing a new standard called *CE-plus* mark. This new standard would involve the inspection of test certificates for products such as toys at the point of import; such inspections at the moment only apply to medical devices, animal/fish products and fruits and vegetables. Clearly, this additional inspection could result in delays in shipments, which would increase the lead time. As lead time increases, the work-in-process and finished goods inventories, and ultimately, the cost would increase. Therefore, this proposed regulation will become a new challenge for many manufacturers if the law passes.

12.3.2 Opportunities Associated with Product Recalls

Despite the above challenges, product recalls can create various new opportunities for companies:.

1. **New business growth.** The growing needs of companies such as Mattel, given their supply chains, to conduct more inspections can stimulate growth for various testing service companies, such as SGS (Switzerland), Underwriter Laboratories (USA) and Bureau Veritas. Sales of safer products and testing kits tend to increase after a recall. For example, immediately after Mattel's recall of some toys with noncompliant levels of lead in paint, various companies started producing and promoting "lead free" toys made in the U.S.[2] At the same time, several companies promoted various lead test kits to concerned parents. Overall, there is certainly a market for toys made in the U.S. that are safer but also more expensive. However, the size of this market remains unclear. On the other hand, German toy-maker PlayMobil produces toys in western Europe and gets them to consumers at prices that are similar to U.S. toymakers like Mattel who outsource their manufacturing to Chinese manufacturers. This means part of new business growth is to consider short supply chains that afford a great deal more visibility.

2. **New technologies.** Besides lead, products can contain noncompliant levels of other regulated or safety-related compounds. As such, there is potential for growth for testing equipment manufacturers such as Elmer Perkins and Agilent.

---

[2] See http://www.thedailygreen.com/ environmental-news/blogs/american/5291?click=main_sr for a listing of "lead free" toys. While this site warned consumers that there is no guarantee that these US-made toys are "truly" lead free, they argued that the chance of having US manufacturers cutting corners is less likely.

Companies also may consider the deployment of RFID throughout the supply chain, if not with the individual product (we already mentioned doing so may not be popular with consumers.) In addition, to deploy RFID technology effectively, companies need to implement information systems such as SAP to capture relevant information for track-and-trace purposes.

3. **New quality standards.** As an essential pre-requisite for conducting business with Western companies, many contract manufacturers and suppliers in Asia are certified with quality standards such as QS9000 (automotive sector), ISO9002 (basic systems), or ISO16949 (environmental). However, many certification processes rely heavily on self-assessment, and third-party verification is sometimes missing, a fact that may trigger the International Organization for Standardization (ISO) to develop new quality standards that do require third-party verification. If this development of new standards happens, it would create new business opportunities for consulting firms and agencies to help various contract manufacturers in the emerging economies to get certified with the new standards in the near future.

12.4 Conclusion

We analyzed and evaluated (a) the way Mattel managed its recalls in 2007; and (b) the company's plans to strengthen product safety following those recalls. We highlighted some key challenges for companies to consider as well as some new business opportunities.

This chapter concludes Part II that focused on ways to mitigate supply chain risks that arise in different contexts. Specifically, we have shown how companies can implement some of the concepts we presented in Part I. More importantly, we use specific cases in Chapters 9, 11 and 12 to highlight the fact the importance of establishing proactive risk mitigation strategies with supply chain partners prior to the occurrence of risk events.

Part III
Review of Quantitative Models for Managing Supply Chain Risk

Chapter 13
Risk Models for Supply Chain Management

Abstract This chapter reviews quantitative models that deal with supply chain risks. While this review is not exhaustive, it provides indicative research literature and quantitative models proposed by supply-chain management researchers. Because most quantitative models in the literature are designed primarily for managing high-frequency-low-impact *operational* risks, we present potential research ideas for managing low-frequency-high-impact *disruption* risks.

13.1 Introduction

After discussing strategic and tactical approaches for managing supply chain risks and actual case studies in Part II, we now turn to reviewing the existing academic research literature that pertains to supply-chain risk management in Part III. Even though the literature reviewed in Chapters 13–15 in this part deals primarily with ways to mitigate "normal" supply chain risks rather than the "abnormal" risks (i.e., disruptions), this review is useful for providing a good starting point for two reasons: (1) this literature can be extended for disruptions, and (2) unlike the general risk management literature that motivated most of Part I of this book, the chapters in this part are firmly rooted in the supply chain literature and practice.

In this chapter, we review primarily quantitative models that deal with supply chain risks. For general approaches to risk management, we refer the reader to the review by Chapman et al. (2002). This chapter (Chapter 13) is not an exhaustive review; rather, it provides indicative research literature and quantitative models proposed by supply-chain management researchers. Chapter 14 deals with modeling of various types of supply-chain flexibility. The final Chapter 15 in this part reviews the use of stochastic programming in the literature to plan for demand uncertainty. Also, we relate various supply chain risk management strategies examined in the literature with actual practices.

In general, we can categorize *supply chain management* efforts and the corresponding risk mitigation models in the literature as follows: (1) supply manage-

M.S. Sodhi, C.S. Tang, *Managing Supply Chain Risk*,
International Series in Operations Research & Management Science 172,
DOI 10.1007/978-1-4614-3238-8_13, © Springer Science+Business Media, LLC 2012

ment, (2) demand management, (3) product management, and (4) information management. For each of these categories, we have supply-chain topics for which we provide indicative literature. Table 13.1 lists the four aforementioned categories and the topics within each along with the corresponding section and subsection numbers in this chapter.

Table 13.1 Topics in the supply chain management literature for managing risks

13.2. Supply Management	13.3. Demand Management	13.4. Product Management	13.5. Information Management
13.2.1 Supply network design	Product rollovers and pricing to:	13.4.1 Postponement	Supply chain visibility; information sharing; vendor-managed inventory, and collaborative planning, forecasting and replenishment for
13.2.2 Supplier relationships	13.3.1 Shift demand across time	13.4.2 Process sequencing	
13.2.3 Supplier selection	13.3.2 Shift demand across markets	13.4.2 Robust strategies for product management	
13.2.4 Supplier order allocation	13.3.3 Shift demand across products		13.5.1 Managing "fashion" products and
13.2.5 Supply contracts	13.3.4 Robust strategies for demand management		13.5.2 Managing "functional" products
13.2.6 Robust strategies for supply management			13.5.3 Robust strategies for information management

Although we focus our review on "robust strategies", most quantitative models presented in this chapter are designed for managing operational risks (i.e., normal risks) rather than disruptive ones. (This is not surprising as these models are for *supply chain management* in general rather than for *supply chain risk management*—see Chapter 16 on researchers' perspectives in this regard.) So, although these quantitative models often provide cost-effective solutions for managing operational risks, they do not address the issue of disruption risks in an explicit manner.

While having robust supply chains is desirable, firms will be more willing to implement "robust" supply chain strategies for mitigating disruption risks if these strategies possess two specific properties: (1) efficiency—the firm could manage operational risks (normal risks) efficiently regardless of the occurrence of major disruptions and (2) resiliency—the firm can sustain its operation during a major disruption and recover quickly afterwards. Christopher (2004), Chopra and Sodhi (2004), and Lee (2004) offer different approaches for establishing resilient supply chains.

Both efficiency and resilience are critical for firms to ensure profitability and business continuity, the latter having elements of supply chain risk management as defined in Section 13.1. Also, when a robust strategy is efficient, most firms can perform cost/benefit analysis or return on investment to justify strategies for improving efficiency under operational risks. As such, the management is more willing to implement a strategy that would enhance the efficiency and resiliency. The notion of efficiency and resiliency is akin to the S.M.A.R.T. supply chain management approach developed by De Waart (2006) for quick and effective response to risk events to ensure cost effectiveness and speedy recovery.

The organization of this chapter is as follows. In Sections 13.2 to 13.5, we review some of the research literature with topics as shown in Table 13.1. In Section 13.6, we discuss managerial attitudes toward risk as these are pertinent to supply chain risk. Section 13.7 concludes this chapter with some suggestions for future research in supply chain risk management.

13.2 Supply Management

To gain cost advantage, many firms outsourced certain non-core functions so as to maintain a focus on their core competence (c.f., Porter, 1985). Since the 1980s, we witnessed a sea change in which firms outsourced their supply chain operations including design, production, logistics, information services, etc. Essentially, supply management deal with six inter-related issues:

1. Supply network design
2. Supplier relationships
3. Supplier selection process (criteria and supplier selection)
4. Supplier order allocation
5. Supply contract
6. Robust strategies for supply management.

13.2.1 Supply Network Design

When designing a global supply chain network, we need to address the following issues:

1. Network configuration: Which available suppliers, manufacturing facilities, distribution centers, and warehouses should be selected?
2. Product assignment: Which facilities (suppliers, manufacturing facilities, distribution centers, etc.) should be responsible for processing which subassemblies, semi-finished products, or finished products?
3. Customer assignment: Which facility at an upstream stage should be responsible for handling the "demand" generated from downstream stages?
4. Production planning: When and how much should each facility produce?

5. Transportation planning: When and which mode of transportation should be used?

Most of the published work in the area of supply network design is based on different deterministic models. For example, by considering the fixed and variable processing cost at each facility, Arntzen et al. (1995) implemented a mixed integer programming model at Digital Equipment Corporation that serves as a planning system for determining optimal decisions related to issues 1, 2, 4, and 5 above. In addition, Camm et al. (1997) develop an integer programming model for Procter and Gamble (P&G) to deal with issues 1 and 3. However, these papers do not deal with risk in any explicit manner.

Some researchers have investigated supply chain network design by capturing certain risk related issues arising from global manufacturing. For instance, Levy (1995) presents a simulation model to examine the impact of demand uncertainty and supplier reliability on the performance of different supply chain network designs (issues 1 and 5). This simulation model has helped a personal computer manufacturer to evaluate the costs and lead times associated with two sourcing alternatives between Singapore and California.

Likewise, Lee and Tang (1998b) develop a stochastic inventory model to examine the tradeoff between the "consignment" and "turnkey" arrangements under demand uncertainty. Their analysis has helped Hewlett Packard (HP) to determine specific arrangements with different contract manufacturers in Singapore and Malaysia. Consignment and turnkey are two common approaches for the contract manufacturers to obtain the requisite parts from various suppliers in outsourced manufacturing. Under consignment, the original equipment manufacturer (e.g., HP) purchases the requisite parts from different suppliers (to enjoy the volume discount), sorts the parts to create kits, and then ships the kits to the corresponding contract manufacturers. However, this arrangement has drawbacks in terms of long lead time and high shipping and handling costs. Turnkey is an alternative arrangement under which the contract manufacturers order the parts directly from suppliers designated by the OEM and then charge the OEM accordingly.

There are also some papers that use stochastic programming to extend supply chain network design under different types of uncertainties pertaining to five issues listed at the beginning of this section. For instance, Huchzermeier and Cohen (1996) develop a modeling framework to show how one can exploit currency exchange rates by shifting production within a global supply chain network. By incorporating network design issues 1, 3, 4 and 5, they formulate the problem as a multi-period stochastic programming problem that aims to maximize the discounted after-tax profit. They also show how flexible global supply chain can provide real options to hedge against exchange rate fluctuations. Likewise, Kouvelis and Rosenblatt (2002) develop a two-stage global supply chain network model that addresses all five issues listed above. More importantly, they consider the case in which government subsidies and tax incentives are present in certain countries for certain products or operations. Their mixed integer programming model provides insights on the effects of financing, taxation, regional trading zones and local content rules on the design of a global supply chain.

13.2.2 Supplier Relationships

As more manufacturers recognized the strategic value of suppliers in the late 1980s, supplier relationships changed from adversarial to cooperative in the U.S. (Helper, 1991). Many firms realized that suppliers could enable a firm to focus on core competence and to reduce cost and product development cycle time at the same time. In addition, various e-markets and information technologies enabled companies to foster different types of relationships with the suppliers, ranging from one-time purchase to virtual integration via information sharing.

Dyer and Ouchi (1993) and Dyer (1996) studied various Japanese and U.S. firms and support the idea of having long-term supplier relationship with fewer strategic suppliers. Tang (1999) has identified four types of supplier relationships: (1) vendor, (2) preferred supplier, (3) exclusive supplier and (4) partner. These four types differ from each other in terms of types of contracts, length of contracts, type of information exchange, pricing scheme, and delivery schedule. By considering the market condition that is measured in terms of the strategic importance level of the part to the buyer and the buyer's bargaining power, Tang recommends different supplier relationship for different market conditions.

Much of the literature reviewed by Tang (1999) focuses on qualitative analysis or strategic analysis rather than operational or short-term opportunities. Closing this gap, Cohen and Agrawal (1999) present an analytical model for evaluating the tradeoff between the flexibility offered by short-term contracts and the improvement opportunities and price certainty associated with long-term contracts. They show analytically that long-term contracts may not always be optimal and provide conditions under which short-term contracts are actually more effective. As firms expand their business globally, their supply chains involve more global partners. For some regional markets, a firm may source locally so as to reduce transportation cost, replenishment lead times and inventory; there may also be tax benefits and low local labor costs. Consequently, firms may source from multiple suppliers. Firms may source from multiple suppliers also to reduce the impact of various operational and disruption risks. Indeed, according to an empirical study conducted by Shin et al. (2000), dual or multiple sourcing is a common business practice.

13.2.3 Supplier Selection Process

Boer et al. (2001) provide a comprehensive review of different methods for selecting suppliers. They divide the supplier selection process into developing of selection criteria, and approving and selecting suppliers. We discuss these in detail below.

Developing supplier selection criteria. Boer et al. (2001) report two decision methods—interpretive structural modeling and expert systems—for forming selection criteria. The interpretive structural modeling technique proposed by Mandal and Deshmukh (1996) that separates dependent criteria from independent criteria:

the independent criteria are important for screening acceptable suppliers while the dependent criteria are critical for final supplier selection. The expert system developed by Vokurka et al. (1996) captures the previous supplier selection process in a knowledge base, which can be used to suggest selection criteria for future supplier selection process. Choi and Hartley (1996) investigate 26 supplier selection criteria used by different partners (automotive assemblers, first-tier suppliers, second-tier suppliers) across the supply chain in their empirical study of the auto industry. These criteria include cost reduction capability, quality improvement capability, and the ability to change production volumes rapidly. By using various multivariate statistical techniques (factor analysis, clustering analysis, multivariate analysis of variance) to analyze the supplier selection criteria reported in 156 surveys, they make the following conclusions:

- The supplier selection criteria are reasonably consistent across the supply chain in the automotive industry. At all levels, commitment to establish cooperative/long-term relationship is an important selection criterion.
- Price is one of the least important criteria, while quality and delivery are important criteria.
- Supplier's technological capability and financial stability are more important criteria for the auto assemblers.

Note that *volume flexibility*, i.e., the ability to change production volumes rapidly, is not considered to be as important as other criteria such as *quality* and *long-term relationship*. Moreover, with disruptions by way of terrorist attacks, hurricanes, earthquakes, and SARS that have occurred since the study by Choi and Hartley (1996), one may also speculate that *business continuity* would become an important supplier selection criterion.

Approving and selecting suppliers. At this stage of the process, the goal is to reduce the set of all potential suppliers to a smaller set of approved suppliers. To do so, the decision maker has to classify all suppliers into approved or disapproved categories. Based on the supplier's performance on the selection criteria, Boer et al. (2001) report the use of the following methods for supplier approval: clustering analysis, data envelopment analysis, and an Artificial Intelligence approach called case-based-reasoning method.

For the final supplier selection out of the approved suppliers list, Boer et al. (2001) report the use of the following decision methods in different settings:

- **Linear weighting models.** By assigning different weights to different criteria, one can compute the overall rating of a supplier by considering the weighted sum of different criteria. In this case, the supplier with the highest rating will be selected.
- **Total cost of ownership.** This method is developed by Ellram (1990) to include all quantifiable costs incurred throughout the life cycle of the item purchased from a supplier. The supplier with the lowest total cost of ownership will be selected.

- **Mathematical programming models.** Most of the methods reported in Boer et al. (2001) are based on various deterministic models including: linear programming, goal programming, data envelopment analysis, etc. The idea is to select supplier(s) with minimum cost.
- **Simulation models.** This method enables the decision maker to capture some of the uncertainties (yield loss, stochastic lead times, etc.) related to supplier selection. By simulating the performance of different suppliers for different criteria under different scenarios, the method can help a decision maker to select a supplier under uncertainty.

There are other quantitative models for supplier selection. For example, Weber and Current (1993) present a mixed integer programming formulation that is intended to capture multiple supplier selection criteria. Current and Weber (1994) formulate the supplier selection problem as a variant of facility location problem. Weber et al. (2000) present an approach for evaluating the number of suppliers to employ by using multi-objective programming and data envelopment analysis. Dahel (2003) extends the model presented in Weber et al. (2000) by incorporating the order quantity decision for each supplier.

While most supplier selection models are deterministic in nature, a few articles specifically address operational risks tied to supplier selection. Tang (1988) presents a supplier selection model that captures the interaction of the supplier's quality and the buyer's quality control (inspection policy). Tagaras and Lee (1996) develop a different supplier selection model that captures different degrees of imperfections in the buyer's manufacturing processes by considering the interaction between the supplier's quality and the buyer's internal manufacturing process. Specifically, they consider that there are two states of the buyer's process: normal or abnormal. When the buyer's process is in the normal state, the output of the process is perfect if the supplier's input is. However, when the buyer's process is in the abnormal state, the output of the process is defective regardless of the supplier's input is or is not. By considering different costs that depend on the output quality, they develop the optimal supplier selection criterion that minimizes the buyer's expected total cost (ordering cost and cost of quality). Kouvelis (1998) presents a supplier selection model that captures the stochastic nature of exchange rate. In his model, the buyer needs to decide the suppliers to be selected and the quantity to be ordered from each of those selected suppliers. As a way to respond to fluctuating exchange rates, the model captures the flexibility for the buyer to shift the order quantity among suppliers dynamically at the expense of switchover costs. When the switchover cost is significantly high, he shows that the buyer may continue to source from suppliers that are more expensive so as to avoid switchover costs.

13.2.4 Supplier Order Allocation

After a set of suppliers is chosen, the buyer needs to determine ways to allocate the order quantity among these selected suppliers. We classify the literature in this area according to different types of operational risks:

1. Uncertain demand
2. Uncertain supply yields
3. Uncertain supply lead times
4. Uncertain supply capacity, and
5. Uncertain supply costs.

Uncertain demand. There is voluminous amount of published works that focus on analytical models for determining optimal order quantity for a *single* supplier under demand uncertainty. For a review of analytical models that deal with a single supplier, the reader is referred to the books by Porteus (2002) and by Zipkin (2000). For models that deal with *multiple* suppliers, Minner (2003) provides a more comprehensive review. When the supply lead times are deterministic, all models assume that the supplier with a shorter lead time charges a lower cost per unit. Due to the complexity of the analysis, most discrete-time models are restricted to two suppliers with lead times that differ by one period. By contrast, Zhang (1996) deals with *three* suppliers with lead times that differ by one and two periods respectively and characterize the optimal ordering policy for each supplier.

To make the analysis of multiple-supplier inventory models more tractable, some researchers consider two supply modes: regular and emergency. The regular supply model is based on a regular supply lead time with finite lead time, while the emergency supply is available instantly. Fukuda (1964) shows that the optimal ordering policy takes on the form of "two order-up-to levels", x and y, where $x < y$. Specifically, the optimal ordering policy can be described as follows: If the inventory at the beginning of a time period z is less than x, then order $(x - z)$ units by using the emergency mode and order $(y - x)$ units by using the regular mode; if $x < z < y$, then order $(y - z)$ units according to the regular mode; otherwise, order nothing. Vlachos and Tagaras (2001) extend Fukuda's model to the case in which the emergency model is capacitated. Scheller-Wolf and Tayur (1999) consider a Markovian periodic review inventory model and show that the optimal ordering policy for the buyer is a modified state-dependent base-stock policy. Specifically, they show that there exists a state-dependent optimal inventory level (target) in each period. In each period, the buyer should first order an amount from the regular supplier so that the inventory position after ordering is as close as possible to the target. The buyer can place an emergency order to fill the gap between the target and the inventory position after ordering from the regular supplier.

Due to the complex analysis of the optimal ordering policies for the multi-supplier case, various researchers restrict their analysis to certain classes of ordering policies. For example, Moinzadeh and Nahmias (1988) analyze an (s_1, s_2, Q_1, Q_2) ordering policy for a continuous time model with regular and emergency supply. Specifically, when the inventory reaches s_1, a regular order of size Q_1 is placed. If the inventory reaches s_2 within the lead time of the regular order, an emergency order Q_2 is placed. Janssens and de Kok (1999) analyze an ordering policy in which the buyer will always order Q units from one supplier in each period, and will order $[S-Q]+$ units from the second supplier so as to bring the inventory position to S. The reader is referred to Minner (2003) for more details.

Instead of focusing on optimal ordering policies, Nagurney et al. (2005) develop a model for analyzing the equilibrium behavior of a three-level supply chain comprising manufacturers, distributors and retailers. By considering uncertain demands at the retailer level, they formulate the problem at each level as a non-linear programming problem. For the retailers, the goal is to determine the optimal order quantity for each retailer based on the wholesale price set by the distributors. However, for the distributors, the goal is to determine the optimal wholesale price based on the manufacturers' price. The manufacturers set their prices to maximize profit adjusted by risk. By considering the first-order conditions of these three inter-related problems, they show how to recast the first order conditions as a set of variational inequalities. Bazaraa et al. (1993) provide details about the relationship between variational inequalities and Nash equilibrium. By exploiting the structure of these variational inequalities, they establish the existence of a unique equilibrium and provide certain characteristics of the equilibrium.

Uncertain supply yields. Consider some single-stage-multiple-period models first. Gerchak, Vickson and Parlar (1988) analyze a finite horizon problem with stationary demand distribution and show that order-up-to policies are not optimal when a buyer receives a random fraction of the order quantity from the supplier. Henig and Gerchak (1990) further show that there exists a critical point for each period such that an order should be placed only when the on-hand inventory at the beginning of the period is below the corresponding critical point. However, the exact order quantity is a complicated function of the system parameters. Agrawal and Nahmias (1998) present a model for evaluating the tradeoff between the fixed costs associated with each selected supplier and the costs associated with yield loss. They show how to determine the optimal number of suppliers with different yields when the demand is known. To limit our focus to supply chain management, we shall highlight some of the models that deal with multiple stages/products. Yano and Lee (1993) provide a thorough review of single stage/period models that deal with lot-sizing models with random yields.

As regards multiple-stage-multiple-period models, Bassok and Akella (1991) consider a two-stage-multiple-period model in which one stage corresponds to raw material ordering and the second stage corresponds to actual production, where yield uncertainty occurs only at the material ordering stage. They show that the existence of two critical points, one for the raw material ordering stage and one for the production stage, and the optimal ordering quantity and the optimal production quantity depends on whether the sum of (on-hand) finished goods and raw materials is larger or smaller than these two critical points, respectively. Because exact analysis of multiple-stage-multiple period models is intractable, Tang (1990) restricts his analysis of a linear control rule for a multi-stage serial production line with uncertain yields at each stage and uncertain demand. This linear control rule intends to "restore" the buffer stock at each stage to its target value in expectation. Hence, this control rule minimizes the expected deviation of the buffer stock levels from their targets. Denardo and Lee (1996) generalize Tang's model by incorporating rework and unreliable machines.

Although, multi-product, multi-stage, and multi-period models are intractable and not much work has been done with such models, there are some exceptions. Akella, Rajagopalan and Singh (1992) study a multi-stage facility with rework that produces multiple parts. Their analysis aims to determine an optimal production rule at each stage that minimizes the total inventory and backorder cost. They assume that the cost function is quadratic, which leads to optimal linear decision rules. Linear decision rules have been analyzed by Gong and Matsuo (1997) as well. Specifically, Gong and Matsuo consider a more general multi-stage facility with re-entrant routings. They formulate a control problem with the objective to minimizing the weighted variance of work-in-process inventory while ensuring that production capacity constraints are satisfied with a pre-specified probability. Their numerical experiments suggest that the linear decision rules perform well when compared with the optimal production policy.

Uncertain lead times. When replenishment lead times are stochastic, most researchers restrict their analyses of multiple supplier models to the case of deterministic demand. When both suppliers have identical lead time distributions (uniform or exponential), Ramasesh et al. (1991) consider an (s, Q) ordering policy where the order quantity Q is split evenly between two suppliers. Due to the complexity of the analysis, the optimal values for the reorder point s and the order quantity Q are determined numerically. By restricting the attention to the (s, Q) ordering policy, Sedarage et al. (1999) extends the model of Ramasesh et al. (1991) by considering more than two suppliers and a non-identical split among these suppliers. Based on the numerical analysis presented in Sedarage et al. (1999), they show that it might be beneficial to order from some suppliers with poor lead time performance in terms of the mean and standard deviation of the lead time. In general, although the exact analysis of multiple suppliers with stochastic lead times is intractable, exact analysis can be obtained for some special cases. For example, Anupindi and Akella (1993) consider a two-supplier model with random demand in which the replenishment lead time of supplier j is equal to one period with probability p_j and two periods with probability $(1 - p_j)$, where $j = 1, 2$. They derive the optimal ordering policy that minimizes the total ordering, holding and backordering costs over a finite horizon. They show that the optimal ordering policy in each period n depends on two critical points x_n and y_n, where $x_n < y_n$, and the on-hand inventory at the beginning period n, z_n. Specifically, order nothing if $z_n \geq y_n$; order from one supplier if $x_n \leq z_n < y_n$; and order from both suppliers if $z_n < x_n$.

Uncertain supply capacity. Most models assume that the supply capacity is unlimited or known. However, unexpected machine breakdowns could affect the supply capacity. Relative little amount of work has been done in the area of uncertain supply capacity. Parlar and Perry (1996) present a continuous time model in which the availability of each of the n suppliers is uncertain because of disruptions like equipment breakdowns, labor strikes, etc. By considering the case that each supplier is either "on" or "off," there are 2^n possible number of states for the whole system. For each of these 2^n states, they analyze a state-specific (s, Q) ordering policy so that the buyer would order Q units when the on-hand inventory reaches s. Ciarallo et al. (1994) develop a discrete time model in which the supply capacity is random with known probability distribution. By considering the total (undiscounted) expected costs (ordering, inventory holding and backordering costs), they show that the objective function is quasi-convex, which implies that an order-up-to policy is optimal. Wang and Gerchak (1996) examine a periodic review model with uncertain supply capacity, uncertain yields and uncertain demand. The objective is to minimize the total discounted expected costs over a finite horizon. They show that the optimal policy possesses the same structure as the optimal policy obtained by Henig and Gerchak (1990) for the case in which only random yield is considered and that the order-up-to policy is optimal.

Uncertain supply cost. While most work focus on demand uncertainty, not much work has been done in the area of uncertain supply cost. For models that examine the issue of uncertain supply cost imposed by an upstream supply chain partner, Gurnani and Tang (1999) analyze a situation in which a retailer has two instants to order a seasonal product from a wholesaler prior to the beginning of a single selling season. They consider the case in which the wholesale price at the second instant and the demand are uncertain; however, the retailer can improve the demand forecast by using market signals observed between the first and second instants. In order to determine the profit-maximizing ordering policy, the retailer needs to evaluate the trade-off between the benefit of having a more accurate forecast and a potentially higher wholesale price at the second instant. By formulating the problem as a 2-period dynamic programming program, they develop an optimal way to allocate the optimal order quantity to be placed at the first and second instants and they provide the conditions under which the retailer should delay his ordering decision until the second instant.

Some researchers develop models for exploiting uncertain currency exchange rates in a global supply chain. Kogut (1985) develops a framework to argue that the benefit of a global supply chain lies in the operational flexibility, which permits a firm to exploit uncertain exchange rates. To examine this issue in a quantitative manner, Kogut and Kulatilaka (1994) develop a stochastic model to examine the value of the flexibility to shift production between two plants located in two different countries. By formulating the problem as a T-period dynamic programming problem and by modeling the exchange rate process as a discrete-time mean reverting stochastic process, they determine the option value of maintaining two manufacturing locations with excess capacity instead of having a single manufacturing location. However, they assume that the capacity of each plant is unlimited so that exactly one plant will be used to produce the required quantity to meet the total demand in each period.

Dasu and Li (1997) generalize Kogut and Kulatilaka's (1994) model by considering the case in which both plants have limited capacity so that both plants will be used to meet the demand in each period. They formulate the problem as an infinite horizon dynamic program with discounting. When the production cost is concave and when the cost of production shifting is linear, they show that the optimal production shifting policy is a two-barrier policy if the exchange rate process satisfies certain conditions. Specifically, under the two-barrier policy, there exists two critical points a and b so that it is optimal to shift the production between two manufacturing locations when the exchange rate is below a or above b. If the exchange rate is between a and b, then it is optimal to keep the same production quantity at each location without any shifting so as to reduce any unnecessary switch-over cost.

However, the models by Kogut and Kulatilaka (1994) and by Dasu and Li (1997) become intractable for more than two countries. For supply chain networks across three or more countries, Huchzermeier and Cohen (1996) present a stochastic dynamic programming problem for evaluating different global manufacturing strategy options. For any given exchange rate in each period, they solve a mixed integer program to determine the optimal production and distribution plan for the entire supply chain network that maximizes the global, after-tax profit. They construct various numerical examples by considering 16 different supply chain network designs, each of which specifies the location of the supplier(s), production plant(s), and market(s). Through these numerical examples, they illustrate the value of a global supply chain network that enables firms to shift its production and distribution plan swiftly as the exchange rates fluctuate.

13.2.5 Supply Contracts

When the partners across a supply chain belong to different firms or divisions, they tend to focus on their own objectives and make their decisions independently. Consequently, locally optimal decisions can cause operational inefficiency and globally suboptimal decision for the entire supply chain. There are two studies highlighting the pitfalls of an *un-integrated* (or decentralized) supply chain. First, when each supply chain partner places their order independently for the case in which the customer demand follows an AR(1) process, Lee et al. (1997c) show this locally optimal ordering decisions will create the "bullwhip" effect that causes operational inefficiency. Second, when each supply chain partner makes their ordering decision by maximizing their own profit for the case and when the customer demand is a deterministic and decreasing function of retail price, Bresnahan and Reiss (1985) show that these locally optimal decisions would result in lower total profit for the entire supply chain.

To improve operational efficiency and/or supply chain coordination, there has been a growing research interest in supply chain contract analysis. Most supply contract models usually deal with a supply chain that consists of one manufacturer (supplier) and one retailer (buyer) who faces customer demand. Even though the economics literature in the area of supply contracts is voluminous, economics researchers usually assume that that the customer demand is either deterministic or stochastic in the sense that demand uncertainty is resolved before the buyer places his order. Tirole (1988) provides a comprehensive review of supply contracts literature in economics.

There are three excellent reviews of supply chain contract analysis by Cachon (2003), by Lariviere (1998), and by Tsay et al. (1998), respectively. These reviews offer different perspectives: Tsay et al. (1998) provide a qualitative overview of various types of contracts when the demand is deterministic and random; Lariviere (1998) shows quantitative analyses of different types of contracts when the demand is uncertain; and Cachon (2003) examines how supply contracts can be used to achieve channel coordination in the sense that each supply chain partner's objective becomes aligned with the supply chain's objective. Since our focus is on supply-chain *risk* management, we focus on a limited set of supply chain contract literature that deals with various types of uncertainties. For this reason, we shall classify the supply chain contract literature according to different risk elements and contract types. Specifically, we shall review different types of supply contracts that can be characterized according to the financial flow and material flow as depicted in Fig. 13.1.

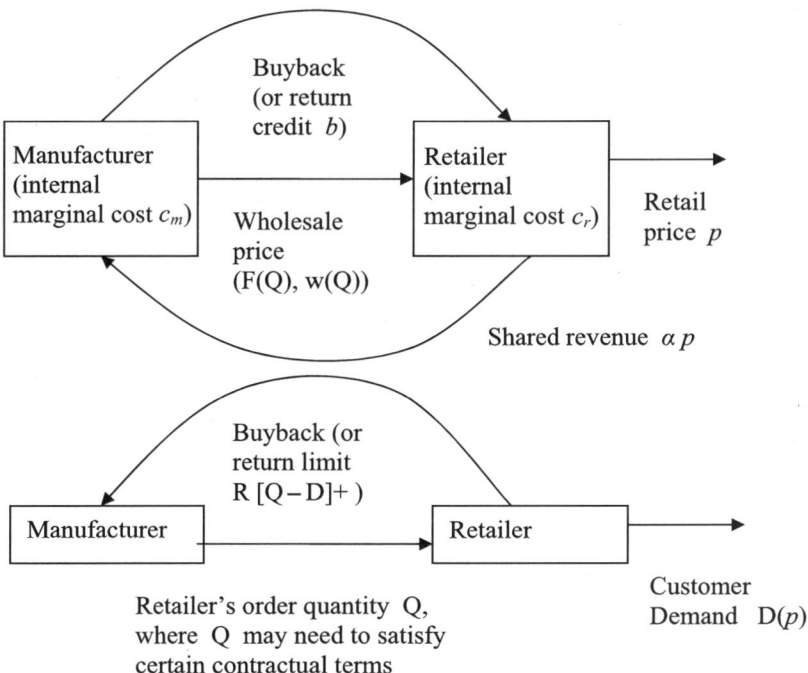

Fig. 13.1 Financial flow and material flow under different supply chain contracts

Wholesale price contracts. Consider the following scenario: the retail price p is fixed, the retailer retains the revenue p and retains the possession of any excess stock that can be salvaged at a price s. Suppose that the manufacturer offers a per unit wholesale price w so that the fixed cost $F(Q) = 0$, and the variable wholesale price $w(Q) = w$. In a single period setting, it is optimal for the retailer to order according to the newsvendor solution based on the corresponding cost structure. Given the retailer's order quantity, the manufacturer needs to determine the optimal w that maximizes his net profit. Lariviere and Porteus (2001) show that the manufacturer's profit function is unimodal when the customer demand distribution $F(x)$ with density function $f(x)$ has an *increasing generalized failure rate* (IGFR); i.e., when $xf(x)/(1 - F(x))$ is increasing in x. Many distributions such as normal, exponential, truncated Normal, Gamma, and Weibull are IGFR. Hence, when the demand distribution is IGFR, one can determine the optimal wholesale price by considering the first order condition. However, Lariviere and Porteus show that a simple price contract w will not achieve channel coordination. Anupindi and Bassok (1999) extend Lariviere and Porteus' single period model to the case in which the retailer faces an infinite succession of identical selling seasons so that it is optimal for the retailer to order up to the newsvendor solution at the beginning of each season. Cachon (2004) generalizes Lariviere and Porteus' single period model to a two-period model with inventory holding and demand updating. Specifically, in Cachon's model, the retailer can place two separate orders at two separate instants before the selling season starts; however, the wholesale price at the second instant is known to be higher. Notice that Cachon's model reduces to Lariviere and Porteus' model when the second order is not allowed. By having the flexibility to place two separate orders, Cachon develops conditions under which channel coordination is achieved.

Next, there are many situations in which a supply chain partner would keep his information private. Corbett and de Groote (2000) consider a situation in which the manufacturer does not know the retailer's holding cost in a deterministic EOQ-type environment. By imposing a prior distribution on the retailer's holding cost, Corbett and de Groote compare various channel coordination schemes in which $F(Q)$ and $w(Q)$ take on different functional forms.

When the demand is deterministic and decreasing linearly in the retail price, Corbett and Tang (1998) examine the case in which the manufacturer does not know the retailer's internal marginal cost c_r study the optimal behavior of each party under different scenarios. By imposing a prior distribution $F(x)$ on the retailer's internal marginal cost c_r and by assuming that the prior distribution $F(x)$ has increasing failure rate; i.e., $f(x)/(1 - F(x))$ is increasing in x, they compare the retailer's and the manufacturer's profits under different scenarios: one-part linear contracts $(F(Q) = 0, w(Q) = w)$, two-part linear contracts $(F(Q) = F \neq 0, w(Q) = w)$, and two-part nonlinear contracts $(F(Q) \neq 0, w(Q) \neq 0)$.

Ha (2001) generalizes Corbett and Tang's (1998) model by analyzing two-part non-linear wholesale price contracts for the case when the demand is stochastic and price-sensitive. Ha shows that channel coordination is not achievable under asymmetric information. When the manufacturer does not know the retailer's fixed ordering cost or the backorder penalty cost, Corbett (2001) examines the benefit of having the manufacturer to own the retailer's inventory (i.e., consignment stock). He shows that consigning stock may not always help the manufacturer.

Babich et al. (2004) analyze supply contracts with supplier default risk. They consider a single product model in which competing risky suppliers compete for business with a retailer. In their model, the suppliers are leaders in a Stackelberg game so that the suppliers would first establish the unit wholesale prices. Then the retailer would determine the order quantity for each supplier by taking demand uncertainty and supplier default uncertainty into consideration. By considering the retailer's discounted expected profit, they show that it is optimal for the competing suppliers to increase their wholesale prices at the equilibrium when the supplier default correlations are low and it is optimal for the retailer to order from suppliers with highly correlated default rates.

Buy-back contracts. In a single-period setting, it is optimal for the retailer to order according to the newsvendor solution. To induce the retailer to order more, it is quite common for the manufacturer to offer a return policy (also known as buy back contracts) so that the manufacturer would "buy back" up to $R\%$ of the retailer's excess inventory $[Q - D]+$ units at a unit rate of b, where $R \leq 100\%$ and $b \leq w$. Therefore, a return policy can be specified by two parameters (R, b). Pasternack (1985) is the first to show that a policy that allows for unlimited returns at partial credit; i.e., $R = 100\%$ and $b < w$, would achieve channel coordination. Moreover, Lariviere (1998) analyze the properties of the manufacturer's and retailer's profits for a class of return policies that coordinate the channel.

Emmons and Gilbert (1998) extend Pasternack's model to the case in which the retailer determines the order quantity Q as well as the retail price p. By considering a specific demand distribution of $D(p)$, they show that return policies or buy back contracts cannot coordinate the channel. However, there exists certain buy back contracts under which both the manufacturer and retailer can obtain higher profits.

Padmanabhan and Png (1997) consider the case in which two competing retailers facing a linear demand curve with an uncertain intercept. Under a full returns policy (i.e., $b = w$), they show that these retailers would increase their order quantities in a competitive environment.

Brown, Chou and Tang (2005) examine a multi-product returns policy in which the retailer can return up to a percentage of the total order quantities; i.e., the allowable return limits are pooled. By comparing the pooled returns policy with the non-pooled returns policy (i.e., the allowable return limits are product-specific), they provide conditions under which the retailer would actually order less under the pooled returns policy.

Revenue-sharing contracts. In retailing, stocking out a product could have a larger impact on the manufacturer's profit because the customer would usually buy a similar product from the retailer. This motivates manufacturer to provide incentive for the retailer to stock more. Clearly, a buy back contract (or a return policy) can serve this purpose; however, the buy-back contract may not be practical in certain situations. For example, in the video rental industry, it is not practical for the video rental stores to return excess inventory of old DVDs to the manufacturer (distributor). This may have triggered the idea for the manufacturer to develop a risk sharing scheme in the form of a revenue sharing contract. The revenue sharing contract can be characterized by the wholesale price w and the portion of the revenue to be shared α. As depicted in Fig. 13.1, the retailer would get a lower wholesale price w upfront but the retailer is required to remit αp for each rental unit to the manufacturer. For instance, as suggested by Mortimer (2004), Blockbuster shared 30–45% of their rental revenue in exchange for a reduced wholesale price of around $8 instead of $65 for each DVD. Tang and Deo (2005) determine the conditions for w and α under which the retailer will obtain a higher profit under the revenue sharing scheme.

In the economics literature, Dana and Spier (2001) show that revenue sharing contracts can be used to coordinate the supply chain, and would induce the retailers to reduce their rental prices under competition. Mortimer (2004) conducted statistical analysis based on the panel data collected at 6,137 video rental stores in the U.S. between 1998 and 2000. She shows that revenue-sharing contracts can enable a retailer to earn more for popular titles or new releases. Pasternack (2002) investigated the effect of a revenue sharing on the optimal order quantity in a newsvendor environment and shows analytically that a revenue-sharing contract can be used to achieve channel coordination. Cachon and Lariviere (2005) show analytically that the revenue sharing contracts are equivalent to buy-back contracts.

Quantity-Based contracts: Quality flexibility and minimum order. To achieve operational efficiency under demand uncertainty, a manufacturer would prefer contracts that would entice retailers to commit their orders in advance while a retailer would prefer contracts that would allow them to adjust their orders when necessary. As a compromise, some manufacturers offer Quantity Flexibility (QF) contracts to their retailers. A QF contract is specified by three parameters: a wholesale price w, an upward adjustment parameter u, where $0 \leq u \leq 1$, and a downward adjustment parameter d, where $0 \leq d \leq 1$. Consider the case in which a retailer placed an order x sometime earlier. Suppose the retailer updates his demand forecast and would like to revise this particular order. Under the QF contract, the retailer can adjust his order to Q by paying w per unit as long as $(1 - d)x \leq Q \leq (1 + u)x$. Notice that the QF contract can be recast as a buy back contract under which the retailer had to buy $(1 + u)x$ units up front but could return or cancel his commitment down to $(1 - d)x$ for a full refund of the wholesale price w. Lariviere (1998) analyzes a QF contract with parameters w, d, and u that coordinates the channel in a single-period setting. Tsay and Lovejoy (1999) provide a detailed analysis of QF contract in a multi-period setting.

When it is costly for a manufacturer to obtain more production capacity, a manufacturer may develop a supply contract to entice each retailer to commit to a minimum quantity in advance. Anupindi and Akella (1997) consider the case in which the retailer is committed to a fixed quantity in each period. In return, the manufacturer offers a discount based on the level of this fixed commitment. They prove that a modified order-up-to policy is an optimal policy for the retailer. Anupindi (1993) examines the case in which the order quantity that a retailer can place in each period is bounded pre-specified lower and upper limits. Bassok and Anupindi (1997) consider the case in which the retailer is committed to order at least $K \cdot N$ units in total over N periods. When demand is independent and identically distributed, they prove that the retailer's optimal order policy in each period is a modified order-up-to policy. As a variation, instead of focusing on the minimum total commitment for each product, Anupindi and Bassok (1998) analyze a multi-product supply contract under which the retailer is committed to a minimum total monetary (i.e., "dollar") value of the products to be purchased over N periods.

For selling seasonal goods, Fisher (1997) and Fisher and Raman (1996) confirm that the early sales data has informational value in the sense that this data can help the retailer to obtain more accurate forecast about the total sales for the whole season. Fisher and Raman show that it is advantageous for the retailer to place a second order after observing the first few weeks of sales data. To ensure that the second order will be replenished within the selling season, the manufacturer needs to impose certain restrictions on the second order quantity. Eppen and Iyer (1997) analyze a "backup agreement" that has been used in the fashion apparel industry. The backup agreement can be characterized by three parameters β, w, k). Prior to the selling season, the retailer commits to Q units for the entire selling season and confirms the first order $(1 - \beta)Q$ at wholesale price w. The retailer can place a second order up to the remaining βQ units (i.e., the backup units) at wholesale price w and receive quick delivery. There is a penalty cost of k for any of the backup units not purchased. Brown and Lee (1997) consider a variant of the backup agreement arising from the semiconductor manufacturing industry.

Uncertain price. While most work focus on demand uncertainty, not much work has been done in the area of uncertain wholesale price. Li and Kouvelis (1999) consider a case in which the wholesale price is a geometric Brownian motion with drift. Facing with uncertain wholesale price, the retailer is required to procure exactly D units by time T, where D is the ultimate demand at time T. Also, an inventory holding cost $h(T - t)$ will be incurred for each unit purchased at time t, where $0 < t < T$. Li and Kouvelis evaluate the cost associated with three different supply contracts. First, in a "time-inflexible contract," the retailer must state up front about the purchase time. In a "time-flexible contract," the retailer may observe price movements and decide dynamically when to buy. They extend their model to the case in which they can procure the item from two manufacturers.

13.2.6 Robust Supply Management Strategies

The multi-supplier strategy is the most common approach for reducing supply chain risks. For example, both Sheffi (2001) and Kleindorfer and Saad (2005) recommend the use of multiple suppliers as a way to manage supply chain operational and disruption risks. For example, as articulated in Huchzermeier and Cohen (1996) and others, using multiple suppliers in multiple countries can enable a firm to manage operation risks such as normal exchange rate fluctuations efficiently. Moreover, doing so can make a supply chain more resilient during a major disruption. For example, as we mentioned in an earlier chapter, when the Indonesia Rupiah devalued by more than 50% in 1997, many Indonesian suppliers were unable to pay for the imported components or materials, and, hence, were unable to produce the finished items for their U.S. customers. However, with a network of 4,000 suppliers throughout Asia, Li and Fung (www.lifung.com), the largest trading company in Hong Kong for consumer-durable goods such as textiles and toys, shifted some production from Indonesia to suppliers in other Asian countries.

In many instances, the buyer does not have the luxury to shift production among different suppliers because of the very limited number of suppliers available in the market. To cultivate additional suppliers, certain supply contracts described in Section 13.2.5 could serve as robust strategies that would make a supply chain more efficient and resilient. For instance, revenue (or risk) sharing contracts are known to be efficient because their use can coordinate the channel partners in the face of uncertain demand (c.f., Pasternack, 2002). In addition, revenue sharing contracts could make a supply chain more resilient. For example, due to uncertain specification of the flu vaccine in any given year, the uncertain market demand, and the price pressure from the U.S. government, there are only two remaining vaccine makers for the U.S. market. This created a shortage of 48 million flu shots in 2004 when Chiron's Liverpool plant was suspended due to bacteria contamination (c.f., Brown, 2004). To make the flu vaccine supply chain more resilient, the U.S. government could consider offering certain risk-sharing contracts to entice more suppliers to enter the flu vaccine market. For instance, the government could share some financial risks with the suppliers by committing to a certain quantity of flu vaccine in advance at a certain price and to buy back the unsold stocks at the end of the flu season at a lower price. With more potential suppliers, the U.S. government would have the flexibility to change their orders from different suppliers quickly when facing major disruptions.

Table 13.2 lists references to the articles mentioned in Section 13.2.

13.3 Demand Management

In this section we focus on articles that emphasize on the use of demand management strategies to "shape" uncertain demand so that a firm can use an inflexible supply to meet the modified demand. In the previous section, we described how manufacturers can use different supply management strategies to mitigate various supply chain operational risks. However, these supply management strategies are ineffective when supply is inflexible. For instance, in the service industry or in the fashion goods manufacturing industry, the is usually fixed. When the supply capacity is fixed, many firms attempt to use different demand management strategies so that they can manipulate demands dynamically to match demand with the fixed

Table 13.2 Summary of supply management articles

Supply Management Aspect	Type of risk	References (in the order of appearance in the subsection)
Supply network design	General	Porter (1985), Arntzen et al. (1995), Camm et al. (1997), Levy (1995), Lee and Tang (1998), Huchzermeier and Cochen (1996), Kouvelis and Rosenblatt (2002),
Supplier relationship	General	Helper (1991), Dyer and Ouchi (1993), Dyer (1996), Shin et al. (2000), Tang (1999), Cohen and Agrawal (1999).
Supplier selection Supplier selection criteria	General	Boer et al. (2001), Mandal and Deshmukh (1996), Vokurka (1996), Choi and Hartley (1996), Ellram (1994)
Supplier approval/selection	General	Boer et al. (2001), Ellram (1994), Weber and Current (1993), Weber (2000), Dahel (2003), Tang (1988), Tagaras and Lee (1996), Kouvelis (1998)
Supply order allocation	Uncertain Demand	Porteus (2002), Zipkin (2000), Minner (2003), Zhang (1996), Fukuda (1964), Vlachos and Tagaras (2001), Scheller-Wolf and Tayur (1999), Moinzadeh and Nahmias (1988), Janssens and de Kok (1999), Nagurney (2005), Bazaraa et al. (1993),
	Uncertain Supply Yields	Gerchak, Vickson, and Parlar (1988), Gerchak (1990), Agrawal and Nahmias (1998), Yano and Lee (1995), Bassok and Akella (1991), Tang (1990), Denardo and Lee (1996), Rajagopalan and Singh (1992), Gong and Matsuo (1997)
	Uncertain Supply Lead Times	Ramasesh et al. (1991), Sedarage et al. (1999), Akella et al. (1993)
	Uncertain Supply Capacity	Parlar and Perry (1996), Ciarallo (1994), Wang and Gerchak (1996), Henig and Gerchak (1990)
	Uncertain Supply Cost	Gurnani and Tang (1999), Kogut (1985), Kogut and Kulatilaka (1994), Li (1997), Kogut and Kulatilaka (1994), Dasu and Li (1997), Huchzermeier and Cohen (1996)
Supply contracts	General	Lee et al. (1997), Bresnahan and Reiss (1985), Cachon (2003), Lariviere (1998), Tsay (1998)
Wholesale price contracts	Uncertain Demand	Lariviere and Porteus (2001), Anupindi and Bassok (1999), Cachon (2002), Corbett and de Groote (2000), Corbett and Tang (1998), Ha (2001), Corbett (2001), Babich et al. (2004)
Buy-back contracts	Uncertain Demand	Lariviere (1998), Emmons and Gilbert (1998), Padmanabhan and Png (1997), Brown, Chou, and Tang (2005),
Revenue sharing contracts	Uncertain Demand	Dana and Spier (2001), Mortimer (2004), Pasternack (2002), Cachon and Lariviere (2005),
Quantity-based contracts	Uncertain Demand	Lariviere (1998), Tsay and Lovejoy (1999), Anupindi and Akella (1997), Anupindi (1993), Bassok and Anupindi (1997), Fisher (1997), Fisher and Raman (1996), Eppen and Iyer (1997), Brown and Lee (1997)
Time-based contracts	Uncertain Price	Li and Kouvelis (1999)
Robust supply management	General	Sheffi (2001), Kleindorfer and Saad (2005), Huchzermeier and Cohen (1996)

supply. We refer the reader to Elmaghraby and Keskinocak (2003), who provide an extensive review of dynamic pricing models and clearance pricing models for selling a fixed number of units over a finite horizon. For literature that deals with coordination of pricing and ordering decisions, we refer the reader to comprehensive reviews by Yano and Gilbert (2004), by Petruzzi and Dada (1999), and by Eliashberg and Steinberg (1993).

Carr and Lovejoy (2000) develop a single-period model for a firm to handle multiple customers with random demand distributions when a firm's supply capacity is fixed. For each customer, they consider the case in which the firm can choose to accept only a fraction of the customer's demand distribution. The objective is to choose different fractions of customer demand distributions so that the firm's expected profit is maximized for a given supply capacity. By analyzing the mean and variance of the total demand generated from different fractions of customer demand distributions, Carr and Lovejoy determine the optimal portfolio of demand distributions.

Van Mieghem and Dada (1999) consider a single product firm that faces a linear demand curve with uncertain intercept and has to decide on its production quantity and price. They consider different strategies including the *price postponement* strategy. Under price postponement, the firm needs to decide on the order quantity in the first period and then determine the price in the second period after observing updated information about the demand. Essentially, the supply is fixed after the first period. Hence, the price postponement strategy enables a firm to use price as a response mechanism to change demand so that the modified demand is better matched with the fixed supply. By formulating the problem as a two-period stochastic dynamic programming problem, Van Mieghem and Dada show that the price postponement is more effective than other strategies being considered.

Besides the demand management strategy examined by Carr and Lovejoy (2000) and Van Mieghem and Dada (1999), it appears that the remaining demand management strategies are designed to generate one or more of the following effects:

1. Shifting demand across time;
2. Shifting demand across markets; and
3. Shifting demand across products.

In the context of supply chain *risk* management, there are also

4. Robust strategies for demand management.

We now review the relevant literature in each of these four categories.

13.3.1 Shifting Demand Across Time

In the service industries such as utilities, airlines and hotels, firms usually set higher prices during peak seasons in order to shift demand to off-peak seasons and to profit from price-inflexible demand during the peak season. This type of pricing mechanism is also known as revenue management or yield management. Offering different prices at different times, it would enable the firm to increase the profit generated from a fixed supply capacity by capturing customers in different segments who are willing to pay different prices for the service offered in different times. For revenue management literature that deals with hotel bookings, we refer the reader to Bitran and Gilbert (1996), to Badinelli (2000), and to the references therein. In most cases, due to uncertain customer arrivals and uncertain cancellations, these models are usually formulated as dynamic programming problems. For revenue management literature that deals with airline reservations, the reader is referred to Dana (1999) and a comprehensive survey provided by Weatherford and Bodily (1992). For revenue management literature that deal with peak-load pricing for managing public utilities, the reader may refer to Crew and Kleindorfer (1986) for a review of economics literature that deals with peak load pricing with uncertain demand. Essentially, many economists have developed various models using different types of demand curves and different types of demand uncertainties to determine the peak-load pricing so that the service provider with fixed capacity can obtain a higher profit. Besides the work by Dana (1999), most economists assumed that the firm knows the time at which peak demand occurs. The reader is referred to a review of revenue management by Talluri and Van Ryzin (2005).

In the context of service marketing, many service firms offer price discount to entice customers to commit their purchase in advance. In many instances, advance-purchase discount can be easily implemented due to new technologies such as smart cards, online payments, electronic money, etc. As articulated in Xie and Shugan (2001), advance-purchase discount can be a win-win strategy for the service provider and their customers. First, advance-purchase discount enables a firm to use this discriminatory pricing mechanism to increase sales by serving different market segments. For example, by considering two-market segments with different reservation values of the service, Dana (1998) shows analytically that it is rational for customers with relatively more certain demands (planned trips) and customers with relatively lower reservation value (leisure travelers) to commit their purchases in advance. This result also implies that customers with less certain demands (unplanned trips) and customers with relative higher reservation value (business travelers) would expect to pay a potentially higher price in the spot market.

Advance-purchase discount enables customers to receive a discount over the spot price or to reserve capacity that may not be available during the spot period. Xie and Shugan (2001) present a two-period model in which the advance-purchase price is announced in the first period but not for the second period; however, the probability distribution of the price for the second period is known to all customers. Since the price in the second period is unknown to the customers and since the reservation price is uncertain for each customer, they would make their purchase in the first period if the surplus obtained from purchasing in advance is higher than the expected surplus obtained from purchasing later. By using backward induction, Xie and Shugan develop the conditions under which the firm should offer advance-purchase discount. By considering an extension in which the prices in both periods are pre-announced, they determine the conditions under which the firm should offer advance-purchase discount.

In the context of supply chain management, we need to address the production planning and inventory control issues that are not addressed in the economics or marketing literature. In most cases in the literature, retailers pre-announce the prices for both periods to their customers. Weng and Parlar (1999) are the first to analyze the benefit of advance-commitment discount. They consider the case in which a retailer offers price discount to entice customers to pre-commit their orders prior to the beginning of the selling season. The advance-commitment discount program can be a win-win solution. First, the customers can enjoy a lower price by pre-committing their orders early. Second, the retailer can benefit from the reduction in demand uncertainty because the advance-commitment discount enable the retailer to convert some uncertain customer demands to pre-committed orders that are known in advance. By considering the demand uncertainty reduction generated by the advance-commitment discount, Weng and Parlar determine the optimal order quantity and the optimal discount rate for the retailer.

Tang et al. (2004) extend Weng and Parlar's (1999) model by considering a more general situation: First, they consider a situation in which the market consists of two customer segments with different purchasing behaviors toward advance-commitment discount. They show how advance-commitment discount would enable the retailer to increase the total expected sales. Second, they consider the case in which the retailer can use the pre-committed orders obtained prior to the beginning of the selling season to improve the accuracy of the forecast of the demand that would occur during the selling season. They show how this improved forecast would enable the retailer to reduce the total expected over-stocking and under-stocking costs. Moreover, they examine various benefits associated with advance-commitment discount programs.

McCardle et al. (2004) extend the Tang et al. (2004) model to the case in which two competing retailers need to decide whether to launch the advance-commitment discount program or not. They show that both retailers would offer the advance-commitment discount program at the equilibrium. However, when there is a fixed cost for implementing this discount program, they develop conditions under which exactly one retailer would offer the discount program at the equilibrium.

The advance-commitment discount program is applicable to non-seasonal products as well. By studying the supply chain operations associated with steel processing, Gilbert and Ballou (1999) present a continuous time model in which the steel distributor offers price discount to customers who pre-commit their orders in advance. By knowing these pre-committed orders earlier, they show that the standard deviation of the demand over the replenishment lead time periods is reduced. By using a traditional approximate cost model for lost sales, they show how advance-commitment discount programs would enable a steel distributor to increase his expected profit and to improve customer service level at the same time. By examining the profits before and after the launch of the advance-commitment discount program, Gilbert and Ballou present an approach for determining the optimal discount price.

These advance-commitment discount models are based on the single product case. In a subsequent paper, Weng and Parlar (2005) examine a situation when a manufacturer produces two products, a standardized product and a make-to-order customized product. The manufacturer offers advance-commitment discount to customers who pre-commit their orders for the standardized product. By considering the case in which the market consists of two segments with different purchasing behaviors toward advance-commitment discount, Weng and Parlar formulate the manufacturer's problem as a stochastic dynamic programming problem. They show that the advance-commitment discount program would enable the manufacturer to increase the total expected demand and to reduce demand uncertainty. When the standardized product is cheaper to produce, they develop conditions under which the manufacturer should offer the advance-commitment discount program.

While the advance-commitment discount program designed to enable a firm to shift customer demand to an earlier period, there is another strategy that would entice customers to shift their demands to a later time. This strategy is called *demand postponement* and is intended to entice some customers to accept shipments at a later time. Iyer et al. (2003) is the first to examine the benefits of the demand postponement strategy. To manage uncertain demand with a fixed supply capacity, they consider the case in which a firm would offer price discount to customers willing to accept late shipments. Essentially, this strategy is akin to the overbooking situation in which an airline may offer incentive to entice some customers to take a later flight. Iyer et al present a two-period model and determine the optimal fraction of customer demands to postpone. In addition, they characterize conditions under which a firm should adopt the demand postponement.

13.3.2 Shifting Demand Across Markets

When selling products with short life cycles in different markets, firms need to manage product rollovers (the process of phasing out old products and introducing new products). As articulated by Billington et al. (1998), different firms have implemented various rollover strategies with different degrees of success. One of the key challenges for managing product rollovers successfully is uncertain demands in different markets. To mitigate the demand risks in different markets, Billington et al. present a "solo-rollover by market" strategy that calls for selling the new product in different markets with non-overlapping selling seasons. The solo-rollover by market strategy is more suitable for situations when there is a natural time delay of the selling season in two different markets. For example, the selling season of ski wear in North America ends in May whereas the selling season in South America begins in June.

Suppose a firm adopts the solo-rollover-by-market strategy. Then the firm has to decide how much to stock for the first market during in first period; how much of the unsold inventory from the first market to transship to the second market at the end of the first period; and how much to stock for the second market at the beginning of the second period. Kouvelis and Gutierrez (1997) examine this stocking and transshipment decisions for two markets with non-overlapping selling seasons. They consider a firm that sells seasonal goods in a primary market in the first period and in the secondary market during the second period. By capturing the possibility of shipping some of the leftover inventory from the primary market to the secondary market at the end of the selling season of the primary market, they present a two-period stochastic dynamic program to determine the optimal production quantity for the corresponding market in each period and the optimal amount of leftover inventory to be shipped from the primary market to the secondary market. Due to the possibility of selling the leftover from the primary market at the second market, they show that the optimal production quantity for the primary market is higher than the case when the secondary market does not exist.

Petruzzi and Dada (2001) extend Kouvelis and Gutierrez's (1997) model to the case in which the firm can use a pricing mechanism to shift some of the demand from the primary market to the secondary market. Specifically, Petruzzi and Dada consider the case that the firm can use information to make better pricing and ordering decisions as follows: First, the firm can choose the selling price as well as the stocking level for the primary market during the first season. Second, the firm can use the actual sales observed in the primary market to improve the accuracy of the forecast of the demand for the secondary market in the second season. Third, given the updated forecast, the firm can determine the transshipment quantity to be shipped from the primary market to the secondary market, the stocking level and the selling price of the product for the secondary market in the second selling season. By formulating the two-period problem as a stochastic programming problem with recourse, Petruzzi and Dada establish the characteristics of the optimal pricing and ordering decisions for both markets.

13.3.3 Shifting Demand Across Products

When selling multiple products in a single market, many marketing researchers have examined various pricing and promotion strategies to entice customers to switch brands or products. The ultimate goal of various marketing strategies is to help a firm to increase market share, sales, or revenue. For example, Raju et al. (1995) present a model to capture the brand switching behavior when a store introduces a store brand to compete with the existing national brands. They show how to determine the optimal retail prices for the national brands and the new store brand so as to maximize the store's revenue. Chong et al. (2001) show how a retailer can obtain higher revenue by adjusting its product assortments and pricing so that the store can offer its customers the right products at the right price. Lilien et al. (1992) provide an extensive review of marketing models that deal with pricing and promotion strategies.

However, in general, these marketing models do not deal with the operational issues arising from supply chain management. In the context of supply chain management, some researchers have developed models by considering the possibility of shifting the supply/demand from one product to another. It seems there are two basic mechanisms that would enable to firm to shift the supply/demand from one product to another. These two mechanisms are

1. *Product substitution*, and
2. *Product bundling*,

as discussed below.

Product substitution. Product substitution can occur in different settings: (1) selling products with similar features, (2) selling products when one dominates another in quality or performance, and (3) using pricing to entice customers to shift demand from one product to another. Let us consider the three in turn.

First, by selling products with similar features, a firm can increase the product substitutability. Chong et al. (2004) show how a firm can increase product substitutability by selecting a specific combination of products with similar attributes/features. Moreover, they show how product substitutability can reduce the variance of the aggregate demand. Rajaram and Tang (2001) present a single-period stochastic model of a firm that sells two substitutable products. Specifically, they consider a situation in which a product with surplus inventory can be used as a substitute for out of stock products. Hence, the demand of one product can be satisfied by the supply for another product. They develop conditions under which product substitutability would enable a firm to reduce the variability of the effective demand for each product. Moreover, they show that the optimal order quantity of each product and the retailer's expected profit increase as product substitutability increases.

Second, consider the case when one product dominates another in terms of quality or performance. For example, in integrated circuit (IC) manufacturing, the output of each production run consists of a random number of chips with different grades measured according to the processing speed. When higher grade chips can be used as substitutes for the lower grade chips, Bitran and Dasu (1992) and Hsu and Bassok (1999) present different models for determining the optimal production quantity at a wafer fabrication facility with random yields.

Third, consider the case when the firm can use pricing to entice customers to shift their demand from one product to another. Parlar and Goyal (1984) consider a case in which the retailer would offer price discount for the old product, say, one-day old doughnut. Clearly, the new and old products are substitutable and the retailer can change the level of product substitution by varying the discount factor. By formulating the problem as a Markov Decision Process and by considering the demands for the old and new products, they determine the optimal order quantity for the new product in each period. Chod and Rudi (2005) examine another situation in which a firm can use differential pricing to entice customers to shift the demand for one product to another. They consider the case in which the firm needs to decide on the production quantity of two similar products in the first period; however, the firm can postpone the pricing decision of the each product until the second period. By extending the model developed by Van Mieghem and Dada (2001), Chod and

Rudi show a firm can obtain a higher profit by delaying the pricing decision until the second period.

Product bundling. In addition to product substitution, a firm can change the demand of the products by developing bundles. There is an increasing number of retail products being bundled together and sold. Examples can be found across a range of products including food (cans of chicken broth), apparel (under garments), cosmetics (shampoo and conditioner), and electronics (computers and printers). When products are sold in bundles, they force the customers to buy all products as a bundle, which will affect the effective demand of the products. Ernst and Kouvelis (1999) examine how product bundles affect the inventory ordering decisions of a firm. Specifically, they consider the case in which the products are sold as a bundle and as individual products. Based on their analysis of a two-product model, they establish the necessary and sufficient conditions for the optimal ordering quantities. They provide insights into the degree of sub-optimality of profits when inventory decisions are made without explicit consideration of demand substitution between the bundles and the individual products. McCardle et al. (2005) present a model for determining optimal bundle prices, order quantities, and profits. By capturing the customer's valuation of individual products, they generate the demand distribution of the product bundle. In addition, they determine how product demand, costs and the relationship of demand between products affect optimal bundle prices and profits. Moreover, they present conditions under which a firm should bundle their products. See Stremersch and Tellis (2002) for a comprehensive review on product bundling literature.

13.3.4 Robust Demand Management Strategies

There are at least two robust demand management strategies already reviewed in this section. First, as described in the previous Section 13.3.3, there are many demand management strategies that would enable a supply chain to shift demand across products. By having the capability to shift demand across products, these strategies can make a supply chain more efficient and resilient. For example, when facing uncertain demand, Chod and Rudi (2005) present a responsive pricing strategy that would enable a firm to increase profit by shifting demand across products. Hence, a responsive pricing strategy would improve supply chain efficiency. In addition, a responsive pricing strategy could improve supply chain resiliency as well. For example, when facing a supply disruption of computer parts from Taiwan after an earthquake, Dell immediately offered special price incentives to entice their online customers to buy computers that utilized components from other countries. The capability to shift customer choice swiftly enabled Dell to improve its earnings in 1999 by 41% even during a supply crunch (c.f., Veverka (1999)).

Second, the demand postponement strategy described in Section 13.3.1 can be a robust demand management strategy that would enhance supply chain efficiency and resiliency (cf. Iyer et al., 2003). Under the demand postponement strategy, a manufacturer may offer price discounts to some retailers to accept late shipments. Essentially, this strategy is akin to the overbooking situation in which an airline may offer incentive to entice a fraction of customers who are willing to take a later flight. By having the capability to shift some of the demands to a later period, it would certain help a firm to manage both operational risks and disruption risks.

Table 13.3 lists references to the articles mentioned in this section.

Table 13.3 Summary of demand management articles

Demand Management Issue	Risk Issues	References (in the order of appearance)
Demand Management	General	Elmaghraby and Keskinocak (2003), Yano and Gilbert (2004), Petruzzi and Dada (1999), Eliashberg and Steinberg (1993), Carr and Lovejoy (2000), Van Mieghem and Dada (1999)
Shifting demand across time	Uncertain demand	Gilbert (1996), Badinelli (2000), Dana (1999), Weatherford and Bodily (1992), Crew and Kleindorfer (1986), Talluri and Van Ryzin (2005), Dana (1998), Xie and Shugan (2001), Weng and Parlar (1999), Tang et al. (2004), McCardle et al. (2004), Weng and Parlar (2005), Iyer et al. (2003),
Shifting demand across markets	Uncertain demand	Billington et al. (1998), Kouvelis and Gutierrez (1997), Petruzzi and Dada (2001),
Shifting demand across products	Uncertain demand	Raju et al. (1995), Chong et al. (2001), Lilien et al. (1992),
Product substitution	Uncertain demand	Chong et al. (2004), Rajaram and Tang (2001), Bitran and Dasu (1992), Hsu and Bassok (1999), Parlar and Goyal (1985), Chod and Rudi (2005), Van Mieghem and Dada (1999)
Product bundling	Uncertain demand	Ernst and Kouvelis (1999), McCardle et al. (2005), Stremersch and Tellis (2002)
Robust demand management strategies	Uncertain supply	Chod and Rudi (2005); Iyer et al. (2003)

13.4 Product Management

To compete for market share, many manufacturers expand their product lines. As reported in Quelch and Kenny (1994), the number of stock keeping units (SKUs) in consumer packaged goods has been increasing at a rate of 16% every year. Marketing research shows that product variety is an effective strategy to increase increasing market share because it enables a firm to serve heterogeneous market segments and to satisfy consumer's variety seeking behavior. However, while product variety may help a firm to increase market share and revenue, product variety can increase manufacturing cost due to increased manufacturing complexity. Moreover, product variety can increase inventory cost due to an increase in demand uncertainty for each product. These two concerns have been illustrated in an empirical study conducted by MacDuffie et al. (1996). They show that the production and inventory costs tend to increase as product variety increases. Therefore, it is critical for a firm to determine an optimal product portfolio that maximizes the firm's profit. The reader is referred to Ramdas (2003) for a comprehensive review of literature in the area of product variety.

To reduce the design and manufacturing costs associated with product variety, firms can increase product variety by developing different variants based on a common platform. For example, in the personal computer industry, different computer models are based on a common platform. Hence, the products would share some common attributes, which make these products mutually substitutable to a certain extent. As discussed in the previous Section 13.3.3, product substitution and product bundling would enable a firm to shift demands across products so that the firm can satisfy more customers without incurring the risk of over-stocking.

However, product substitutability is a key challenge for researchers to develop analytical models to evaluate market share, revenue, and manufacturing cost associated with different product portfolio. As articulated in Ulrich et al. (1998), there is no explicit analytical model for determining an optimal product portfolio with substitutable products.

Still, various researchers have examined product variety issues using different approaches; Ho and Tang (1998) review of articles in the area of marketing, operations management and economics that deal with product variety. For example, Ulrich et al. (1998) study the mountain bikes industry and suggest that firms need to take their internal capabilities such as process technology, distribution channels, product architecture, supply chain network, etc., into consideration when making product variety decision. Krishnan and Kekre (1998) develop a regression model to examine the impact of functional features on software development cost. Martin et al. (1998) present a method for examining the impact of product variety on replenishment lead time. Moreover, by considering the attribute levels of different products associated with a product portfolio, Chong et al. (2004) develop a logit model for determining the mean and variance of the sales associated with different compositions of a product portfolio. Caro and Gallien (2005) present a multi-armed bandit model for selecting an optimal product portfolio of fashion items that maximizes the expected profit over a finite horizon.

In the context of supply chain risk management, the key concern is to determine ways to reduce inventory cost associated with a given portfolio of products. Based on the classical inventory theory (c.f., Porteus, 2002 and Zipkin, 2000), it is well known that the average inventory level associated with the order-up-to policy depends on mean and the standard deviation of the demand over the replenishment lead time. Therefore, to develop cost-effective product variety strategies, researchers

have developed different approaches for reducing the standard deviation of the demand over the replenishment lead time. For instance, as explained in Section 13.3.1,
we can reduce the demand uncertainty over the replenishment lead time periods by
using pricing mechanisms such as advance-commitment discount, peak load pricing, etc. In this section, we shall review articles based on three specific product
management strategies: postponement, process sequencing, and product substitution as well as robust strategies. We have already discussed product substitution in
Section 13.3.3 so we shall limit our discussion in this section to

1. Postponement,
2. Process sequencing, and
3. Robust strategies for product management.

13.4.1 Postponement

Consider a manufacturing system that produces two end-products. The system has
N processing stages, where stage 0 is a "dummy" stage. As depicted in the Fig. 13.2,
the first k stages are common to both end-products and after this stage the products
are differentiated in the sense that they may require different operations or different
components. We call stage k as the "point of differentiation." Lee and Tang (1997)
describe how delayed product differentiation can be achieved via standardization of
components and subassemblies, modular design, postponement of operations, and
re-sequencing of operations. Recall that stage 0 is a dummy stage; hence, there is
no postponement if $k = 0$. Let T be the total lead time of the entire manufacturing
process, and $L(k)$ be the lead time from stage 0 to stage k.

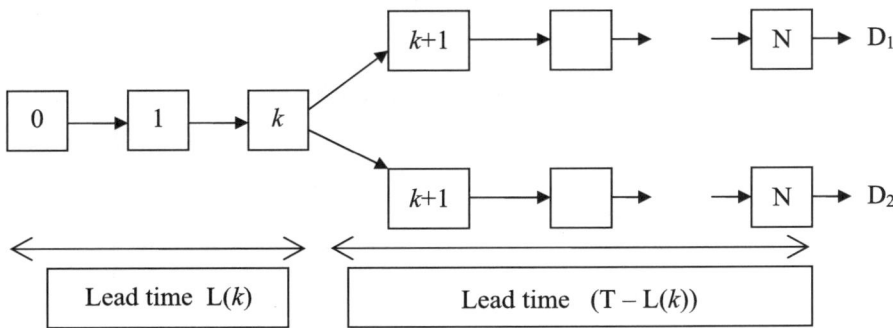

Fig. 13.2 A system with point of differentiation at stage k

The postponement models can be classified according to (a) operating modes
(make-to-stock and make-to-order) and (b) demand forecasts (no forecast updating
and with forecast updating). Although this two-by-two classification suggests four

categories, we are not aware of models that deal with make-to-order systems with forecast updating. The remaining three are discussed at length below:

Make-To-Order systems without forecast updating. Lee (1996) develops a theoretical analysis of the postponement strategy. When there is no demand forecast updating, he examines the benefits of postponement in a make-to-order (MTO) system and a make-to-stock (MTS) system. In the MTO system, work-in-process inventory is held only at stage k and each end product is customized on demand. Depending on the availability of the inventory at stage k and the processing capacity at stages $(k + 1)$ through (N), the time it takes to respond to a customer is uncertain. In an MTO system, it is common to measure the system performance according to the mean response time and the probability of the response time being less than a target response time. Using these two performance measures, Lee shows that the optimal order up level S is decreasing in k. Hence, one can reduce the inventory level by delaying the point of differentiation. While the base stock level S is decreasing in k, the inventory holding cost rate is likely to be increasing in stage k; i.e., it is more costly to hold inventory at a later stage. By considering the tradeoff between lower inventory level and higher inventory holding rate, Lee provides conditions under which postponement is beneficial.

In Lee's (1996) model, the end products are differentiated according to a single feature. As such, all end products can be customized from a single point of differentiation. However, when the end products are differentiated according to multiple features, there could be multiple points of differentiation. This observation motivates Swaminathan and Tayur (1998a) and (1998b) to define the semi-finished products held at different points of differentiation as "vanilla boxes." Essentially, the firm will stock these vanilla boxes and then customize different types of vanilla boxes into different end products on demand. By considering the capacity for the customization process and different demand scenarios, Swaminathan and Tayur formulate the problem as a stochastic programming problem with recourse. By examining the structure of the stochastic programming, they develop a solution methodology for determining the optimal configuration of vanilla boxes that minimizes the expected stock-out cost and the inventory holding cost of vanilla boxes.

Make-To-Stock Systems without forecast updating. For a make-to-stock system, Lee (1996) consider the case in which only finished product inventory is held; i.e., inventory is held after stage N. Conceptually speaking, this system is akin to the single-depot, multi-warehouse distribution system examined by Eppen and Schrage (1981). By assuming that the demand distributions for the end-products are independently normal across time but may be correlated within a time period, Lee applies the approximate analysis developed by Eppen and Schrage to determine the base-stock level and the average inventory level for each end-product. Moreover, Lee shows that the finished product inventory level for each end-product is decreasing in the point of differentiation k.

As articulated by Lee (1996), the postponement strategy can be implemented in a MTO or a MTS system. This observation motivates Su et al. (2005) to develop a model to compare the total supply chain costs associated with postponement in a

MTO and a MTS system. Their analysis shows that the MTO system is more cost effective when the number of end products exceeds a certain threshold level.

When the end products are differentiated according to different features, the corresponding manufacturing process can have multiple points of differentiation. Garg and Tang (1997) extend the model presented by Lee (1996) by considering a system with multiple points of differentiation. Since system with multiple points of differentiation is akin to a multi-echelon distribution system, Garg and Tang (1997) first extend Eppen and Schrage's (1981) two-echelon model to a three-echelon model. Then they show that postponement at each of the differentiation points would result in inventory savings. Instead of relying on the approximate analysis developed by Eppen and Schrage to evaluate different postponement strategies, Aviv and Federgruen (1998b) show how one can develop an exact analysis of the model presented in Garg and Tang (1997).

Lee and Tang (1997) examine a system that can keep work-in-process inventory at every single stage. They develop a stochastic inventory model by capturing the investment cost per period for redesigning the products and/or processes, the unit processing cost, and the inventory holding cost at each stage. Their analysis is based on a decomposition scheme in which the manufacturing process is decomposed into N independent stages, each of which will follow an order-up-to level policy. The decomposition scheme enables them to approximate the system-wide cost function associated with the point of differentiation k. By examining the underlying property of this explicit cost function, they develop conditions under which no postponement is optimal; i.e., the optimal point of differentiation is stage 0. Also, they discuss the conditions under which postponement is beneficial.

In the postponement literature, most researchers assume that the production capacity is unlimited. To examine how production capacity can affect the value of postponement, Gupta and Benjaafar (2004) develop a queuing model for examining the benefits of postponement in a MTO and a MTS system with limited production capacity. When the production capacity for stages 1 through k (point of differentiation) is limited, Aviv and Federgruen (2001a) present a multi-product inventory model for the case when the product demand is random and periodical. They show that the underlying inventory model can be formulated as a Markov Decision Process, and that delayed product differentiation is always beneficial even when the system has limited capacity.

Make-To-Stock Systems with forecast updating. In the postponement literature, most researchers assume that the product demands in each period are random, but they are independent across time and their distributions are known. As a result of these assumptions, the benefit of postponement is derived from "risk pooling" in the sense that all stages before the point of differentiation (stage k) would plan according to the "aggregate demand" instead of individual product demand. Besides risk pooling, postponement enables a firm to delay the product differentiation so that the production quantity decision for the final products can be made in a later period of time. When the timing of this decision is delayed, the firm can use the actual demands observed in earlier period to obtain more accurate forecasts of future demands. To explore further about the benefit of postponement with forecast updating, Whang and Lee (1998) extend the make-to-stock model presented in Lee (1996) by considering the case in which the demand $D_i(t)$ of end-product i in period t possesses the following form:

$$D_i(t) = \mu_i + \sum_{j=1}^{t} \varepsilon_{ij} \quad \text{where } i = 1, 2, \ldots, n, \; t = 1, 2, \ldots, T, \text{ and } \varepsilon_{ij} \approx N(a_{ij}, \sigma_{ij}^2)$$

Whang and Lee assume that the parameters a_{ij} and σ_{ij} are known for the normally distributed error term. This demand distribution is a form of random walk that enables one to capture a series of random shocks (economic trends, random noises, etc.). As time goes on, the decision maker can use the shocks observed in earlier periods (i.e., some of the ε_{ij}'s are now known) to develop a more accurate forecast of $D_i(t)$. By incorporating the capability to obtain more accurate demand forecast as time goes on, Whang and Lee show that one can obtain substantial reduction in the end-product inventory by using more accurate forecasts. In addition, they show analytically that significant inventory savings can occur even when the point of differentiation k occurs in the early stage.

Aviv and Federgruen (2001b) consider a more general demand distribution in which the parameters of the demand are unknown. In their model, they consider the case in which the decision maker would update the demand forecast in a Bayesian manner. They show analytically that the standard deviation of the total demand over the lead time periods decreases over time. Furthermore, they show that this standard deviation decreases with the point of differentiation k. This implies that it is more beneficial to update the demand forecast when postponement occurs in a later stage. The reader is referred to Garg and Lee (1998), Aviv and Federgruen (1998b), and Yang et al. (2004) for comprehensive reviews of the research literature that examines different postponement-related issues.

13.4.2 Process Sequencing

As noted in Section 13.4.1, postponement is an effective way to reduce variability in a supply chain. Lee and Tang (1998a) suggest that variability can also be reduced by reversing the sequence of manufacturing processes in a supply chain. Their suggestion is motivated by the re-engineering effort at Benetton. In the woolen garment industry, virtually all manufacturers will use the dye-first-knit-later sequence; i.e., dye the yarns into different colors first and then knit the colored yarns into different finished products. However, as a way to reduce inventory, Benetton pioneered the knit-first-dye-later process by reversing the "dyeing" and "knitting" stages (c.f., Dapiran, 1992). Intuitively speaking, the knit-first-dye-later strategy would be beneficial when there is only one style of woolen sweaters with multiple colors. This is because it would result in delaying product differentiation after the "knitting" stage. However, when there are multiple styles and multiple colors, it is not clear which

strategy is better. To determine the conditions under which a particular process sequence is better, Lee and Tang (1998a) develop a model of a production system that
produces products with 2 features (A and B), each of which has 2 choices (1 and 2).
As depicted in Fig. 13.3, the product demands $(X_{11}, X_{12}, X_{21}, X_{22})$ are assumed to
be multinomially distributed with parameters $(N; \theta_{11}, \theta_{12}, \theta_{21}, \theta_{22})$, where the total
demand N is normally distributed with mean μ and standard deviation σ, and θ_{ij}
corresponds to the probability that the customer will choose choice i of feature A
and choice j of feature B. By considering p as the probability that a customer will
purchase a product with choice 1 of feature A and by examining the conditional
probabilities: $\mathrm{Prob}(B1 \mid A1) = f(p)$ and $\mathrm{Prob}(B1 \mid A2) = g(p)$, one can express θ_{ij}
in terms of p, $f(p)$ and $g(p)$.

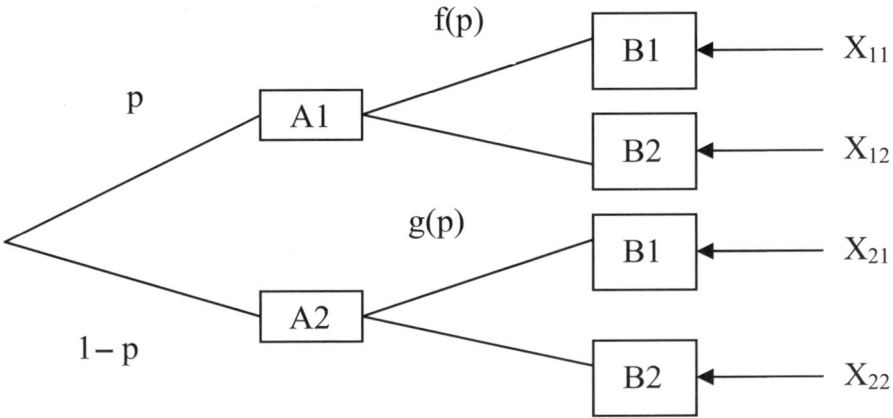

Fig. 13.3 A two-stage system with 2 features and 2 choices

Lee and Tang argue that the total expected cost associated with the intermediate
products is proportional to the sum of the variances of demands in a period. As such,
they show that the process sequence A-B has a smaller variance than the process
sequence B–A if

$$(\mu - \sigma^2)\{p(1-p) - [pf(p) + (1-p)g(p)]\{1 - [pf(p) + (1-p)g(p)]\}\} < 0$$

This result has the following implications. Consider the special case in which
$f(p) = g(p) = q$, where q is independent of p. If the product demand is stable
(i.e., when $\mu > \sigma^2$), then the process sequence A–B has a lower variance when fea-
ture (attribute) A is less variable than attribute B (i.e., when $|0.5 - p| > |0.5 - q|$).
However, the reverse is true when $\mu < \sigma^2$.

By considering the sum of the standard deviations as an alternative measure,
Kapuscinski and Tayur (1999) conduct their analysis associated with the special
case. They show that it is optimal to process the attribute with less variability first,
regardless of the values of μ and σ^2.

Other researchers have generalized Lee and Tang's (1998) model in other ways. Federgruen (1998) develops a general definition of when attribute A is more variable than attribute B in terms of θ_{ij}. He shows a more general conditions under which the process sequence A–B has a smaller variance than the process sequence B–A. Jain and Paul (2001) generalize Lee and Tang's model by incorporating two important characteristics of fashion goods markets, namely, heterogeneity among customers and unpredictability of customer preferences. Yeh and Yang (2003) develop a simulation model by incorporating additional factors such as lead times, ordering policies, and inventory holding cost. By using the data obtained from a garment manufacturer, they show how their simulation model can be used to select a process sequence that minimizes the total expected cost.

13.4.3 Robust Product Management Strategies

Among the product management strategies reviewed in this section, the postponement strategy described in Section 13.4.1 is a robust strategy for enhancing the efficiency and the resiliency of a supply chain. As reported in Lee (1996), postponement is an effective strategy for improving supply chain efficiency when facing uncertain demands for different products. In addition, the postponement strategy can increase supply chain resiliency. For example, as discussed in an earlier chapter, after Philip's semiconductor plant was damaged in a fire in 2000, Nokia was facing a serious supply disruption of radio frequency chips. Since Nokia's cell phones were based on the modular design concept, Nokia was able to postpone the insertion of these radio frequency chips until the end of the assembly process. Due to this postponement strategy, Nokia was able to reconfigure the design of their basic phones so that the modified phones could accept slightly different chips from other suppliers. Consequently, Nokia satisfied customer demand smoothly and obtained a stronger market position. The reader is referred to Hopkins (2005) for details.

Table 13.4 lists the references in this section.

13.5 Information Management

Fisher (1997) classifies most consumer products as fashion products or functional products. Basically, fashion products usually have shorter life cycles and higher levels of demand uncertainties than the functional products. Therefore, different information management strategies would be needed to manage for different types of products especially in the presence of supply chain risks. For this reason, we classify the work in this section according to the product types: (1) fashion products and (2) functional products. In addition, we consider (3) robust strategies for information management.

Table 13.4 Summary of product management articles

Product Management Issue	Risk Issue	References (in the order of appearance)
Product Management	General	Quelch and Kenny (1994), MacDuffie et al. (1996), Ramdas (2003), Ulrich et al. (1998), Krishnan and Kekre (1998), Martin et al. (1998), Chong et al. (2004), Ho and Tang (1998), Caro and Gallien (2005), Porteus (2002), Zipkin (2000)
Postponement	General	Lee and Tang (1997),
Make to order systems without forecast updating	Uncertain demand	Gunasekaran and Ngai (2005), Lee (1996), Tayur et al (1998a), Tayur et al. (1998b)
Make to stock systems without forecast updating	Uncertain demand	Lee (1996), Eppen and Schrage (1981), Su et al. (2005), Garg and Tang (1997), Aviv and Federgruen (1998b), Lee and Tang (1997), Gupta and Benjaafar (2004), Aviv and Federgruen (2001a)
Make to stock systems with forecast updating	Uncertain demand	Whang and Lee (1998), Lee (1996), Aviv and Federgruen (2001b), Garg and Lee (1998), Aviv and Federgruen (1998b), Yang et al. (2004)
Process Sequencing	Uncertain demand	Lee and Tang (1998), Kapuscinski and Tayur (1999), Federgruen (1998), Jain and Paul (2001), Tayur (1999), Yeh and Yang (2003)
Robust strategies	Uncertain supply	

13.5.1 Information Management Strategies for Managing Fashion Products

As articulated in Section 13.4, reducing the standard deviation of the demand over the replenishment lead time would result in inventory reduction for the entire supply chain. When managing products with short life cycles, short replenishment lead times could enable a retailer to place more than one order over the selling season. For example, various researchers have considered the situation in which a retailer can place two orders over the selling season. Specifically, the retailer can place one order prior to the beginning of the selling season and another order during the selling season. In the fashion goods industry, this type of replenishment system is called the "quick response" system. Clearly, the second order provides a great opportunity for the retailer to obtain more accurate demand forecast by using the actual sales data. As mentioned before, Fisher and Raman (1996) develop a two-period stochastic dynamic programming model with demand forecast updating to analyze the quick response system. By implementing their model at a skiwear company called Obermeyer, they illustrate how the quick response system would enable Obermeyer to achieve a higher customer service level with a lower level of inventory. The reader is referred to Raman (1998) for a review of quantitative models of quick response systems.

Gurnani and Tang (1999) analyze a similar quick response system except that the unit cost for the second order is uncertain. By formulating the problem as a two-period dynamic programming problem, they show that an order-up-to level policy is optimal. Instead of focusing on the retailer's perspective, Iyer and Bergen (1997) and Iyer (1998) analyze the impact of a quick-response system on both retailers' and manufacturers' inventories. They show that, when the customer service level is at least 0.5, the quick-response system is not Pareto in the sense that the retailer would obtain a higher expected profit and the manufacturer would achieve a lower expected profit. They develop conditions under which a quick-response system is beneficial to both the retailers and manufacturers. Donohue (2000) considers a variant of the quick-response system in which the retailer can place their orders in two modes: low cost with long lead time and high cost with short lead time. By formulating the problem as a two-period dynamic programming problem, she derives an optimal ordering policy and an optimal contract that coordinates the supply chain.

All quick response models assume that the manufacturer can always fill the second orders placed by the retailers. However, as articulated in the Benetton case prepared by Signorelli and Heskett (1984), manufacturers may not be able to guarantee complete fulfillment of the second order. Smith et al. (2002) investigate the retailer's optimal order quantities, the retailer's profit, and the manufacturer's profit for the case when the manufacturer can only fulfill the second order partially. By considering a stylized model, they show that the manufacturer should provide either complete fulfillment or no fulfillment of the second orders when the underlying demand distribution is either uniform or exponential. Specifically, they show analytically that partial fulfillment of the second orders is never optimal for the manufacturer.

13.5.2 Information Management Strategies for Managing Functional Products

When managing products with long life cycles, market information is critical for generating accurate demand forecasts. However, since wholesalers, distributors, manufacturers, and suppliers are farther remove from the consumer market, they usually do not have first-hand market information such as point of sales data, customers' preferences, and customer response to various pricing and promotion strategies. Instead, upstream supply chain partners usually generate their demand forecasts based on the orders placed by their downstream partners. Planning according to the orders placed by the downstream partners would create a phenomenon termed the "bullwhip effect" as coined by Procter and Gamble. Essentially, the bullwhip effect depicts the phenomenon in which the orders exhibit an increase in variability up the supply chain, even when the actual customer demands were fairly stable over time (c.f., Sterman, 1989). The increase in variability of the orders up the supply chain can cause many problems for the upstream partners including higher inventory, lower customer service level, inefficient use of production and transportation capacities, etc. In order to mitigate the bullwhip effect, one needs to identify the root causes.

Lee et al. (1997b) is the first to show that the bullwhip effect can occur even when every supply chain partners operate optimally and rationally. The bullwhip effect has also been shown independently by Bagahana and Cohen (1998). To establish the existence of the bullwhip effect, Lee et al. develop a 2-level supply chain that consists of a retailer and a manufacturer. They assume that the retailer "knows" that the underlying demand process follows an auto-regressive process AR(1) so that the demand in period t, denoted by D_t, is equal to:

$$D_t = d + \rho D_{t-1} + \varepsilon_t.$$

Notice that d represents the base demand level, ρ represents the correlation of demands in successive periods, where $|\rho| < 1$, and ε_t represents the error term that is normally distributed with mean 0 and standard deviation σ. Lee et al. (1997b) consider the case in which the retailer would act rationally by following an order-up-to level policy and by placing an order Q_t in period t. To show that the bullwhip effect occurs, they prove that $\mathrm{Var}(Q_t) \geq \mathrm{Var}(D_t)$.

Gilbert (2005) generalizes Lee et al. model by considering a more general demand process than the AR(1) process that is known as the Autoregressive Integrated Moving Average (ARIMA) time-series. When the underlying demand process is an ARIMA process, Gilbert shows that the order quantity Q_t associated with the order-up-to level policy is also an ARIMA process. Li et al. (2005) develop a simulation model for the case when the demand process is an ARIMA process. By varying the values of the parameters associated with the ARIMA process, they show that the bullwhip effect does not always occur. More importantly, they discover an "anti-bullwhip effect" that would occur for certain values of the parameters by showing that $\mathrm{Var}(Q_t) \leq \mathrm{Var}(D_t)$; i.e., the variance of the order quantity is lower than the variance of the demand.

While Lee et al. (1997b) show that the bullwhip effect will occur when the retailer has knowledge about the demand distribution, Chen et al. (1998), (2000a) and (2000b) investigate the occurrence of the bullwhip effect for the case when the retailer does not know the underlying demand process follows an AR(1) process; however, the retailer would use a moving average or an exponential smoothing method to forecast future demands. They show that the bullwhip effect will occur and that the bullwhip effect will be larger when the retailer uses an exponential smoothing forecast instead of a moving average forecast. Zhang (2004) extends the work of Chen et al. by examining the impact of different forecasting methods on the bullwhip effect. Sodhi and Tang (2011) show the bullwhip effect, at its core, is due to demand characteristics and leadtime, with information distortion by way of batch size etc. contributing to an incremental effect, which they quantify for an arborescent supply chain.

To mitigate the bullwhip effect, Lee et al. (1997c) identify four root causes of the bullwhip effect: demand forecasting, batch ordering, supply shortage, and price variations. In addition, they propose strategies for mitigating the bullwhip effect including:

1. Information sharing,
2. Vendor managed inventory, and
3. Collaborative forecasting and replenishment planning.

Below we review articles that examine these three strategies.

Information sharing. Lee et al. (2000) study the benefits of information sharing in a two-level supply chain. They consider the case in which the retailer has the information about the underlying demand distribution (i.e., an AR(1) process) and the retailer would order according to an order-up-to policy in each period.

When there is no information sharing, the manufacturer has the information about the underlying demand distribution and the retailer's ordering policy; however, the manufacturer does not have the information about the actual demand realized in period t (i.e., the manufacturer does not know the realization of the error term ε_t in period t).

When there is information sharing, the retailer would share the information about the actual demand realized in period t as well. By assuming that there exists a reliable exogenous source of inventory, information sharing has no impact on the retailer because the retailer's orders are always received in full. By examining the inventory level and the relevant costs incurred by the retailer and the manufacturer, Lee et al. show analytically that information sharing is beneficial to the manufacturer, not the retailer. Moreover, information sharing is most beneficial to the manufacturer especially when the correlation coefficient ρ is high. Also, in order to entice retailer to share demand information with the manufacturer, Lee et al. suggest various mechanisms including price discount and replenishment lead time reduction.

Cheng and Wu (2005) extend Lee et al.'s model to the multi-retailer case and they conclude that information sharing would enable the manufacturer to reduce both the inventory level and the total expected cost. Lee et al. commented that information sharing would be less valuable to the manufacturer if it uses the historical stream of orders from the retailer to forecast demand. Raghunathan (2001) confirms this analytically for the case when the underlying demand is an AR(1) process. Gaur et al. (2005) extend Raghunathan's model to the case in which the demand process is a more general process than the AR(1) process, namely, the AR(p) process for $p \geq 1$ and the autoregressive moving-average process ARMA process.

By assuming that the underlying demand is independent and identically distributed, Gavirneni et al. (1999) develop a model to examine the benefits of information sharing for the case in which the manufacturer has limited production capacity. In their model, the retailer has the information about the underlying demand distribution and the retailer would order according to an (s, S) policy. Under the (s, S) policy, the retailer would place an order in a period only when the inventory level drops below s. When there is no information sharing, the manufacturer has the information about the underlying demand distribution and the retailer's ordering policy; however, the manufacturer does not have the information about the retailer's inventory level. When there is information sharing, the retailer would share the information about the actual inventory level with the manufacturer in each period. They show that information sharing is beneficial to the manufacturer especially when the manufacturer's production capacity is higher or when the demand uncertainty level is moderate. Cachon and Fisher (1997) and (2000) analyze the benefits of information sharing for the N-retailer case in which the manufacturer has limited production capacity. By assuming that each retailer implements a (R, nQ) policy, they show analytically that information sharing is beneficial to the retailer and the manufacturer. In addition, Cachon and Fisher (2000) show numerically that lead time reduction will be more beneficial than information sharing.

Zhao et al. (2002) develop a simulation model to examine the impact of forecasting methods such as moving average, exponential smoothing, and Winters' method, etc., on the value of information sharing in a supply chain that has 1 manufacturer and N retailers. They show that the cost savings for the entire supply chain are more substantial when the retailers share information about future orders with the manufacturer than the case in which the retailers share information about the customer demand.

While many companies reported that sharing information (such as customer demand, inventory level, or demand forecast) among supply chain partners is beneficial, there are several obstacles for supply chain partners to share private information. For instance, retailers are reluctant to share information with the manufacturer because of fear (lower bargaining power, information leakage, etc.). Besides fear, there are other problems associated with forecast sharing in practice. Terwiesch et al. (2005) articulate that when a retailer revises his forecasts (or soft orders) frequently before placing a firm order, the manufacturer may ignore the revisions. Also, when a manufacturer is unable to fulfill the firm order in one period, the retailer may inflate his soft orders in future periods to ensure sufficient supply. As such, this could lead to a "lose-lose" situation. By using the data collected from a semiconductor company, Terwiesch et al. (2005) show empirically that the manufacturer would penalize the retailer for unreliable forecasts by delaying the fulfillment of forecasted orders. Also, they show that the retailer would inflate their orders resulting in excessive order cancellations. Therefore, both manufacturer and retailer would lose when sharing forecast information.

Vendor Managed Inventory. As articulated in Lee et al. (2000), information sharing is beneficial to the manufacturer, not to the retailer. As such, many manufacturers develop various initiatives to entice the retailer to share demand information with the manufacturer. Besides offering price discount, various manufacturers launched an initiative called Vendor Managed Inventory (VMI). Under the VMI initiative, the retailers delegate the ordering and replenishment planning decisions to the manufacturer. In return, the manufacturer gains direct information access regarding customer demand and retailers' inventory positions. To ensure the retailer achieves higher customer service levels with lower inventory costs, the manufacturer either owns the inventory at the retailer's warehouse subject to a minimum inventory level or issues some form of promises that the inventory at the retailer's warehouse will stay within certain pre-specified limits.

Under the VMI initiative, the retailer can reduce the overhead and operating costs associated with replenishment planning, while enjoying certain guaranteed service levels. Even though the manufacturer takes on the burden to manage the retailer's inventory under the VMI initiative, the manufacturer can derive the following benefits: (1) reduced bullwhip effect due to direct information access regarding customer demands and (2) reduced production/logistics/transportation cost due to coordinated production/replenishment plans for all retailers. Disney and Towill (2003) develop a simulation model to analyze the bullwhip effect under the VMI initiative. Their simulation results confirm that VMI can reduce the bullwhip effect by 50 percent. Clearly, reducing the bullwhip effect and coordinated planning would enable the manufacturer to reduce inventory. Johnson et al. (1999) examine the performance of VMI in different settings: (a) the manufacturer has limited capacity and (b) some retailers adopt the VMI scheme while the remainders adopt the information sharing scheme. By considering the case that VMI would enable the manufacturer to coordinate the replenishment plan by consolidating the customer demands (instead of orders placed by the retailers), they show that VMI would reduce inventories for the manufacturer and the retailer.

Aviv and Federgruen (1998a) develop an analytical model to evaluate the retailer's and the manufacturer's operating cost under an information-sharing scheme and an VMI initiative. Under both systems, the manufacturer has information about customer demand. However, the replenishment plans are determined by the retailers under the information sharing scheme, while the manufacturer decides on the timing and magnitude of the replenishment shipments to the retailers. Therefore, under the information sharing scheme, the effective demand process faced by the manufacturer is essentially the superposition of orders placed by the retailers in an uncoordinated manner. By considering that the underlying demand distribution is normal and by using the fact that the manufacturer has the authority to coordinate the customer demands under the VMI initiative, Aviv and Federgruen show that the manufacturer can reduce the production and inventory costs under both systems. To examine further about the benefit of the VMI initiative under which the manufacturer has the authority to determine the delivery schedule and quantity for each retailer, Cetinkaya and Lee (2000) presents an analytical model for determining an optimal coordinated replenishment and delivery plan for different retailers located in a given geographical region. By assuming that the demands at the retailers are independent Poisson processes, they compute an optimal replenishment quantity and delivery schedule that minimizes the total production, transportation and inventory carrying costs while meeting certain customer service levels.

Besides the analytical models that examine the benefits of the VMI initiative, there are other studies using simulation models. The reader is referred to Sahin and Robinson (2005) and the references therein. Several retailers and manufacturers reported successful implementations of VMI. For example, Clark and Hammond (1997) show that the VMI initiated by Campbell Soup provided a win-win situation for Campbell Soup and the retailers. For additional examples of successful implementations of VMI, please see Aviv and Federgruen (1998a), Cetinkaya and Lee (2000), and the references therein.

Collaborative forecasting. Under the information sharing scheme or the VMI initiative, not much collaborative effort is needed. To induce collaboration between the retailers and the manufacturers, Voluntary Inter-industry Commerce Standards (VICS) association developed an initiative called Collaborative Planning, Forecasting and Replenishment (CPFR). Under this initiative, both parties would develop mutually agreeable demand forecasts jointly. To develop mutually agreeable demand forecasts, the manufacturer would generate an initial demand forecast based on his market intelligence on products, and the retailer would create her own initial demand forecast based on customer's response to pricing and promotion decisions. Both parties would share their initial demand forecasts and would reconcile the differences in their forecast to obtain a common forecast. Once both parties agree on the common demand forecasts, the retailer would develop a replenishment plan and the manufacturer would develop a production plan independently. See www.cpfr.org for details.

The crux of CFPR is collaborative forecasting. Aviv (2010c) is the first to develop a framework for modeling the collaborative forecasting process between a retailer and a manufacturer. To specify the demand process when there is no collaborative forecast, he specifies the demand process based on an individual party p's perspective, where $p = r, m$, where r denotes the retailer and m denotes the manufacturer. Specifically, when there is no collaborative effort, the underlying demand process D_t from party p's perspective is given by:

$$D_t = d + \psi_t^p + \varepsilon_t^p, \quad \text{for } p = r, m,$$

where d represents the base demand level, ψ_t^p represents the cumulative forecast adjustment made by party p in past periods up to the beginning of period t, and ε_t^p represents the residual forecast error of party p's forecasting method.

By considering the correlation between ψ_t^m and ψ_t^r, one can capture the correlation between the forecast adjustments made by the retailer and the manufacturer. Under the collaborative forecasting initiative, Aviv assumes that the retailer and the manufacturer would select the best forecast adjustment ψ_t so that the forecast error is minimized. Based on this specific construct, he computes the optimal collaborative forecast adjustment in each period. For the retailer, he computes the variance of the total demand over the replenishment lead time with no collaborative forecasting and that with collaborative forecasting. For the manufacturer who needs to satisfy the order placed by the retailer, he computes the mean and the variance of the total aggregate order quantity to be placed by the retailer under no collaborative forecast and under collaborative forecast. Even when these quantities can be expressed in closed form expressions, it is intractable to evaluate the benefit of collaborative forecast analytically. As a surrogate, Aviv develops an aggregate supply chain performance measure that is based on the variance of the whole system: the sum of the variance of the total demand over the retailer's replenishment lead time and the variance of the total order quantity over the manufacturer's replenishment lead time. He shows analytically that collaborative forecast would reduce the system-wide variance. In a subsequent paper, Aviv (2002) extends this analysis to auto-correlated demand. Specifically, he considers the case in which the demand process possesses the following form:

$$D_t = d + \rho D_{t-1} + \psi_t^p + \varepsilon_t^p, \quad \text{for } p = r, m.$$

Aviv (2005) further extends the analysis to the case in which the manufacturer operates in an environment that calls for production smoothing.

As articulated by Aviv (2001), it is very difficult to evaluate the benefit of CPFR analytically even for a two-level supply chain. Therefore, many of the comparisons are conducted numerically. While these numerical examples provide some insights, there is no guarantee that this insight is applicable to a realistic supply chain. This observation has motivated Boone et al. (2002) to develop a simulation model to compare the performance of the CPFR initiative with the performance of a traditional replenishment policy based on a reorder point. By using the data collected from a Fortune-500 company and by using a simple process to generate demand forecast, their simulation model suggests that CPFR would increase customer service level and reduce inventories for both the manufacturer and the retailer. Aviv (2004) provides a comprehensive review of CPFR literature.

13.5.3 Robust Information Management Strategies

As reported in this section thus far, strategies based on information sharing, vendor managed inventory, or collaborative forecasting and replenishment planning would increase "supply chain visibility" in the sense that the upstream partners have access to information regarding the demand and inventory position at downstream stages. As supply chain visibility improves, each supply chain partner can generate more accurate forecast of future demands and better coordination. We have cited various articles that show how these strategies would enable a supply chain to become more responsive to customer demand with less inventory and lower cost. Hence, the information management strategies reported in this section would increase supply chain efficiency.

However, there are few articles on how these information management strategies would increase supply chain resiliency. Still, we have reasons to believe that the CPFR strategy can enable a supply chain to develop a production planning system that would improve resiliency. While Aviv (2005) discuss the mechanism for supply chain partners to generate a common demand forecast in a collaborative manner, we are not aware of specific models in the literature that deal with the collaborative replenishment planning. We envision a more complete CPFR system may improve supply chain resiliency. For example, consider a CPFR system in which all supply chain partners generate a common demand forecast, share inventory information, and adopt a common ordering rule that is based on the "proportional restoration rule" developed by Denardo and Tang (1992). Specifically, under the proportional restoration rule, the retailer would order Q_t^r and the manufacturer would order Q_t^m in period t, where:

$$Q_t^r = d + (T^r - I_t^r)\alpha^r \quad \text{and} \quad Q_t^m = d + [(T^m - I_t^m) + (T^r - I_t^r)]\alpha^m.$$

Notice that d represents the common demand forecast, T^p represents the "target" inventory position for party p, I_t^p represents the inventory held at party p at the beginning of period t, and $0 < \alpha^p \leq 1$ represents party p's restoration factor, where $p = r, m$ (i.e., retailer and manufacturer, respectively). Denardo and Tang (1992) use numerical examples to show that this ordering rule is efficient. In addition, in a later paper, Denardo and Tang (1996) show analytically that this ordering rule would "restore" the inventory level at each stage to its target even when the demand forecast d is inaccurate. Thus, one can conclude that such a CPFR system would improve supply chain efficiency and resiliency.

Table 13.5 lists references in this section.

13.6 Managerial Attitudes

Given the prominence of supply-chain risk in the business press, it is worthwhile asking what role managers play in tolerating or even engendering supply-chain risk. Consider managerial attitudes towards risk in general and towards initiatives for managing supply-chain disruptions in particular.

Managers' attitude towards risks. Sharpira (1986) and March and Sharpira (1987) study managers' attitude towards risks and conclude that:

- Managers are insensitive to estimates of the probabilities of possible outcomes.
- Managers tend to focus on critical performance targets, which affect the way they manage risk.
- Managers make a sharp distinction between taking risks and gambling.

The first conclusion can be explained by the fact that managers do not trust, do not understand, or simply do not much use precise probability estimates. This is consistent with the observations reported in De Waart (2006) and the results obtained by other researchers (c.f., Kunreuther, 1976 and Fischoff et al., 1981). Since managers are insensitive to probability estimates, March and Sharpira (1986) noted that managers are more likely to define risk in terms of the magnitude of loss such as "maximum exposure" or "worst case" instead of expected loss. The second conclusion is based on the observation that most managers are measured by a set of performance targets. March and Sharpira (1986) argue that these performance targets would cause the managers to become more risk averse (or risk prone) when their performance is above (or below) certain target. Finally, the third conclusion

Table 13.5 Information management references

Information Management Aspect	Risk Issue	References (in the order of appearance)
Information Management	General	Fisher (1997)
Managing Products with *Short* Life Cycles	General	Fisher and Raman (1996), Gurnani and Tang (1999), Iyer and Bergen (1997), Iyer (1998), Donohue (2000), Signorelli and Heskett (1984), Smith (2002)
Managing Products with *Long* Life Cycles	General	Sterman (1989), Lee et al. (1997b), Bagahana and Cohen (1998), Gilbert (2005), Li et al. (2005), Chen (1998), (2000a) (2000b), Zhang (2004), Sodhi and Tang (2011), Lee (1997c)
Information sharing	Uncertain demand	Lee et al. (2000), Cheng and Wu (2005), Raghunathan (2001), Gaur et al. (2005), Gavirneni et al. (1999), Cachon and Fisher (1997), Cachon and Fisher (2000), Zhao et al. (2002), Terwiesch et al. (2005)
Vendor managed inventory	Uncertain demand	Lee et al. (2000), Disney and Towill (2003), Johnson et al. (1999), Aviv and Federgruen (1998a), Cetinkaya and Lee (2000), Sahin and Robinson (2005), Clark and Hammond (1997)
Collaborative forecasting	Uncertain demand	Aviv (2001), Aviv (2002), Boone et al (2002), Aviv (2004)
Robust Information Management Strategies	Uncertain demand	Denardo and Tang (1992; 1996)

is driven by the fact that companies tend to reward managers for obtaining "good outcomes" but not necessarily for making "good decisions."

Managers' attitude towards initiatives for managing supply chain disruption risks. According to various major case studies conducted by Closs and McGarrell (2004), Rice and Caniato (2003) and Zsidisin et al. (2001) and (2004b):

- Most companies recognize the importance of risk assessment programs and use different methods, ranging from formal quantitative models to informal qualitative plans, to assess supply chain risks. However, most companies invested little time or resources for mitigating supply chain risks.
- Due to few data points, good estimates of the probability of the occurrence of any particular disruption and accurate measure of potential impact of each disaster are difficult to obtain. This makes it difficult for firms to perform cost/benefit analysis or return on investment analysis to justify certain risk reduction programs or contingency plans.

- Firms tend to underestimate disruption risk in the absence of accurate supply chain risk assessment. As reported in Kunreuther (1976), many managers tend to ignore possible events that are unlikely. This may explain why few firms take commensurable actions to mitigate supply chain disruption risks in a proactive manner. As articulated in Repenning and Sterman (2001), firms rarely invest in improvement programs in a proactive manner because "nobody gets credit for fixing problems that never happened."

13.7 Conclusions

In this chapter we have reviewed various quantitative models for managing supply chain risks. We found that these quantitative models are designed primarily for managing operational risks, not disruption risks. However, some of these strategies have been adopted by practitioners for managing risk because these strategies are robust: they can make a supply chain become not only more efficient in terms of handling operational risks but also more resilient in terms of managing disruption risks.

As there are not many models for managing disruption risks, we present six potential ideas for future research:

1. **Demand and supply stochastic processes.** Virtually all models reviewed in this paper are based on the assumption that the demand or the supply process is stationary. To model various types of disruptions mathematically, one may need to extend the analysis to deal with non-stationary demand or supply process. For instance, one may consider modeling the demand or the supply process as a "jump" process to capture the characteristics of major disruptions.

2. **Objective function.** The performance measures of the models reviewed in this paper are primarily based on the expected cost or profit. The expected cost or profit is an appropriate measure for evaluating different strategies for managing operational risks. When dealing with disruption risks that rarely happen, one may need to consider alternative objectives besides the expected cost / profit. For instance, Sharpira (1986) and March and Sharpira (1987) articulated that managers tend to focus on performance targets. Hence, when developing strategies for managing supply chain disruption risks, one may consider using certain performance targets such as recovery time after a disruption. The reader is referred to Brown and Tang (2005) and the references therein regarding various alternative performance targets in the context of single-period inventory models.

3. **Supply management strategies.** When developing supply management strategies for managing disruption risks, both academics and practitioners suggest the idea of "back-up" suppliers. To capture the dynamics of shifting the orders to these back-up suppliers when a major disruption occurs, one need to develop a model for analyzing dynamic supply configurations of suppliers including contract manufacturers, transportation providers, and distribution channels.

4. **Demand management strategies.** Among the demand management strategies presented in Section 13.3, it appears that dynamic pricing / revenue manage-

ment has great potential for managing disruption risks because a firm can deploy this strategy quickly after a disruption occurs. In addition, revenue management looks promising especially after successful implementations of different revenue management systems in the airline industry for managing operational risks.

5. **Product management strategies.** When selling products on line, e-tailers can change their product assortments dynamically according to the supply and demand of different products. This idea can be extended to brick and mortar retailers for managing disruption risks. Chong et al. (2001) show that store manager can manipulate customer's product choice and customer's demand by reconfiguring the set of products on display, the location of each product and the number of facings of each product. They suggest that one can utilize dynamic assortment planning to entice customers to purchase certain products that are widely available (when other products are in short supply).

6. **Information management strategies.** Among the information management strategies described in Section 13.5, we think the Collaborative Planning, Forecasting and Replenishment (CPFR) strategy is promising because it fosters a tighter coordination and stronger collaboration among supply chain partners. While Aviv (2005) develops a mechanism for generating collaborative forecasts, there is no model that captures the collaborating replenishing planning. It is conceivable that the value of a more complete CPFR system is much higher than a system that is solely based on collaborative forecasting.

Chapter 14
Modeling the Value of Flexibility

Abstract In this chapter, we present different stylized quantitative models to quantify the benefit of adding flexibility to the supply chain in terms of mitigating supply chain risk. Our results suggest that a firm can obtain significant value from implementing various flexibility strategies with only a limited amount of flexibility. In other words, to reduce supply chain risks, companies need to make their supply chains only slightly more flexible. Even without hard data, our analysis can increase a firms confidence for making small investments implementing the flexibility strategies described in Chapter 7.

14.1 Introduction

In this chapter, we recall the flexibility strategies described in Chapter 7 and ask the question: What is the *benefit* of adding flexibility to the supply chain? To answer this, we quantify the *value of flexibility* in the supply chain by analyzing different stylized models. Our results are unambiguous in that companies can obtain significant benefits (in terms of supply chain risk mitigation) from having only low levels of flexibility. This has implications for managers in that they need not reconfigure their supply chains completely to make them robust: they just need to add some flexibility.

We do not consider the *cost* for implementing additional flexibility in our model for two reasons. First, this kind of implementation cost is situation-specific. Second, to determine the optimal level of flexibility for a specific situation, one can always examine the trade-off between the situation-specific implementation cost and the value obtained at different levels of flexibility, and it is this value that is the focus of this chapter.

The models presented in this chapter indicate that a firm can obtain significant value from implementing these flexibility strategies with only a limited amount of flexibility. Even without reliable data and absolutely accurate cost and benefit analysis, our analysis can increase a firm's confidence for implementing the flexibility strategies described in Chapter 7 with only a small investment.

We consider risks in three categories in the supply-chain context: *supply risk*, *process risks*, and *demand risks*. As we discussed in Section 3 of Chapter 2, supply risks include the risks associated with supply cost, supply quality, and supply commitment. Process risks include the quality, time and capacity risks associated with in-bound and out-bound logistics and in-house operations. Demand risks include the risks associated with demand uncertainty.

This chapter is organized as follows. Sections 14.2–14.6 present the stylized models that analytically quantify the value of flexibility: Section 14.2 examines the value of multiple suppliers; Section 14.3 inspects the value of flexible supply contracts; Section 14.4 explores the value of flexible manufacturing; Section 14.5 investigates the value of postponement; and Section 14.6 studies the value of responsive pricing. Some of these analyses are motivated by models presented by other researchers in the literature. Section 14.7 concludes this chapter.

14.2 Supply-Cost Risk and the Value of Flexibility via Multiple Suppliers

Consider a situation in which a manufacturer has five pre-qualified suppliers with uncertain supply costs. In any time period, the unit cost of supplier j, denoted by C_j ($j = 1, 2, \ldots, 5$), is equal to $5, $10, or $15 with equal probability. In each period, the manufacturer always orders from the supplier who offers the *lowest* unit cost and each supplier has adequate capacity to handle the manufacturer's demand in each period.

What is the value to the manufacturer by having five qualified suppliers? Suppose, instead, that the manufacturer has only one qualified supplier. Then, based on our assumptions, the *expected unit cost* associated with sourcing from this supplier only, denoted by UC(1), is given as: $UC(1) = (1/3)(5 + 10 + 15) = \$10$. Next, consider the case in which the manufacturer can source from two suppliers and the manufacturer selects the supplier with a lower unit cost. The corresponding expected unit cost associated with sourcing from two potential suppliers, denoted by UC(2), is $UC(2) = E[\min\{C_1, C_2\}]$. By enumerating all possible pricing scenarios by the two suppliers, it can be shown that UC(2) = $7.8. Similarly, UC(3) = $6.6, UC(4) = $5.9, and UC(5) = $5.6. Thus the expected unit cost to the manufacturer drops from $10 to $5.6 as the number of qualified suppliers increases from 1 to 5 in this example.

Let the value of flexibility, $V(n)$, be defined as the percentage decrease in the expected unit cost by ordering from n suppliers instead of 1 supplier, i.e., $V(n) = (UC(1) - UC(n))/UC(1)$. Figure 14.1 shows that the percentage of savings in the expected unit cost $V(n)$ is increasing and concave in n. The figure also implies that significant savings in the expected unit cost can occur when a firm orders from only two or three suppliers.

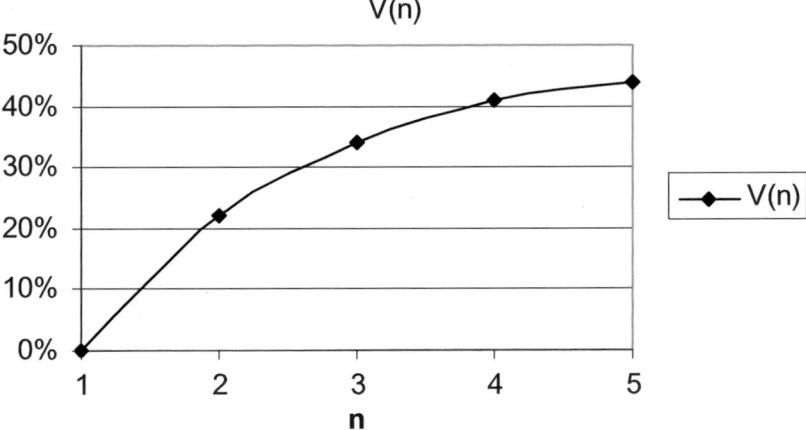

Fig. 14.1 The power of flexibility via multiple suppliers

Below, we generalize the above results by analytically establishing that the percentage of savings in the expected unit cost $V(n)$ is increasing and concave in n for three common probability distributions: Bernoulli, Uniform and Exponential.

Consider a situation in which a manufacturer has n suppliers with uncertain (and independently and identically distributed) costs. We first consider the case in which supplier costs follow the same Bernoulli distribution. We then consider uniformly and exponentially distributed costs.

In the Bernoulli case, the unit cost charged by supplier i is equal to C_i, where C_i equals c_h with probability p, and equals c_l with probability $(1-p)$, where $c_h > c_l$. Suppose that the manufacturer is committed to source from supplier 1 exclusively; i.e., $n = 1$. Then the expected unit cost, denoted by UC(1), is given as: UC$(1) = E(C_1) = p \cdot c_h + (1-p) \cdot c_l = c_l + p(c_h - c_l)$. Next, consider the case when the manufacturer sources from two suppliers 1 and 2. Because the manufacturer orders from the supplier with the lower realized unit cost, the corresponding expected unit cost is denoted by UC(2), where: UC$(2) = E(\min\{C_1, C_2\})$. By observing that $\min\{C_1, C_2\}) = c_h$ for the case when $C_1 = C_2 = c_h$,

$$\text{UC}(2) = E(\min\{C_1, C_2\}) = p^2 \cdot c_h + (1 - p^2) \cdot c_l = c_l + p^2(c_h - c_l).$$

By using the same argument, we can show that when the manufacturer sources from n suppliers, the corresponding expected unit cost is denoted by UC$(n) = c_l + p^n(c_h - c_l)$. Using simple calculus, we can also show that the expected unit cost UC(n) is decreasing and convex in n. In general, UC(n) is given by the expected value of the minimum order statistic and, therefore, UC(n) depends on the probability distribution of the unit price charged by any of the suppliers.

Suppose that the unit cost C_i is *uniformly* distributed over $[a, b]$. Then we can use standard order-statistic results to show that UC$(n) = E(\min\{C_1, C_2, \cdots, C_n\}) = a + (b - a)/(n + 1)$. Alternatively, suppose that the unit cost C_i is *exponentially* distributed with rate λ. Then we can use standard order-statistic results to show that

$$\mathrm{UC}(n) = \mathrm{E}(\min\{C_1, C_2, \ldots, C_n\}) = \frac{1}{n\lambda}.$$

For both these probability distributions, $\mathrm{UC}(n)$ is decreasing and convex in n.

To state the result formally, as defined earlier in this section, let $V(n)$ be the relative savings in the expected unit cost that is obtained by n suppliers instead of 1 supplier, i.e., $V(n) = (\mathrm{UC}(1) - \mathrm{UC}(n))/\mathrm{UC}(1)$. Since $\mathrm{UC}(n)$ is decreasing and convex in n, we have:

Theorem 14.1. *The relatives savings in the expected unit cost, $V(n)$, is increasing and concave in n for Bernoulli, Uniform and Exponential cost distributions.*

The above theorem suggests that significant savings in the expected unit cost can be obtained by using only a small number of suppliers. This illustrates the value to a company that introduces only a small amount of flexibility by way of multiple suppliers to mitigate the supply-cost risk.

14.3 Supply-Commitment Risk: the Value of Flexibility via Flexible Supply Contracts

Consider a supply chain comprising a supplier, a manufacturer, and a retailer. The supply cost is c per unit, the wholesale price is p per unit, and all unsold units have 0 salvage value. We consider a two-period model in which the retailer places his order only at the *end* of period 1. However, due to the supply lead time, the manufacturer needs to place an order with the supplier at the *beginning* of period 1, i.e., prior to the order placed by the retailer. This ordering process is similar to that described in the Sport Obermeyer case prepared by Hammond and Raman (1995).

At the beginning of period 1, the manufacturer estimates that the retailer would order $D = a + \varepsilon$ at the end of period 1, where ε corresponds to the uncertain market condition to be realized in period 1. Based on the information about c, p, and D, the manufacturer orders x units at the beginning of period 1. Under the flexible supply contract, the manufacturer is allowed to modify this order from x units to y units after receiving the actual order from the retailer at the end of period 1. Consider the case when the retailer orders $d = a + e$ at the end of period 1, where e is the realized value of ε. Under the u-flexible contract, the modified order y must satisfy: $x(1-u) \leq y \leq x(1+u)$, where $u \geq 0$ represents the "allowable adjustment" in percentage. Thus, the parameter u represents the flexibility level of the u-flexible contract.

14.3.1 Uniformly Distributed Market Shocks

To illustrate the impact of u on the manufacturer's expected profit, we set $p = 2$, $c = 1$, $a = 100$ and assume that the market shock ε is uniformly distributed between -50 and 50, i.e., the probability density function is $1/100$. Given u, x, and e, the manufacturer can determine the optimal modified order y that maximizes the manufacturer's profit:

$$\pi(u,x,e) = \max_{y}[p\min\{y, a+e\} - cy], \quad \text{subject to } x(1-u) \leq y \leq x(1+u).$$

Since $p = 2$, $c = 1$, $a = 100$, the manufacturer's optimal profit $\pi(u,x,e)$ can be expressed as:

$$
\begin{aligned}
\pi(u,x,e) &= x(1+u) & \text{if} & & e &> x(1+u) - 100 \\
\pi(u,x,e) &= 100 + e & \text{if} & \quad x(1-u) - 100 < e &\leq x(1+u) - 100 \\
\pi(u,x,e) &= 2(100+e) - x(1-u) & \text{if} & & e &\leq x(1-u) - 100
\end{aligned}
$$

Since e is the realized value of ε that is uniformly distributed between -50 and 50, we can compute $\pi(u,x)$, the (ex-ante) expectation of $\pi(u,x,e)$:

$$
\begin{aligned}
\pi(u,x) &= E_e(\pi(u,x,e)) \\
&= \int_{-50}^{x(1-u)-100} [2(100+e) - x(1-u)]\frac{1}{100}\,de \\
&\quad + \int_{x(1-u)-100}^{x(1+u)-100} (100+e)\frac{1}{100}\,de \\
&\quad + \int_{x(1+u)-100}^{50} x(1+u)\frac{1}{100}\,de \\
&= \frac{1}{100}(-x^2u^2 - x^2 + 200x + 100xu - 2500)
\end{aligned}
$$

By considering the first order condition, we can show that the optimal initial order quantity x^* that maximizes the manufacturer's expected profit $\pi(u,x)$ is given as $x^* = (200 + 100u)/(2(1+u))$. (When $u = 0$, $x^* = 100$. This corresponds to the newsvendor solution with $F(x^*) = (p-c)/p = 0.5$, where F is the probability distribution of D.) Substitute x^* into $\pi(u,x)$, the manufacturer's optimal expected profit under the u-supply contract, denoted by $\pi(u) = \pi(u,x^*) = (100 + 100u + 25u^2)/(1+u^2) - 25$. (Notice that $\pi(0) = 75$ under the 0-flexible contract; i.e., when $u = 0$.)

Let the value of flexibility, $V(u)$, be defined as the percentage increase in the manufacturer's optimal expected profit over the 0-flexible contract, where:

$$V(u) = \frac{\pi(u) - \pi(0)}{\pi(0)} = \frac{\dfrac{100 + 100u + 25u^2}{1+u^2} - 100}{75}.$$

It can be easily shown that $V(u)$ is increasing and concave in u. When we vary u from 0 to 50%, Figure 14.2 illustrates that $V(u)$ is increasing and concave in u. Also, significant benefits associated with the u-flexible contract can be obtained when u is relatively small. We note that although Tsay and Lovejoy (1999) have numerically observed that there are diminishing returns to flexibility in these quantity flexible contracts, our work establishes this analytically, albeit in a simpler setting.

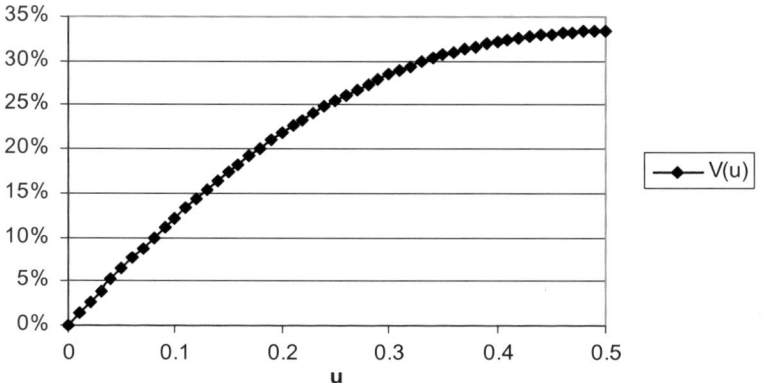

Fig. 14.2 The power of flexibility via flexible supply contract

14.3.2 General Market Shocks

We now analyze the general model in which the market shock ε follows a general probability distribution and the parameters p, c, and a are not restricted to any particular value. This description is based on a stylized model of a flexible supply contract that serves our purpose of analytically investigating the value of supply flexibility. For a more general model of quantity-flexible (QF) contracts, we refer the reader to Tsay and Lovejoy (1999).

Consider a manufacturer that orders a product from a supplier and sells the product to a retailer at unit price p. A simple QF contract can be specified by two parameters: the unit cost c and an adjustment parameter u. Consider the case in which the manufacturer placed an order x sometime earlier. As the manufacturer obtained more accurate demand information, he would like to revise this particular order x. Under the QF contract, the manufacturer is allowed to adjust his order from x to y as long as y satisfies: $(1-u)x \leq y \leq (1+u)x$. Therefore, the QF contract provides the 'flexibility' for the manufacturer to adjust his order as more accurate market condition becomes available. In this case, the manufacturer is considered to engage in a u-flexibility contract when $u > 0$, and is consider to engage in a 0-flexibility contract when $u = 0$.

To quantify the value of supply flexibility via the QF supply contract, we consider a stylized two-period model in which the retailer places her order at the end of period 1. As described earlier, at the beginning of period 1, the manufacturer estimates that the retailer's order to be realized at the end of period 1 is given as: $D(1) = a + \varepsilon$, where ε corresponds to the uncertain market condition to be realized in period 1 and $F(\cdot)$ denotes the cumulative distribution function for ε.

Due to supply lead time, the manufacturer needs to place an initial order quantity x at the beginning of period 1 prior to the realization of ε. At the end of period 1, the manufacturer observes the realization of $\varepsilon = e$ and can then revise his order from x to y. Let us first consider the second-stage problem of determining the optimal revised order y^* given a market realization $\varepsilon = e$, an initial order quantity x, and a flexibility parameter u. This second-stage problem can be formulated as the following linear program (A):

$$\pi(u,x,e) = \max_{y,s}\{ps - cy\}$$

$$s \le y,$$

$$s \le a + e,$$

$$x(1-u) \le y \le x(1+u),$$

where $\pi(u,x,e)$ is the optimal profit associated with the u-flexibility contract for a given initial order of x and market realization e. By considering the optimal solution to the above program, it is easy to show that the optimal profit $\pi(u,x,e)$ can be expressed as:

$$\pi(u,x,e) = \begin{cases} (p-c)x(1+u) & \text{if } e > x(1+u) - a \\ (p-c)(a+e) & \text{if } x(1-u) - a \le e \le x(1+u) - a \\ p(a+e) - cx(1-u) & \text{if } e < x(1-u) - a \end{cases}$$

For an initial order of x, the 'ex-ante' expected profit associated with the u-flexibility contract is then equal to:

$$\Pi(u,x) = \int_{x(1+u)-a}^{\infty} (p-c)x(1+u)\,dF(e)$$

$$+ \int_{x(1-u)-a}^{x(1+u)-a} (p-c)(a+c)\,dF(e)$$

$$+ \int_{x(1-u)-a}^{-\infty} (p(a+e) - cx(1-u))\,dF(e)$$

One can show that $\Pi(u,x)$ is concave in x and that the optimal $x^*(u)$ is the unique x that solves:

$$(p-c)(1+u)(1 - F(x(1+u) - a)) - c(1-u)F(x(1-u) - a) = 0$$

For the 0-flexibility case; i.e., when $u = 0$, $x^*(0)$ is the traditional newsvendor solution, where:

$$x^*(0) = a + F^{-1}\left(\frac{p-c}{p}\right).$$

We denote the optimal 'ex-ante' expected profit as $\Pi(u) = \Pi(u, x^*(u))$. Let $V(u)$ denote the relative increase in expected profit of a u-flexibility contract over that of a 0-flexibility contract, i.e., $V(u) = (\Pi(u) - \Pi(0))/\Pi(0)$.

For the case of uniformly distributed market shocks, it is possible to develop a closed-form solution for $x^*(u)$ and, hence, $\Pi(u)$ and $V(u)$ (see Section 14.3.1.). For a general distribution of market shocks, however, a closed-form solution for $\Pi(u)$ does not exist. However, we can first approximate $\Pi(u)$ by setting $x(u) = x^*(0) = a + F^{-1}((p-c)/p)$, that is, $\Pi(u) \approx \Pi(u, x^*(0))$. Then we can approximate $V(u)$ by $V(u) \approx V(u, x^*(0)) = (\Pi(u, x^*(0)) - \Pi(0))/\Pi(0)$. To validate this approximation scheme, we have numerically solved for the optimal $x^*(u)$ for the case of normally distributed market shocks, and our results show that $V(u, x^*(0))$ is very close to $V(u)$.

Using calculus, we can prove the following:

Theorem 14.2. *The (approximated) relative increase in expected profit $V(u, x^*(0))$ is increasing and concave in u.*

The proof is provided by Tang and Tomlin (2008). Our numerical work (omitted) also validated that $V(u)$ is increasing and concave in u. Hence, one can conclude that the manufacturer can obtain significant value with only a small amount of supplier flexibility; i.e., when u is small. This illustrates the power of limited flexibility to mitigate the supply-commitment risk.

14.4 Process risk: the Value of Flexibility via Flexible Manufacturing

Jordan and Graves (1995) focus on the benefits of process flexibility for managing demand risk and show that a firm can extract most of the benefit associated with flexibility by implementing a manufacturing process with only limited flexibility. Their analysis is based on a single-stage manufacturing process, which has been extended to a multiple-stage process by Graves and Tomlin (2003). Please refer to Jordan and Graves (1995) for an in-depth analysis of a model in which different plants are capable of producing different number of products.

However, process flexibility can also help mitigate the process risk associated with fluctuating capacities. The following stylized model is a modification of a special case presented in Jordan and Graves (1995) in that our model focuses on process risks in terms of uncertainty of plant capacity rather than uncertainty of demand.

14.4.1 (Discrete) Uniformly Distributed Plant Capacity

Consider a firm that sells 4 different products (1, 2, 3, and 4), each with a demand of $D_1 = D_2 = D_3 = D_4 = 100$ units. The firm owns 4 different plants; the capacity of each plant j ($j = 1, 2, 3, 4$), denoted by C_j, is equal to 50, 100, or 150 units with equal probability 1/3; i.e., C_j follows a discrete uniform distribution. In this setting, there is no (expected) redundant capacity in the sense that the expected total aggregate capacity of all 4 plants is 400 units, which is equal to the total aggregate demand of all 4 products.

We focus on the following system configurations. A system is considered to possess "h-flexibility" when each plant has the capability of producing exactly h products and when the system is configured as follows: When $h = 1$, each plant j is capable of producing product j only, where $j = 1, 2, 3, 4$. Hence, 1-flexibility system corresponds to the system with no flexibility. Figure 14.3 depicts the h-flexibility system for $h = 1, 2, 3, 4$. Clearly, the 4-flexibility system corresponds to the system with total flexibility.

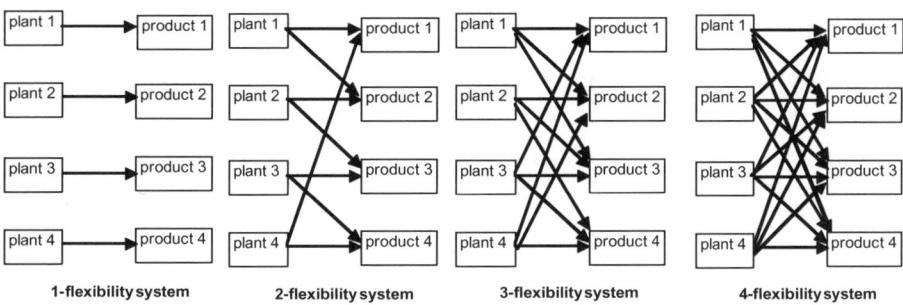

Fig. 14.3 h-flexibility manufacturing systems

We now compute the effective sales associated with each system as depicted in Fig. 14.3. For each h-flexibility system, let $A(h)$ be the set of arcs so that $(i, j) \in A(h)$ only if plant j is capable of producing product i. In this case, for any arc (i, j), let X_{ij} be the number of units of product i to be produced by plant j. For any given realization of plant capacities, say, $\mathbf{C} = \mathbf{c} = (c_1, c_2, c_3, c_4)$) the effective sales associated with the h-flexibility system, denoted by $S(h; \mathbf{c})$, corresponds to the optimal objective function of the following network flow problem (P), where:

$$S(h; \mathbf{c}) = \max_{X_{ij} \geq 0} \sum_{i=1}^{i=4} \sum_{j=1}^{j=4} X_{ij}$$

$$\text{s.t.} \quad \sum_{i:(i,j)\in A(h)} X_{ij} \leq c_j \ \forall j; \quad \sum_{j:(i,j)\in A(h)} X_{ij} \leq D_i \ \forall i.$$

Let $S(h)$ be the expected effective sales associated with the h-flexibility system, where $S(h) = E(S(h;\mathbf{c}))$. It is easy to observe from Fig. 14.3 that

$$S(1) = \sum_{j=1}^{4} E(\min\{C_j, D_j\}) = \sum_{j=1}^{4} E(\min\{C_j, 100\}), \quad \text{and}$$

$$S(4) = E\left(\min\left\{\sum_{j=1}^{4} C_j, \sum_{i=1}^{4} D_i\right\}\right) = E\left(\min\left\{\sum_{j=1}^{4} C_j, 400\right\}\right).$$

By considering the probability of each of the 81 possible plant capacity scenarios, one can show that $S(1) = 333.33$, $S(2) = 367.9$, $S(3) = 367.9$, $S(4) = 367.9$. Note that the expected total shortfall of demand for a h-flexible system is equal to $E(\sum D_i) - S(h)$. Therefore, the demand shortfall is an opposite measure of effective sales of a h-flexibility system. This result implies that the 2-flexibility system yields the same capability as the 4-flexibility system.

Define the value of flexibility, $V(h)$, as the percentage increase in the expected sales associated with a h-flexible system over the 1-flexibility system (i.e., the system with no flexibility). Specifically, we define $V(h) = (S(h) - S(1))/S(1)$. Figure 14.4 shows that the percentage increase in the expected sales $V(h)$ is increasing and concave in h. Figure 14.4 implies that significant increase in the effective sales can occur even with a system with limited process flexibility. Therefore, to reduce process risks, it is sufficient to operate a manufacturing system with limited flexibility. This illustrates the power of process flexibility via flexible manufacturing process.

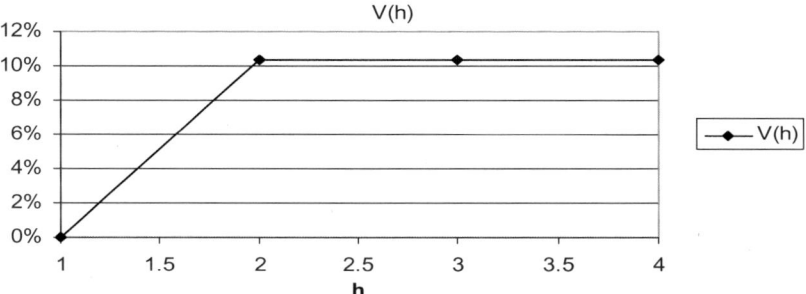

Fig. 14.4 The power of flexibility via process flexibility

14.4.2 Normally Distributed Plant Capacity

We now analyze the benefit of process flexibility for the case in which plant capacities are *normally distributed* by considering demand shortfall (i.e., unfilled demand)

as a measure. We show analytically that the percentage reduction in demand short-fall is indeed increasing and concave in h.

Consider a system with n plants and n products. The capacity of each plant j, denoted by C_j, is i.i.d. *normally* distributed with a mean μ and a standard deviation σ. Each product i has a deterministic demand equal to μ. In a manner directly analogous to Jordan and Graves (1995), one can show that the expected demand "shortfall" (i.e., the quantity of demand not filled) is equal to $E[\max_M\{\sum_{i\in M}\mu - \sum_{j\in P(M)} C_j\}]$, where M is any subset of products and $P(M)$ is the subset of plants that can produce at least one of the products in M. Jordan and Graves (1995) proposed an "efficiency" metric $\Pi(M^*)$ that reflects the probability that the shortfall of a limited-flexibility system exceeds the shortfall of a totally-flexible system (i.e., one in which every plant can produce every product.) They established that $\Pi(M^*)$ is a good predictor of the "efficiency" of a flexibility configuration, that is, a limited-flexibility system with a small value of $\Pi(M^*)$ gives a similar expected shortfall to that of a totally-flexible system. We use the same efficiency metric here, but adapt if for the case of uncertain capacities rather than uncertain demands. In this case,

$$\Pi(M^*) = \max_M \text{Prob}\left\{\left(\sum i \in M\mu - \sum_{j\in P(M)} C_j\right) > \max\left\{0, n\mu - \sum_{j=1}^{n} C_j\right\}\right\}$$

To illustrate the value of process flexibility, we consider the h-flexibility system configurations proposed by Jordan and Graves (1995). In the h-flexibility configuration, each plant has the capability to produce exactly h products in the following manner. When $h = 1$, the system has no flexibility as each plant i is capable of producing product i only. Hence, 1-flexibility system corresponds to a system with no flexibility. In the 2-flexibility system, the system has some flexibility in the sense that plant 1 can produce products 1 and 2, plant 2 can produce products 2 and 3, ..., and plant n can produce products n and 1. In general, the level of flexibility can be specified by the parameter h. Applying the analysis of Jordan and Graves (1995, p. 591) to our case, we obtain

$$\Pi(h) = \left(1 - \Phi\left(\frac{(h-1)\mu}{\sigma\sqrt{0.5n}}\right)\right) \cdot \Phi\left(-\frac{(h-1)\mu}{\sigma\sqrt{0.5n}}\right)$$
$$= \left[\Phi\left(-\frac{(h-1)\mu}{\sigma\sqrt{0.5n}}\right)\right]^2$$

for $1 \le h \le n/2$. By using the properties of a standard normal distribution $\Phi(z)$, one can use simple calculus to show that $\Pi(h)$ is decreasing and convex in h. (Jordan ad Graves (1995) provided some numerical results (Figure 9 on p. 591) to illustrate this point.) In this case, the percentage increase in efficiency associated with a h-flexible system over the 1-flexibility system can be defined by $V(h)$, where: $V(h) = (\Pi(1) - \Pi(h))/\Pi(1)$. Because the function $\Pi(h)$ is decreasing and convex in h, we can derive the following:

Theorem 14.3. *The percentage increase in efficiency associated with a h-flexible system, $V(h)$, is increasing and concave in h.*

The above theorem suggests that significant system efficiency occurs in systems with limited flexibility; i.e., when h is small. This illustrates the power of limited flexibility to mitigate process risk.

14.5 Demand Risk: the Value of Flexibility via Postponement

The following description is a simplified version of the model presented by Lee and Whang (1998). A firm produces two end-products by using a two-stage production process. The firm adopts a τ-postponement strategy when it takes τ time periods to produce a generic semi-finished product at the first stage and $T - \tau$ time periods to customize these generic products into two different end-products. Figure 14.5 depicts a process under the τ-postponement strategy. Since the generic product is flexible, the production process is more flexible as τ increases.

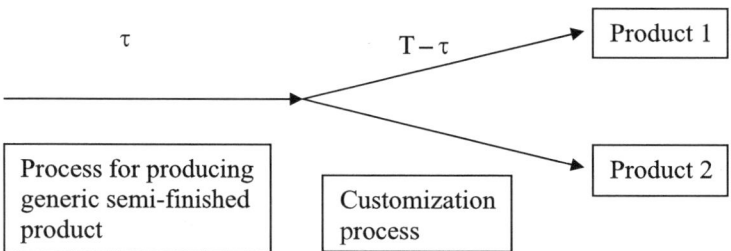

Fig. 14.5 A manufacturing process associated with the τ-postponement strategy

Let $D_i(t)$ denote the demand of product i to be realized t periods in the future, where $i = 1, 2$. Lee and Whang (1998) considered two demand models, namely, (a) the independent and identically distributed (IID) model $D_i(t) = \mu_i + \varepsilon_{it}$; and (b) the random walk (RW) model: $D_i(t) = \mu_i + \varepsilon_{i1} + \varepsilon_{i2} + \cdots + \varepsilon_{i,t-1} + \varepsilon_{it}$, where $i = 1, 2; t = 1, \ldots, T$; and ε_{it} are independently and identically (i.i.d.) *normally* distributed random variables with mean 0 and standard deviation σ.

For any τ-postponement strategy, let $S^{\text{IID}}(\tau)$ and $S^{\text{RW}}(\tau)$ be the optimal order-up-to level for end-product i associated with the IID and the RW demand models, respectively. Similarly, let $V^{\text{IID}}(\tau)$ and $V^{\text{RW}}(\tau)$ be the percentage savings of safety stock over the 'no postponement' strategy associated with the IID and the RW demand models, respectively. By considering a situation in which the sales and inventory review take place every time period, the inventory is held only in finished goods form, and the entire production system is managed according to a periodic review order-up-to system, Lee and Whang (1998) proved the following result for the two-product case (proof omitted):

$$V^{\text{IID}}(\tau) = \frac{S^{\text{IID}}(0) - S^{\text{IID}}(\tau)}{S^{\text{IID}}(0)} = 1 - \sqrt{0.5\left(1 - \frac{T - \tau + 1}{T + 1}\right)}$$

$$V^{\text{RW}}(\tau) = \frac{S^{\text{RW}}(0) - S^{\text{RW}}(\tau)}{S^{\text{RW}}(0)}$$

$$= 1 - \sqrt{0.5\left(1 - \frac{(T - \tau + 1)(T - \tau + 2)(2T - 2\tau + 3)}{(T + 1)(T + 2)(2T + 3)}\right)}$$

It can be easily shown that $V^{\text{IID}}(\tau)$ is increasing and convex in τ while $V^{\text{RW}}(\tau)$ is increasing and concave in τ.

To illustrate the impact of the τ-postponement strategy on the percentage savings of safety stock, let us consider a numerical example in which $T = 50$. Figure 14.6 shows that the percentage savings of safety stock (i.e., $V^{\text{IID}}(\tau)$ and $V^{\text{RW}}(\tau)$) increase as τ increases. Specifically, in the RW demand model, significant savings can occur at the early state of the production process (i.e., even when τ is small). This illustrates two important points: (i) the power of flexibility via product postponement depends on the underlying demand structure; and (ii) significant benefits associated with product flexibility can be obtained by postponing the product differentiation at the early stage of the production process (i.e., when τ is small).

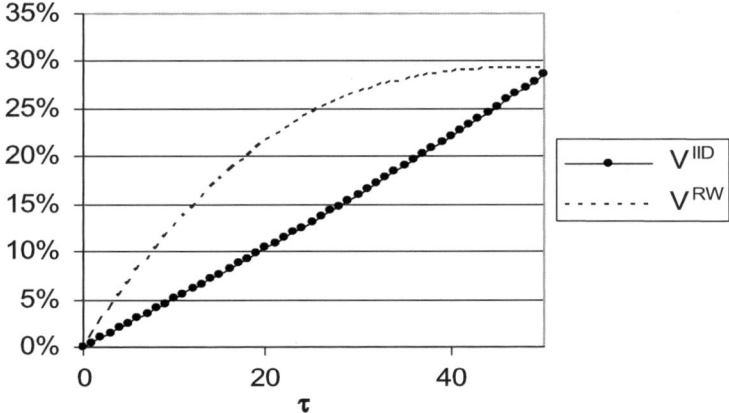

Fig. 14.6 The power of flexibility via postponement

14.6 Demand Risk: the Value of Flexibility via Responsive Pricing

We now present a stylized model to illustrate the benefit of postponing the pricing decision. This model is motivated by the work of Van Mieghem and Dada (2001) in which they examine the joint effect of production postponement and price post-ponement. Whereas Van Mieghem and Dada (1999) focus on the benefit of price postponement to a single-product firm, we consider the benefit to a two-product firm. Furthermore, Van Mieghem and Dada (1999) consider two extreme pricing times: the price is set either before any uncertainty is resolved or after all uncertainty is resolved. We examine a model in which uncertainty reduces over time and consider a range of pricing times. The later the firm must commit to prices, the more flexible it is. While Van Mieghem and Dada (1999) model additional decisions, e.g., production and capacity, we choose to focus exclusively on the timing of the pricing decision so as to provide insight into the value of timing flexibility.

Consider a manufacturer that sells two substitutable products (1 and 2) through a retailer over a selling season that starts after period T. At the beginning of period 1, the manufacturer estimates that the total demand for product i over the selling season is given by $D_i(1)$, where

$$D_i(1) = a_i + S_{i1} + S_{i2} + \cdots + S_{i,T-1} + S_{i,T} - bp_i + \delta(p_j - p_i), \quad i,j = 1,2; \quad j \neq i.$$

In this case, a_i represents the "primary demand" of product i, S_{it} represents the "shock" to the primary demand of product i that occurs in period t, b measures price sensitivity, and δ measures product substitutability. In our model, we assume that the shock S_{it} follows an auto-regressive process of order one (i.e., AR(1)) process so that $S_{it} = \rho_i S_{i,t-1} + \varepsilon_{it}$ for $i = 1,2$, and $t = 1,2,\ldots,T$, where $0 \leq \rho_i < 1$, and ε_{i1}, ε_{i2}, $\ldots$, $\varepsilon_{i,T-1}$, $\varepsilon_{i,T}$ are i.i.d. normal random variables with mean 0 and standard deviation σ_i. Without loss of generality, we set $S_{i0} = 0$ for $i = 1,2$.

To keep things simple, we consider the case in which the manufacturer and the retailer are both owned and controlled by a single firm. We also assume that the manufacturer has the capacity to meet the actual demand of each product over the selling season that starts after period T. In this integrated supply chain, the unit cost of each product i is given as c and we only need to decide on p_i; i.e., the retail price of each product i. (The wholesale price is determined mutually by the manufacturer and the retailer.)

Suppose that the firm has the flexibility to set and announce p_{it}; i.e., the retail price of each product i, at the end of period t, where $t = 1,\ldots,T$. Once the retail price is announced, we assume that the firm is committed to sell each product at p_{it} during the selling season that starts after period T. This implies that the firm must announce the actual retail price no later than the end of period T. Clearly, the firm would benefit from postponing the pricing decision because it would allow the firm to gain more accurate information about the market demand before setting the actual retail price. To formalize this thinking, we say that the firm employs the t-postponement strategy when the actual retail price is determined at the end of period t. Timing flexibility increases as t increases.

Suppose a firm adopts the t-postponement strategy by setting the retail price p_{it} at the end of period t. Then the value of S_{ik}, has been realized: $S_{ik} = s_{ik}$ for $k = 1,2,\ldots,t$. For any given $\mathbf{s} = \{s_{ik}, i = 1,2, \text{ and } k = 1,2,\ldots,t\}$, the conditional seasonal demand can be expressed as: $D_i(t)|\mathbf{s} = a_i + s_{i1} + s_{i2} + \ldots + s_{i,t-1} + s_{it} +$

$S_{i,t+1} + \ldots + S_{i,T-1} + S_{i,T} - bp_{it} + \delta(p_{jt} - p_{it})$. In this case, the optimal retail price p_{it} that maximizes the manufacturer's expected profit can be determined by solving the following problem (P):

$$\pi(t,s) = \max_{\{p_{1t}, p_{2t}\}} (p_{1t} - c) \, E(D_1(t)|\mathbf{s}) + (p_{2t} - c) \, E(D_2(t)|\mathbf{s}).$$

To solve problem (P), we need to determine $E(D_i(t)|\mathbf{s})$ for $i = 1, 2$. In preparation, we first use the AR(1) process to express $S_{i,t+k}$ for any given $\mathbf{s} = \{s_{ik}, i = 1, 2,$ and $k = 1, 2, \ldots, t\}$, getting:

$$S_{i,t+k} = \rho_i^k s_{i,t} + \rho_i^{k-1} \varepsilon_{i,t+1} + \rho_i^{k-2} \varepsilon_{i,t+2} + \cdots + \rho_i \varepsilon_{i,t+k-1} + \varepsilon_{i,t+k}$$
$$\text{for } i = 1, 2 \text{ and } k = 1, 2, \ldots, T - t.$$

By using this expression for $S_{i,t+k}$ for $k = 1, 2, \ldots T - t$, we can express $D_i(t)|\mathbf{s}$ as follows:

$$D_i(t)|\mathbf{s} = a_i + s_{i1} + s_{i2} + \ldots + s_{i,t-1} + \left(\frac{1 - \rho_i^{T-t+1}}{1 - \rho_i} \right) s_{it} + \left(\frac{1 - \rho_i^{T-t}}{1 - \rho_i} \right) \varepsilon_{i,t+1}$$
$$+ \left(\frac{1 - \rho_i^{T-t-1}}{1 - \rho_i} \right) \varepsilon_{i,t+2} + \cdots + \left(\frac{1 - \rho_i^2}{1 - \rho_i} \right) \varepsilon_{i,T-1}$$
$$+ \varepsilon_{i,T} - bp_{it} + \delta(p_{jt} - p_{it}) \quad \text{for } i = 1, 2.$$

Since ε_{ik} are i.i.d. normal random variable with mean 0, it is easy to show that:

$$E(D_i(t)|\mathbf{s}) = a_{it} - bp_{it} + \delta(p_{jt} - p_{it}),$$
$$\text{where} \quad a_{it} = a_i + s_{i1} + s_{i2} + \cdots + s_{it} + \left(\frac{1 - \rho_i^{T-t+1}}{1 - \rho_i} \right) s_{it} \text{ for } i = 1, 2$$

Given $E(D_i(t)|\mathbf{s})$, we can determine the first order conditions associated with problem (P) and show that:

$$\pi(t,s) = \frac{(a_{1t} + a_{2t} - 2bc)^2}{8b} + \frac{(a_{1t} - a_{2t})^2}{8(b + 2\delta)}.$$

We now determine the 'ex-ante' expected profit associated with the t-postponement strategy, which is equal to $\pi(t) = E_s(\pi(t,\mathbf{s}))$. In preparation, we first use the AR(1) process to express $S_{ik}, k = 1, 2, \ldots, t$, as:

$$S_{ik} = \rho_i^k S_{i0} + \rho_i^{k-1} \varepsilon_{i1} + \rho_i^{k-2} \varepsilon_{i2} + \cdots + \rho_i \varepsilon_{i,k-1} + \varepsilon_{ik} \text{ for } i = 1, 2 \text{ and } k = 1, 2, \ldots, t.$$

We can then express a_{it} in the 'ex-ante' form in terms of S_{i0} and ε_{ik}, where:

$$a_{it} = a_i + \rho_i \left(\frac{1 - \rho_i^T}{1 - \rho_i} \right) S_{i0} + \left(\frac{1 - \rho_i^T}{1 - \rho_i} \right) \varepsilon_{i1}$$

$$+ \left(\frac{1 - \rho_i^{T-1}}{1 - \rho_i} \right) \varepsilon_{i2} + \cdots + \left(\frac{1 - \rho_i^{T-t+1}}{1 - \rho_i} \right) \varepsilon_{it}.$$

By using the fact that $S_{i0} = 0$, that $\varepsilon_{i1}, \varepsilon_{i2}, \ldots, \varepsilon_{i,T-1}, \varepsilon_{i,T}$ are i.i.d. normal random variables with mean 0 and standard deviation σ_i, and the fact that $E(X^2) = Var(X) + [E(X)]^2$ for any random variable X, one can show that:

$$\pi(t) = \left(\frac{1}{8b} + \frac{1}{8(b+2\delta)} \right)$$

$$\cdot \sum_{i=1}^{2} \sigma_i^2 \left[\left(\frac{1 - \rho_i^T}{1 - \rho_i} \right)^2 + \left(\frac{1 - \rho_i^{T-1}}{1 - \rho_i} \right)^2 + \cdots + \left(\frac{1 - \rho_i^{T-t+1}}{1 - \rho_i} \right)^2 \right]$$

$$+ \frac{(a_1 + a_2 - 2bc)^2}{8b} + \frac{(a_1 - a_2)^2}{8(b+2\delta)}.$$

By expanding the terms in $\pi(t)$, one can show that:

$$\pi(t) = \left(\frac{1}{8b} + \frac{1}{8(b+2\delta)} \right)$$

$$\cdot \sum_{i=1}^{2} \sigma_i^2 \left(\frac{1}{1 - \rho_i} \right)^2 \left[t - 2\rho_i^{T-t+1} \left(\frac{1 - \rho_i^t}{1 - \rho_i} \right) + \rho_i^{2(T-t+1)} \left(\frac{1 - \rho_i^{2t}}{1 - \rho_i^2} \right) \right]$$

$$+ \frac{(a_1 + a_2 - 2bc)^2}{8b} + \frac{(a_1 - a_2)^2}{8(b+2\delta)}.$$

By examining the first- and second-order differences of $\pi(t)$ with respect to t, we can conclude that the expected profit associated with the t-postponement strategy is increasing and concave in t for $0 < \rho_i < 1$ and is increasing linearly in t for $\rho_i = 0$.

Let the value of flexibility, $V(t)$, be the percentage increase in the manufacturer's expected profit associated with the t-postponement strategy over the 0-postponement strategy, where $V(t) = (\pi(t) - \pi(0))/\pi(0)$. For $0 < \rho_i < 1$, $\pi(t)$ is increasing and concave in t and, so, $V(t)$ is also increasing and concave in t. Therefore, delaying the pricing decision is beneficial because it allows the manufacturer to set a more profitable price after observing some market signals for each product. In addition, significant benefit can be obtained by delaying the pricing decisions for a few periods (i.e., when t is relatively small). To illustrate, we set $a_1 = a_2 = 100$, $b = 2$, $c = 5$, $\delta = 3$, $\sigma = 5$, $\rho_1 = \rho_2 = 0.7$, $T = 20$, and we vary t from 1 to 20. As shown in Figure 14.7, the percentage increase in the manufacturer's expected profit associated with the t-postponement strategy over the 0-postponement strategy; i.e., $V(t)$, is increasing and concave in t.

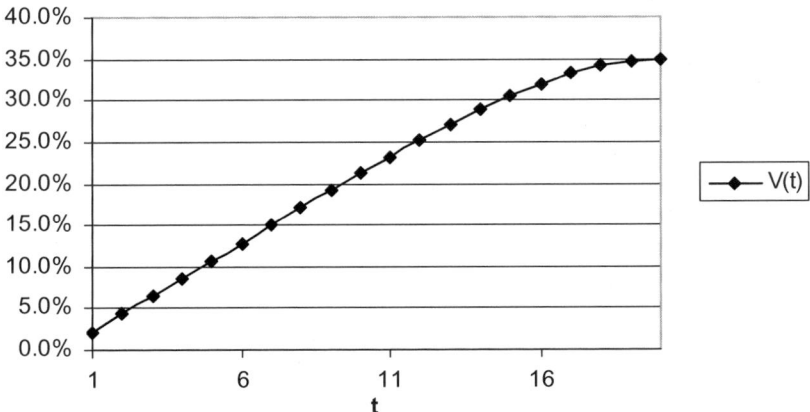

Fig. 14.7 The power of flexibility via responsive pricing

14.7 Conclusions

Following up on the flexible strategies in Chapter 7 as "defensive" mechanisms for mitigating supply chain risks, in this chapter we have shown that a firm can obtain significant benefit by investing in only a small amount of flexibility in its supply chain.

We have examined the benefits of different flexibility strategies in the context of supply chain risk management. By employing stylized models, we have shown analytically that a firm does not need to invest in a high degree of flexibility to mitigate supply, process and demand risks, because most of the benefits are obtained at low levels of flexibility. While some of the stylized models are based on work presented in the literature, our interest is to provide a structure to illustrate the power of flexibility for reducing supply chain risks. We hope our results in this chapter—see also Tang (2006a)—are compelling for implementing some of these flexibility strategies. In many real-life settings, exact cost-benefit analyses of flexibility investments are not feasible due to limitations of data availability. Note that only a limited amount of flexibility is needed to mitigate risk—this should encourage firms to build flexibility into their supply chains. Of course, when implementing a particular strategy in a particular context, a firm needs to establish a structured evaluation process that includes risk identification, risk assessment, decision analysis, mitigation and contingency planning.

Before closing, we recognize that value of the flexibility strategies in Chapter 7 lies not only in risk mitigation as discussed in this chapter but can also be used as "proactive" mechanisms for firms to compete. Consider three successful examples:

One, in the late 80s, Yamaha developed low cost and high quality motorcycles to compete with Honda. To compete, Honda improved its process flexibility so that Honda can introduce new models of motorcycles frequently. This flexibility process strategy enabled Honda to defeat Yamaha in the motorcycle market. The reader is referred to Stalk and Hout (1990) for details.

Two, Zara, the Spanish fashion company, has earned its reputation as the "Fast Fashion" company. Specifically, Zara used the flexible process strategy to speed up the design and the production process so that the company can change its complete fashion collection within 2-3 weeks. Consequently, Zara has become Europe's most profitable fashion company with double digit growth rate annually for the last 10 years. See Ferdows et al. (2004) for details.

Three, flexible pricing strategies via dynamic pricing have revolutionized the airline industry in the 90s. Specifically, when selling limited seats on an airplane with uncertain demand, airlines always adjust their ticket price dynamically so as to meet uncertain demand with limited supply. Cook (1998) reported that dynamic pricing has generated "almost $1 billion of incremental annual revenue" at American Airlines. In the context of e-tailing, dynamic pricing can certainly increase online traffic. For instance, Lands' End Overstock site (http://www.landsend.com) has generated additional traffic after they introduced the "on the counter" event. Specifically, Lands' End puts a new group of products every Saturday for sale at a reduced price. The price of each item is then reduced by 25% if it is not sold by Monday, 50% by Wednesday, and 75% by Friday. With this pre-announced markdown price schedule, many online shoppers would need to monitor the sales of these items so as to time their purchase accordingly. As online traffic increases, the total sales can increase as well.

Chapter 15
Stochastic Programming for Supply Chain Planning Under Demand Uncertainty

Abstract In this chapter we focus on stochastic programming for making optimal supply chain planning decisions under uncertainty. In particular, we extend deterministic linear programming for supply-chain planning (SCP) by using stochastic programming to incorporate the issues of demand risk and liquidity risk. Because the resulting stochastic linear programming model is similar to that of Asset-Liability Management (ALM) and because the literature using stochastic programming for ALM is extensive, we survey various modeling and solution choices developed in this literature and discuss their applicability to supply chain planning. This survey forms a basis for making modeling/solution choices in research and in practice to manage the risks of unmet demand, excess inventory and inadequate cash liquidity when demand is uncertain.

15.1 Introduction

In this chapter we consider stochastic programming for making optimal supply chain planning decisions under uncertainty. In particular, we extend the linear programming (LP) model of deterministic supply-chain planning to take demand uncertainty and cash flows into account for the medium term. Because the resulting stochastic LP model is similar to that of Asset-Liability Management (ALM) and because the literature using stochastic programming is extensive, we survey various modeling and solution choices developed in the ALM literature and discuss their applicability to supply chain planning in this chapter. This survey forms a basis for making modeling/solution choices in research and in practice to manage the risks of unmet demand, excess inventory and inadequate cash liquidity when demand is uncertain.

Companies usually manage supply chain risks either at the strategic (long term) or at the tactical (medium term) level (Sodhi and Lee, 2007). In the medium term—typically 12–26 weeks for consumer electronics companies but up to 24 months for

M.S. Sodhi, C.S. Tang, *Managing Supply Chain Risk*,   259
International Series in Operations Research & Management Science 172,
DOI 10.1007/978-1-4614-3238-8_15, © Springer Science+Business Media, LLC 2012

petrochemical companies—supply-chain planning should incorporate the risks of unmet demand, excess inventory and even liquidity.

We extend deterministic linear programming for supply-chain planning (SCP) by using stochastic programming for demand uncertainty to consider unmet demand and excess inventory and by incorporating cash flows to consider liquidity risk as well. By noting that procurement and production decisions to match supply and demand in (centralized) SCP under different demand scenarios are similar to investment decisions in interest-rate-based securities to match the future cash flows from these securities to the financial institution's future liabilities under different interest-rate scenarios, we look to the ALM literature for modeling perspectives. Indeed, stochastic programming has been extensively used by researchers and practitioners for ALM (Sodhi, 2005b) and we can benefit from the experience of ALM researchers.

Our survey of the ALM (and supply chain) literature from the viewpoint of modeling and solution choices for stochastic programming for SCP prepares the groundwork for researchers and practitioners to consider the use of stochastic programming for managing demand, inventory and liquidity risks associated with SCP for the medium-term. Still, our focus is on stochastic programming and on demand uncertainty: We do not survey heuristics, dynamic programming algorithms, or optimal policies from the inventory management literature. We provide a simple model motivated by a Japanese consumer electronics company for illustration rather than for an instance of a real-life application. We do not address the credit risk of not being able to collect payments from customers, something quite important for liquidity risk. Our survey of the literature, as with the previous Chapters 13 and 14, is indicative rather than exhaustive. Finally, we do not consider decentralized supply chains or adaptation of financial instruments like options that may have application for some of the supply-chain contracts like the QF contracts in Chapter 14.

We motivate the use of stochastic programming as a modeling choice for SCP under uncertainty in Section 15.2. Section 15.3 presents a stochastic program formulation for an idealized supply chain under uncertainty. Section 15.4 presents various modeling choices and Section 15.5 presents solution choices, both motivated by the ALM literature, before the conclusion in Section 15.6.

15.2 Motivation

For the *long* term, SCP with multiple uncertain factors is too broad to tackle with mathematical programming alone and may require scenario planning (Sodhi, 2003). However, modeling solutions have been attempted in at least three ways: (1) through multiple runs of deterministic models (Geoffrion, 1976; Geoffrion and Powers, 1995; Cohen and Lee, 1988), (2) through simulation of deterministic models (Iassinovski et al., 2003), or (3) through stochastic programming (Eppen et al., 1989; Alonso-Ayuso et al., 2003; Swaminathan and Tayur, 1999). By contrast, for the *short* term, operational supply chain risks such as one-off delays in delivery of raw materials are usually dealt with operationally rather than through modeling solutions.

For the *medium* term, the deterministic mathematical programming literature provides models for planning production and distribution based on a single demand forecast. For example, Arntzen et al. (1995) and Camm et al. (1997) developed mixed integer programming models for SCP at Digital Equipment Corporation and for Procter & Gamble, respectively. These deterministic models are solved by using linear programming or heuristics solutions offered by vendors like SAP, i2 and Manugistics (Sodhi, 2001). Typically, a rolling horizon is used so that the forecast as well as the plan are updated every week within the company's enterprise resource planning (ERP) system supported by an advanced planning system (APS) (Sodhi, 2000; 2001). Sometimes, "what-if" forecasts are used in different runs to capture demand uncertainty. However, deterministic modeling even with multiple what-if forecasts is not adequate for managing supply chain risks because such modeling does not take consider the risk pertaining to unmet demand, excess inventory, or inadequate liquidity.

Unmet demand risk (understocking) and inventory risk (overstocking) can be managed by having the "appropriate" level of inventory determined by certain rules, heuristics, or algorithms developed in the inventory-management literature (c.f., Mula et al., 2006; Tarim et al., 2004; and Zipkin 2000). Most stochastic inventory management models focus on determining the optimal ordering and production quantities that minimize the long-term expected cost associated with inventory and unmet demand. However, these stochastic inventory models are inadequate in the sense that they do not consider risks associated with a firm's cash flows for acquiring excessive raw materials when demand is unexpectedly high, and for disposing excessive finished goods (write-offs) when demand is unexpectedly low (as in the case of Cisco's $2 billion write-off in 2001). Researchers have considered multi-echelon inventory theory as a way to analyze SCP under uncertainty (e.g., Porteus 2002; Minner 2003). However, this approach may not be appropriate for dealing with products with short life cycles and it is intractable for complex supply chains with multiple products (Tang 2006a).

Another approach has been to consider uncertainty in the context of *decentralized* supply chains. Starting with the seminal work of Clark and Scarf (1960), researchers have considered, for instance, *contracts* (e.g., Cachon 2003; Martínez-de-Albéniz and Simchi-Levi 2005), *channel coordination* (e.g., Gan, Sethi and Yan 2004, 2005), *incentives* using real options (e.g., Miller and Park 2005; Kleindorfer and Wu 2003; Huchzermeier and Cohen 1996; Kogut and Kulatilaka 1994), *capacity investment* (e.g., Birge 2000; Lederer and Mehta 2005), and *R&D investment* (Huchzermeier and Loch 2001). While these decentralized SCP models are not designed for centralized planning for complex supply chains, they may well have implications for such planning and mathematical modeling.

Stochastic programming can be a useful choice for modeling SCP for the medium term when demand is uncertain. It has been applied to electric power generation, telecommunication network planning, and financial planning; c.f., Sen (2001) for the medium term. While recent advances in computational technology encourage more development in stochastic programming, many advantages of stochastic programming as explained by Birge and Louveaux (1997) have not translated into widespread use in practice except primarily in financial applications like ALM.

Stochastic programming for ALM goes back to the late 1960s and its use has been proposed by, among others, Bradley and Crane (1972), Klaassen (1998), Consigli and Dempster (1998), Dattatreya and Fabozzi (1995), Dert (1999), Sodhi (2005b), Zenios and Ziemba (2005), and Ziemba and Mulvey (1999).

The use of stochastic programming for SCP is showing increasing promise. Dormer et al. (2005) illustrate how Xpress-SP—a stochastic programming suite—can be used to solve certain supply chain management problems. Leung et al. (2005) and Sodhi (2005a) present different stochastic programming models for production planning in a supply chain. However, these models focus on production planning and do not deal with liquidity risk stemming from uncertain demand. As such, we seek to encourage the development of stochastic programming models for managing demand, inventory and liquidity risks.

As further motivation, consider the fulfillment process at the digital video camera division of a Japanese consumer-electronics company (Sodhi, 2005a). The company has several regional offices to deal with regional demand from customers that include electronics retail chains with multiple stores, and to coordinate with the company's headquarters for orders and fulfillment. Every week, each customer provides a weekly "order" of a 26-week rolling horizon at the SKU-level to the designated regional office. Each regional office aggregates its customers' orders by week and then sends the aggregate weekly orders to headquarters. Headquarters allocates supplies to these aggregate orders and ships against the current week's aggregate orders from a central warehouse either to customers' warehouses directly or to regional warehouses for final shipment to customers.

To match demand and supply in this company, the planning problem can be modeled as a multi-period network flow model in which the central warehouse is the "source" node and the customers' and regional offices are the "sink" nodes. Sodhi (2005a) presents a *deterministic* linear programming model as well as a *stochastic* programming extension of this problem that is associated with a single-product, three customers, and a planning horizon of $T = 10$ weeks. The deterministic model has 150 decision variables and 100 constraints while the stochastic model, using 2^{10} ($= 1024$) scenarios, has about 1,500 decision variables and 10,000 constraints. The former solves in a fraction of a second and the latter takes about 10 seconds on the same computer. One benefit of the stochastic programming model is that, even with infinite capacity, the model does not recommend fulfilling all of the customer's orders owing to a high risk of excess inventory. *This underscores our belief that we cannot manage demand and supply risks by simply running deterministic models with a rolling horizon because doing so does not allow us to trade off unmet demand and excessive inventory.*

While stochastic programming is an appropriate modeling choice, the computational requirement is a big challenge. For example, suppose we increase the planning horizon T from 10 weeks to 26 weeks. Then the stochastic program for the single-product example described earlier would have over 100 million decision variables and close to 200 million constraints. Therefore, in order to use stochastic programming as a "practical" modeling choice for SCP under uncertainty, one needs to make modeling choices carefully.

15.3 A Stochastic Program for Supply-Chain Planning

15.3.1 A Stochastic Program for Asset Liability Management

Before we formulate a model for SCP under uncertainty, let us review Asset Liability Management (ALM) and examine a model (A) for ALM with interest rate uncertainty (Sodhi 2005b). Let ζ_t be the interest-rate scenario that ends at time t and the scenarios evolve according to a "binary event tree" over time. There are *six* sets of decision variables, the two operational ones being the amount of security i purchased (x) or sold (y). Each security in the portfolio generates cash in each period under each interest-rate scenario, and this cash can be used to meet the bank's liabilities in that period. Securities not sold are carried over as "inventory," and the total value of all securities in the portfolio at the end of the decision horizon T, discounted to the current time is what is sought to be maximized. These six sets of decision variables are:

x_{i,ζ_t} = Amount purchased of (original) principal of security i
y_{i,ζ_t} = Amount sold of principal of security i
h_{i,ζ_t} = Holdings of principal of security i *after* trades
b_{ζ_t} = Amount borrowed at current short interest rate ρ_{ζ_t}, plus a premium Δ
l_{ζ_t} = Lending amount at the current short interest rate ρ_{ζ_t}
L_{ζ_t} = Liability under scenario ζ_t

The parameters that accompany the above decision variables are:

κ_{i,ζ_t} = Cash generated from the principal and interest per unit of security i
π_{i,ζ_t} = Ex-dividend computed price at time t for principal for each unit of security i (same as market price when $t = 0$)
ρ_{ζ_t} = Single-period interest rate; $1 at time $t \equiv (1 + \rho_{\zeta_t})$ in $t + 1$
δ_{ζ_t} = Present value of a cash flow of $1 at time period t in scenario ζ_t
Δ = Premium paid over the short interest rate for single-period borrowing
T_i = Transaction cost per trade dollar associated with trading of security i
T = Planning horizon
l_{-1} = Single-period lending due in current time period ($t = 0$)
b_{-1} = Single-period borrowing from last period due in current time period
$h_{i,-1}$ = Holding of security i at beginning of current time period

The stochastic program associated with this ALM model can be formulated as problem (A):

$$(A) \quad \max \ 2^{-T} \sum \zeta_T \delta_{\zeta_T} \left\{ \sum_i \left[\pi_{i,\zeta_T} h_{i,\zeta_T} \right] + l_{\zeta_T} - b_{\zeta_T} \right\}, \quad \text{subject to:}$$

$$\sum_i \kappa_{i,\zeta_t} h_{i,\zeta_{t-1}} + l_{\zeta_{t-1}} (1 + \rho_{\zeta_{t-1}}) + b_{\zeta_t} + \sum_i (1 - T_i) \pi_{i,\zeta_t} y_{i,\zeta_t}$$

$$\sum_i (1 + T_i) \pi_{i,\zeta_t} x_{i,\zeta_t} - l_{\zeta_t} - b_{\zeta_{t-1}} (1 + \rho_{\zeta_{t-1}} + \Delta) = L_{\zeta_t} \quad \forall \zeta_t$$

$$h_{i,\zeta_t} + y_{i,\zeta_t} - x_{i,\zeta_t} - h_{i,\zeta_{t-1}} = 0 \quad \forall i, \forall \zeta_t$$

$$x_{i,\zeta_t}, y_{i,\zeta_t}, h_{i,\zeta_t}, l_{\zeta_t}, b_{\zeta_t}, = 0 \quad \forall i, \forall \zeta_t$$

for all $t = 0, \dots, T$ (for all three sets of constraints).

There are two sets of constraints besides the non-negativity ones: cash balance and inventory holdings. The first set specifies the net cash that needs to be raised in order to match the liability L_{ζ_t} under scenario ζ_t. Cash comes from: (a) the holding of security i from period $t-1$; (b) the principal and interest from lending in period $t-1$; (c) the amount borrowed in period t; and (d) the sale of securities in period t. Cash is spent on: (a) purchasing securities in period t (along with transaction costs); (b) lending in period t; and (c) the principal plus interest associated with the amount borrowed in period $t-1$. The second set corresponds to the inventory balancing equation of each security i in period t.

15.3.2 A Stochastic Program for Supply Chain Planning

Consider the electronics company example illustrated in Section 15.2 along with consideration of its cash flows. Take an idealized situation with a manufacturer with a single plant and a single warehouse that faces customer demand. The plant purchases inputs from a supplier, converts these inputs into finished goods, and transports all finished goods to the warehouse. The procurement, conversion, and transportation lead times are known; however, future demand for finished goods is uncertain. Our focus is on the planning of materials and cash flows over a planning horizon T (Fig. 15.1). Material flows comprise inputs from the supplier to the plant and finished goods from the plant to the warehouse and then onto the customers. Information flows are orders from the manufacturer to the supplier. Cash flows include the financial transactions between the bank and the manufacturer including the (delayed) payments from the manufacturer to the supplier and from the customers to the manufacturer.

To formulate this SCP model as a stochastic program S, we need to model uncertain demand by specifying various demand scenarios with certain probabilities. (We shall discuss ways to model other types of uncertainties in Section 15.4.1.) Let ζ_t be a demand scenario for period t and ζ_{t-1} its "parent" scenario so that the evolution of demand scenario over time follows a "binary" event tree (c.f., Luenberger, 1998) in which each scenario ζ_t emerges with two "child" scenarios ζ_{t+1} with equal probability. Thus, the model with T periods has 2^T scenarios. Note that a binary event tree for generating demand scenarios is only one way of representing uncertainty

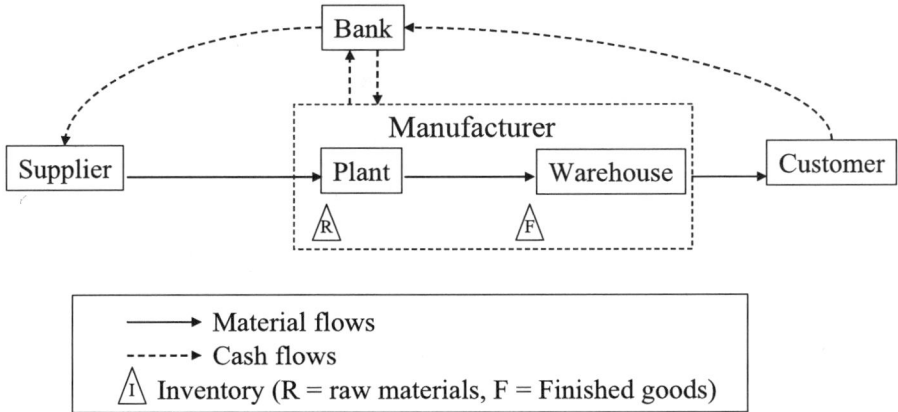

Fig. 15.1 Material and cash flows in a supply chain

with a single source of uncertainty that is motivated by the way uncertain interest rate is modelled in the ALM model. We describe a stochastic process for generating a binary event tree and the demand scenarios such a tree entails in Section 15.3.4.

In the SCP model, there is the planning horizon T as before, and another horizon T_E that exceeds T by at least the sum of the lead times of the two material flows and the lead time for cash settlement on the sell side. The extended horizon enables us to handle the awkward end conditions created by lead times. Our SCP model has *nine* sets of ***decision variables***, the three operational ones being the amount of an input i purchased (x), the amount of a finished product produced (y), and the amount of the product sold (z). These nine set of decision variables are:

$x_{i,t} =$ Amount of input i to be purchased at the beginning of period t (independent of scenario)

$y_{j,t} =$ Amount of product j to be produced at the beginning of period t (independent of scenario)

$z_{j,\zeta_t} =$ Amount of product j to be sold in period t

$h_{i,\zeta_t} =$ Inventory level of input i at the end of the period t

$g_{j,\zeta_t} =$ Inventory level of product j at the end of the period t

$b_{\zeta_t} =$ Amount borrowed in period t at interest rate $\rho_t + \Delta$

$b_{\zeta_t}^p =$ Single-period borrowing over a limit B at a higher rate $\rho_t + \gamma\Delta$, where $\gamma > 1$. (Hence, risk aversion to excessive borrowing is captured by the parameter γ.)

$l_{\zeta_t} =$ Amount of cash in account during period t earning interest at rate ρ_t

$u_{j,\zeta_t} =$ Unmet demand for product j at the end of period t

The first two sets of variables, x and y, represent procurement and production plans, respectively, for the next so many periods. Unlike the other decision variables, these are independent of demand scenario ζ_t because we have to make these plans now for a certain time horizon. Alternatively, we could have modeled these variables as scenario-independent variables up to a certain time horizon and from then on as scenario-dependent. However, there seems little lost by having these variables scenario-independent and much gained by having the number of x and y decision variables grow linearly rather than exponentially with the number of periods. Although procurement and production plans are usually obtained using deterministic modeling in MRP, ERP, or APS technology, the solution produced is different from that obtained by a stochastic model because the latter hedges these decisions for risk that a deterministic model cannot.

Our stochastic programming model has the following **parameters**:

$\kappa_{j,\zeta_t} =$ Selling price for product j(net of any selling costs)
$\pi_{i,\zeta_t} =$ Unit price of input i in period t

Note: Both selling price of product j and purchasing price of input i are assumed to be known functions of demand. Therefore, these two parameters are scenario-dependent (or, we can model these more compactly as node-dependent in relation to the demand event tree).

$\rho_t =$ Single-period interest rate at time t; \$1 at time $t \equiv \$(1+\rho_t)$ at time $t+1$
$\delta_t =$ Present value of a cash flow of \$1 at time period t
$\alpha_{ji} =$ Amount of input i needed to make one unit of product j (i.e., bill of materials)
$\beta =$ Fraction of unmet demand that is backordered for next period, $0 \le \beta \le 1$
$D_{j,\zeta_t} =$ Demand for finished product j in scenario ζ_t
$\Delta =$ Premium paid over the short rate for single-period borrowing; strictly positive
$L_s^M =$ Material-flow lead time for supplier to fulfill order (i.e., supply lead time)
$L_p^M =$ Material-flow lead time for manufacturer to convert raw materials to finished goods (i.e., production lead time)
$L_s^C =$ Cash-flow lead time for manufacturer to pay supplier following receipt of supplies (i.e., account payable lead time)
$L_p^C =$ Cash-flow lead time for customer to pay manufacturer following sale of finished goods (i.e., account receivable lead time)
$T =$ Planning horizon (with periods in weeks or months)
$T_E =$ Extended horizon with $T_E \ge T + L_s^M + L_p^M + L_p^C$
$B =$ Borrowing limit
$C_t =$ Joint production capacity across all finished products in period t.

We also need parameters to reflect the inventories at hand or already "in the pipeline" for the coming weeks. These **initial values** are denoted by the following:

$l_{-1} =$ Initial amount of cash in account at the beginning of current time period ($t=0$)
$b_{-1} =$ Amount borrowed in the last period that is due in the current time period
$h_{i,-1} =$ Inventory level of input i at the end of last period
$g_{j,-1} =$ Inventory level of product j at the end of last period

$x_{i,-L_s^M}, \ldots, x_{i,-1} =$ Outstanding orders of input i (to be received)

$y_{j,-L_p^M}, \ldots, y_{j,-1} =$ Work in process inventory of product j (to be completed)

$z_{j,-L_p^C}, \ldots, z_{j,-1} =$ Amounts (in units) of product j in corresponding to accounts receivable.

There are 2^T scenarios for the binary event tree that we have chosen as our way of representing demand uncertainty. Each scenario ζ_t occurs with probability 2^{-T} and the expected present value of the net cash at period T is $2^{-T} \delta_T \sum \zeta_T [l_{\zeta_T} - b_{\zeta_T}]$. Suppose our objective is to maximize this quantity (there may be other objective functions as we discuss later in Section 15.4.4). Then we can formulate this SCP model as the following stochastic program (S):

$$(S) \qquad \max 2^{-T} \delta_T \sum_{\zeta_T} [l_{\zeta_T} - b_{\zeta_T}], \qquad \text{subject to:} \qquad (15.1)$$

$$\sum_j \kappa_{j,\zeta_{t-L_p^C}} z_{j,\zeta_{t-L_p^C}} + l_{\zeta_{t-1}} (1 + \rho_{t-1})$$

$$+ b_{\zeta_t} + b_{\zeta_t}^p - \sum_i \pi_{i,\zeta_{t-L_s^C}} x_{i,\zeta_{t-L_s^C}}$$

$$-l_{\zeta_t} - b_{\zeta_{t-1}} (1 + \rho_{t-1} + \Delta) - b_{\zeta_{t-1}}^p (1 + \rho_{t-1} + \gamma\Delta) = 0 \quad \forall \zeta_t \qquad (15.2)$$

$$h_{i,\zeta_t} + \sum_j y_{j,t} \alpha_{ji} - x_{i,t-L_s^M} - h_{i,\zeta_{t-1}} = 0 \quad \forall i, \forall \zeta_t \qquad (15.3)$$

$$g_{j,\zeta_t} + z_{j,\zeta_t} - y_{j,t-L_p^M} - g_{j,\zeta_{t-1}} = 0 \quad \forall j, \forall \zeta_t \qquad (15.4)$$

$$z_{j,\zeta_t} + u_{j,\zeta_t} - \beta u_{j,\zeta_{t-1}} = D_{j,\zeta_t} \quad \forall j, \forall \zeta_t \qquad (15.5)$$

$$b_{\zeta_t} \leq B \quad \forall \zeta_t \qquad (15.6)$$

$$\sum_j y_{j,t} \leq C_t \quad \forall t \qquad (15.7)$$

$$x_{i,t}, y_{j,t}, z_{j,\zeta_t}, h_{i,\zeta_t}, g_{j,\zeta_t}, l_{\zeta_t}, b_{\zeta_t}, b_{\zeta_t}^p, u_{j,\zeta_t} \geq 0 \quad \forall j, \forall i, \forall \zeta_t \qquad (15.8)$$

for all $t = 0, \ldots, T_E$ for all constraints (15.2)–(15.8).

Note that the objective function discounts the value of cash at time T while the time periods in the constraints go up to the extended horizon T_E. Constraints (15.2) represents the balance of cash flows for each demand scenario as follows similar to the constraints for ALM. Cash flows in from accounts receivable based on the sale of the $z_{j,\zeta_{t-L_p^C}}$ units of product j earlier due in period t; from the principal and interest associated with the cash in account in period $t - 1$; and from amounts borrowed at the base and at the higher rate in term t. Cash flows out to the accounts payable of the $x_{i,t-L_s^C}$ units of input i purchased earlier that is due in period t; to the cash in account in period t; and to the principal and interest associated with the amounts borrowed in period $t - 1$ at the two different rates. As with the ALM model, constraints (15.3) and (15.4) correspond to inventory balancing at the beginning and at the end of each period in each scenario: in this case we have inputs i and product j, respectively instead of securities. The remaining constraints are different from those in the ALM model. Constraint (15.5) ensures that sales do not exceed demand (plus the fraction of backordered demand). Constraint (15.6) specifies the "soft" borrowing limit B

above which it is more expensive to borrow. Constraint 15.7) specifies the joint production capacity constraint.

Notice that stochastic program (S) is feasible and bounded. Feasibility can be established by setting the decision variables x and y to zero for all t. Boundedness can be established by noting that cash is generated only from sales, which are limited by the demand. Moreover, production is limited by the joint production capacity.

15.3.3 Comparison of SCP Model with the ALM Model

The two models (A) and (S) presented in Section 15.3.1 and 15.3.2 are quite similar. The model (A) has interest-rate scenarios and (S) has demand scenarios. Both have "material" flows: (S) has physical units being ordered or produced while (A) has the number of securities being purchased or sold. As already indicated, the constraints (15.2)–(15.4) for balancing cash and inventory in (S) have their counterpart for (A). We can impose a soft limit B on single-period borrowing (i.e., constraint (15.6) for program (A) as well.

However, the backorder constraint (15.5) and the joint-production capacity constraint (15.7) are relevant to SCP only. Moreover, the variables for SCP are slightly different in the sense that the purchase decision vector x_i and the production decision vector y_j in (S) do not depend on the scenarios owing to lead times; however, the buy/sell decision of each security i in (A) is scenario-specific for ALM. Hence, the number of decision variables associated with x and y in program (S) grows only linearly, while the number of decision variables grows exponentially with the number of periods for the ALM model. This makes it easier to solve (S).

A fundamental difference lies in the generation of scenarios. In ALM, the scenarios for interest rate should, as a set, (1) match the yield curve, (2) provide the same period-on-period returns for all securities to prevent arbitrage, and (3) discounted cash flows of securities should match market prices for a chosen set of securities like treasuries. Such a restriction does not apply to (S). However, this difference applies only to the generation of scenarios and not to the models themselves or to solution techniques for these models.

There are operational differences that do affect modeling. For ALM, the volume of buying and selling is unrestricted (at least in the model A) whereas for SCP, selling is limited by the demand. Production capacity or joint-production capacity in (S) is another difference as is conversion from inputs to finished product. Finally, there are lead times for material and cash flows in (S)—such lead times do not have an equivalent in ALM as buy/sell decisions of securities are immediately executed.

Overall, the programs (S) and (A) are similar in terms of types of decision variables, types of constraints, and the coefficient matrix structure of the constraints. Therefore, we can discuss the applicability of modeling and solution choices for ALM to SCP.

15.3.4 Demand-Scenario Generation: An Example

For the SCP model (S), the demand for product j in period t under scenario ζ_t (i.e., D_{j,ζ_t}) can be generated according to a binary event tree (i.e., each scenario ζ_t in period t has two child scenarios in period $t+1$) as follows. Consider modifying an autoregressive model of first order, i.e., AR(1), in the following manner (Sodhi 2005a): Given an unbiased forecast $\mu_j[t]$, the $D_{i,\zeta}$ follows the stochastic process $D_{j,\zeta_0} - \mu_j[0] = 0$, and

$$D_{j,\zeta_t} - \mu_j[t] = \theta_j(D_{j,\zeta_{t-1}} - \mu_j[t-1]) + \varepsilon_t$$

for $t \geq 1$, where θ_j, $|\theta_j| \leq 1$, is the product-dependent auto-correlation coefficient. Notice that the above process for $D_{j,\zeta}$ will be reduced to the standard AR(1) process when $\mu_j[t]$ is a constant over time. However, to model certain effects such as season-related effect or product-lifecycle effect, we can make $\mu_j[t]$ depend on time. To be consistent with the findings of Lee, So, and Tang (2000) regarding the consumer-packaged goods and the electronics industries, we can assume $\theta_j > 0$. The error ε_t is i.i.d. across time periods and equals v or $-v$ with equal probability for some positive scalar v; hence, $E(\varepsilon_t) = 0$ and $\mathrm{Var}(\varepsilon_t) = v^2/4$ for all t. We shall discuss other mechanisms in the literature for generating demand scenarios in Section 15.4.1.

15.3.5 Risk Measures Consideration

Besides the optimal solution to program (S), the decision maker should evaluate the goodness of the solution by considering various risk measures. We propose three risk measures pertaining to SCP under uncertainty, adapting the Value-at-Risk (VaR) measures from financial risk management. The first risk measure is called Demand-at-Risk (DaR)—a measure of unmet demand at any given period t. For example, by considering the distribution of the unmet demand of product j in period t (i.e., u_{j,ζ_t}), let $\mathrm{DaR}_t(p)$ be the "critical value" that satisfies: $\mathrm{Prob}\{u_{j,\zeta_t} > \mathrm{DaR}_t(p)\} = p$; i.e., there is a probability p that the unmet consumer demand u_{j,ζ_t} exceed $\mathrm{DaR}_t(p)$ in period t. Using a similar setup, we can create a measure called Inventory-at-Risk (IaR) that measures the inventory level g_{j,ζ_t} exceeding a "certain threshold" in any given period (i.e., $\mathrm{Prob}\{g_{j,\zeta_t} > \mathrm{IaR}_t(p)\} = p$.) Finally, we can define a measure called Borrowing-at-Risk (BaR) that measures $b_{\zeta_t}^p$ (the extent of excessive borrowing above the limit B); i.e., $\mathrm{Prob}\{b_{\zeta_t}^p > \mathrm{BaR}_t(p)\} = p$.

These three risk measures can be computed from the output of our stochastic program (S) because we know the probability distributions unmet demand, excess inventory and liquidity for any time period corresponding to the optimal solution. Note that we can reduce the DaR at the expense of increasing IaR or BaR or vice versa. So, one needs to balance these risk measures. However, all three risk measures can be reduced if uncertainty or the lead times were to be reduced.

15.4 Modeling Choices

To model SCP, we look for guidance in the ALM literature for modeling choices including: representation of uncertainty, time periods and length of the decision horizon, the objective function, and constraints. Modeling choices are important not only to convince managers of the efficacy of the model but also for solution considerations.

15.4.1 Representation of Uncertainty

The SCP model (S) relies on an auxiliary demand model to provide demand scenarios that evolve in the form of a binary event tree with either of any pair of child nodes being equally probable. Only demand uncertainty is taken into account similar to the ALM model (A) where only interest-rate uncertainty is considered. However, our model can incorporate other uncertain factors if we can handle a (much) larger number of scenarios computationally. Scenarios can be used to capture one or more sources of uncertainty. We have considered demand scenarios with only one underlying source of uncertainty in (S) but we could alternatively consider correlated demand of multiple products, uncertain production yields, unreliable delivery times and/or uncertain supply, at least in principle, using a (very) large number of scenarios.

The binary event tree, as described in Section 15.3.4, is a simple way for generating demand scenarios. However, in the ALM model, the binary event tree for interest rate uncertainty is built using the yield curve and the prices of treasury options by assuming a "risk-neutral" world with "risk-neutral" probabilities as opposed to objective or subjective ones. All 2^T scenarios are included in the ALM model so that the number of periods T cannot be very large in practice (c.f., Black et al., 1990; Ho and Lee, 1986; and Heath et al., 1990).

To model demand uncertainty, we need to first determine the underlying factors that affect product demand (e.g., market conditions, product novelty, etc.) and then use an appropriate stochastic process to generate scenarios for the demand of multiple products for multiple time periods. To do so, we can adopt some of the research development in ALM. Specifically, for interest-rate scenario generation using multiple factors, Mulvey (1996) uses the two-factor model of Brennan and Schwartz to generate a sample of interest-rate scenarios. Hull (2009) discusses different interest-rate models using both one- and two-factor. Bradley and Crane's (1972) generate interest-rate scenarios by using an n-ary tree with any number of underlying factors.

While our current model, as depicted in program (S), assumes that the demands for different products in each time period are independent of each other, we may need to construct more realistic scenarios in which these demands are correlated. ALM researchers have created event trees that use the covariance between across asset classes such as stocks, bonds and real estate although not with individual assets. Kouwenberg (2001) and Gaivorinski and de Lange (2000) create event trees for the ALM model with selected asset *classes* (stocks, bonds, real estate, etc.). They match the means and the covariance matrix of the joint continuous probabilities obtained from history (or future expectations) of returns to those obtainable from the event tree as recommended by Hoyland and Wallace (2001) by solving a nonlinear problem that penalizes differences in the two sets of means and covariance matrices. Consigli and Dempster (1998) use UK data from 1924–1991 comprising third-order autoregressive equations to generate annuals returns (and hence probability distributions) for similar asset classes: ordinary shares, fixed-interest irredeemable bonds, bank deposits, index-linked securities, and real estate. Dert (1999) follows a similar approach for Dutch pension funds with a "vector" autoregressive model for wage inflation, price inflation, cash, stocks, property, bonds, and GNP. Pflug et al. (2000) use principal component analysis on historical data to extract factors of uncertainty that drive asset returns and interest rates and use these to create scenario trees by matching statistical properties. Mulvey and Shetty (2004) generate ALM scenarios with multiple economic factors.

15.4.2 Decision variables

Our SCP model as formulated as program (S) has scenario-specific decision variables z, h, g, l, and b that correspond to the current scenario ζ_0 at $t = 0$ and to the future scenarios ζ_t, $0 < t \leq T_E$. However, in contrast to the ALM model, the planning decision variables x and y can be modeled as scenario-independent as explained earlier. If there are specific *time-fence* restrictions on revising the planned values in subsequent periods, these restrictions can be easily incorporated into our model (S) using initial values for x and y up to the time fence.

15.4.3 Time Periods and the Decision Horizon

Our model is motivated by a SCP problem arising from such industries as consumer electronics or consumer-packaged goods where it makes sense to consider time periods in weeks. If the time periods are "too long", we will have fewer time periods but will lose the granularity needed for operations. If the time periods are "too short", then the number of time periods and consequently the number of scenarios, constraints, and variables will increase, impacting solution tractability. For instance, when $T = 6$ periods, the number of scenarios is only 128 but when $T = 26$ periods, the number of scenarios exceeds 67 million. Therefore, we need to consider the length of the time-interval carefully.

In the ALM literature, some researchers have proposed the use of successive time periods of increasing length to reduce the number of periods T (e.g., Cariño et al., 1994; 1998ab). The idea of increasing length of time periods successively is supported by commercial supply chain software such as that from SAP or i2 as is the case with commercial ALM software, e.g., that from PeopleSoft, now part of Oracle. For a period of six months in our SCP model, we could have $T = 6$ with two periods of one week, one period of a fortnight, two periods of one month each, and one period of one quarter although doing so makes things trickier in a rolling horizon model with lead times. Therefore we may consider 20–30 equal periods of weeks in the electronics industry and months in the petrochemical industry.

15.4.4 Objective function

The objective of our stochastic program (S) is to maximize the expected present value of the net cash (cash minus borrowed amount) in period T. We can easily include the value of the assets in period T (as in the ALM model) by adding the asset value of the inputs and finished goods inventory at liquidation prices as well as the present value of accounts receivable.

Besides maximizing expected present value, *minimizing cost* has been used in many deterministic tactical SCP models with penalties for not meeting demand and costs for holding inventory. When the purchasing price for inputs and the selling price for finished goods are *not* available (unlike the assumption for (S)), we can change the objective function in our model (S) from maximizing the expected present value of net cash to minimizing the total penalty associated with over-stocking and under-stocking. This situation occurs in the ALM model as well. While it is common to maximize the expected present value of net cash position, Dert (1999) has considered a different ALM model in which the goal is to minimize the expected costs of funding a defined-benefit pension fund.

Minimizing the probability of an adverse event—running out of cash, not meeting demand, or having excessive inventory—has been captured in our risk measures such as BaR, DaR, and IaR as defined in the previous section. Similar risk measures have been proposed for ALM and other financial models as a way to control risk. For tactical supply chain management this could translate into the likelihood of more than a certain quantity of backorders or surplus respectively (refer to chance constraints below as well).

For model (S), the linear objective function is justified when the decision maker is actually risk-neutral or when his risk-aversion is captured implicitly by penalizing any borrowings above a threshold value B (or a series of such thresholds with increasingly higher penalties) from the expected present value. Modeling risk aversion explicitly is challenging. One objective function is to *maximize risk-averse utility* $u(\cdot)$ as a function of the *subjective* probability distribution $f(\cdot)$ of "wealth" over terminal scenarios ζ_T (i.e., w_{ζ_T}). In this case, the objective function becomes: $\max u(f(w_{\zeta_T}))$. A simplified version is the expected utility function $\sum_{\zeta_T} \phi_{\zeta_T} u_1(w_{\zeta_T})$ where ϕ_{ζ_T} is the probability of the scenario (Klaassen, 1997). Any utility-based objective entails nonlinear functions, e.g., expected log value of excess horizon return (Worzel et al., 1994). Kallberg and Ziemba (1983) discuss the relevance of different forms of concave utility functions and show that they all give similar results if the average risk aversion is the same. Others attempted to avoid nonlinearity by adding constraints to model risk-aversion (Bradley and Crane 1972; Kusy and Ziemba 1986; Cariño, et al., 1994, 1998a, 1998b). Such constraints—soft or hard—can be used with a linear objective function (e.g., Kusy and Ziemba (1986).

15.4.5 Constraints

One modeling choice consideration is how to use (S) to generate output that can be used as input or guidance for a more detailed but deterministic model. Thus, we can argue for a "simple" model (S) that suppresses many practical constraints but incorporating uncertainty, and then accompany this simple model by examining a detailed model with all the constraints but without incorporating uncertainty.

For stochastic programming models, the so-called *non-anticipatory constraints* arise from the requirement that for scenarios sharing past events up to a certain time, the decisions must be the same at this time (e.g., Birge and Louveaux, 1997). In (S), we have used scenarios in the form of a scenario tree so the non-anticipatory constraints are implicitly embedded in the event tree.

In the development of the ALM model, some researchers (e.g., Dert, 1999; and Drijver et al., 2000) use chance constraints to limit the probability of not meeting the liabilities, i.e., under-funding the pension fund. For SCP, not meeting the liabilities is analogous to not meeting demand. However, according to Kusy and Ziemba (1986), such ALM models with chance constraints tend to have theoretical problems in multi-period situations. Therefore, one needs to be cautious about imposing additional chance constraints in our SCP model.

15.5 Solution-Technique Choices

Our SCP model (S) is simple to understand. However, solving it directly is computationally challenging because it has more than 100 million constraints and a similar number of decision variables for a single-product SCP problem with a 26-period planning horizon. Still, for a single product, the distribution-only model (i.e., no production) solves in only about 5 seconds for a 10-period problem on a 1.2 GHz Pentium laptop with 512 Meg memory using XPRESS-MP^{TM} software (Sodhi 2005a). This observation suggests that there is hope to solve such models directly, possibly taking into account its special network structure and using solution methods as proposed, for instance, by Mulvey and Vladimirou (1992). In addition to seeking direct methods for solving (S), we should consider other solution techniques including decomposition, sampling of scenarios, aggregation, and combined simulation/optimization methods that have been proven to be effective for solving program (A) for the ALM model. Because (S) has a coefficient matrix similar to that of program (A) for the ALM model, the solution techniques for (A) may be effective for solving (S).

15.5.1 Decomposition

When the number of inputs and/or the number of finished goods are large, our stochastic program (S) could become intractable. However, one can always decompose the joint-production capacity constraint (7) using Benders decomposition by finished product (c.f., Benders 1962). This decomposition approach will enable us to decompose problem (S) into sub-problems, each of which corresponds to a single finished product j. As noted in Sodhi (2005a), this single-product sub-problem can be solved efficiently. In this case, we could think of using this heuristic decomposition approach as a practical way to manage demand risk in SCP. When solving the stochastic program (A) for the ALM model, Benders decomposition has generated some promising results: Gondzio and Kouwenberg (2001) use Benders decomposition and an efficient model generator to solve problems arising from a Dutch pension funds application with $T = 6$ and up to 13^6 (≥ 4.8 million) scenarios on a parallel computer with 16 processors. Still, to temper our optimism, Mulvey and Shetty (2004) describe challenges in solving for even a modest number of scenarios (4096) on a 128-processor machine; they describe the use of interior-point and parallel Cholesky methods as well as new ways to reduce the number of floating point operations. Dert (1999) refers to an iterative heuristic for a chance-constrained ALM model that tackles only one or two time periods per iteration. Thus, despite the merits of decomposition, the large number of scenarios means that researchers have to approximate uncertainty in some way for optimal solution.

Besides using Benders decomposition for solving (S), we should consider applying some of the following decomposition schemes that have been proven to be effective for solving program (A). For example, Bradley and Crane (1972) obtain sub-problems using decomposition that can be solved efficiently. Kusy and Ziemba (1986) use the algorithm developed by Wets (1983) for a stochastic linear program (LP) with fixed recourse, restricting uncertainty to that of deposit flows. Birge (1982) provides a solution method to tackle the large size of multistage stochastic LP's in general. Building upon the work of Kusy and Ziemba (1986) and that of Birge (1982), Cariño et al. (1994, 1998ab) use Bender's decomposition to solve problems that have up to 6 periods, 7 asset classes, and 256 scenarios for a Japanese insurance company. Mulvey and Vladimirou (1992) adopt a generalized network structure for a multi-period asset allocation problem with different classes of bonds, equities, and real estate. They solve problems up to 8 periods, 15 asset classes, and 100 scenarios using the progressive hedging algorithm developed by Rockafellar and Wets (1991).

There are other ways to decompose the problem especially for parallel computation. One approach is to relax the afore-mentioned non-anticipatory constraints so that we can treat the scenarios independently from each other and thus decompose the stochastic model. All the scenarios can be solved in parallel if a parallel computer were available. This approach enables one to exploit parallel computing for solving multi-stage stochastic problems as in program (S). Nielsen and Zenios (1993) report significant savings in computational effort by using parallel algorithms for solving stochastic programs developed by others including Rockafellar and Wets (1991) and Mulvey and Ruszcynski (1995).

15.5.2 Aggregation

Recall from Section 15.4 that the number of demand scenarios generated from a binary event tree grows exponentially in the number of periods. As a way to reduce the number of scenarios, aggregation appears to be a reasonable way to approximate the uncertainty itself. In SCP models, it is common to aggregate finished goods with similar demand patterns, production lead times, or product costs, etc. as an "aggregated product" when dealing with thousands of finished goods. Hence, we can consider using aggregation as a way to reduce the size of our stochastic program (S). Besides aggregating finished goods, one may consider aggregating customers with similar demand patterns or suppliers with similar supply costs or lead times. Two innovative aggregation schemes for reducing the size of stochastic program (A) for the ALM model have been proposed and analyzed by Klaassen (1998). These two aggregation schemes call for aggregating different scenarios into an "aggregated scenario" and for aggregating different time periods into an "aggregated time period." Owing to the similarity between stochastic programs (S) and (A), it is likely these aggregation schemes are applicable to our SCP model.

Birge (1985) obtains a single aggregated scenario by aggregating the rows and columns of the LP coefficient matrix for the stochastic model using weights corresponding to the probability distribution of the random variables to determine error bounds for the expected value problem. His aggregation scheme and the corresponding results apply to the case when only the right hand side of the constraints is stochastic. This aggregation may be applicable to our SCP problem when only the demand D_{j,ζ_t} of each product j on the right hand side of constraint (5) is stochastic. For the case when all coefficients in the constraint matrix are stochastic as well, one may consider using the aggregation technique developed by Wright (1994) that generalizes Birge's (1985) work. Moreover, our stochastic program (S) for SCP under demand uncertainty is essentially a stochastic LP that has a finite number of (discrete) scenarios and is therefore an ordinary LP. Therefore, we can aggregate scenarios by aggregating columns first (Zipkin, 1980a) and then aggregating rows later (Zipkin, 1980b). See Sodhi and Tang (2011) for such an aggregation to support a sales-and-operations planning process.

15.5.3 Sampling Scenarios

The restrictions that apply to the set of scenarios for ALM do not apply to the set of scenarios for SCP as we explained earlier. Therefore, we do not need to use all scenarios from a binary event tree for our SCP model. For instance, we could use simulation to generate a sample of scenarios using one or more factors of uncertainty. However, we could get poor quality solutions depending on how the demand scenarios were generated (Kouwenberg, 2001). Moreover, the solution in each run can be quite different from the previous one with the same input because the number of randomly-generated scenarios may be only a tiny percentage of a large population of very diverse scenarios. This makes comparison of different runs with different inputs (e.g., forecasts) difficult. To reduce the variance of scenarios generated in different runs, Mulvey and Thorlacius (1999) among others use antithetic sampling scenarios for different interest rates, exchange rates, stock returns, and inflation rates.

Besides sampling scenarios, one can combine simulation and optimization to get a better representation of uncertainty (e.g., Seshadri et al., 1998; Sen, 2001).

15.6 Conclusion

We have presented a stochastic programming formulation for a simple SCP model that involves the purchase of inputs from a supplier, their conversion to finished goods at a single plant, and the eventual stocking and selling of these goods from a single warehouse facing uncertain demand. We have also compared the models in these two domains, drawing out similarities and differences. We have argued that the ALM literature leads the SCP literature in its use of stochastic programming and is therefore a useful guide for researchers and practitioners wishing to develop stochastic programming models for SCP.

One conclusion from this survey is that no matter how fast computers become or how much solver technology improves over time, stochastic programming will remain as much of an art as it is a science. As such, modeling choices and solution choices have to be made carefully, matching these choices to the objectives and being able to compromise on the objectives. We hope our survey will help researchers and practitioners to do so.

There is also a cautionary lesson we need to draw from the ALM literature. In practice, there is much emphasis on Monte Carlo simulation to "stress test" an existing or a proposed ALM portfolio. Regardless of the sophistication of scenarios, the simulation-only approach is simplistic compared to dynamic stochastic optimization. Still, it may satisfy managers because simulation is more easily understood by them than stochastic optimization. It is quite conceivable then that for tactical SCP practice too, popular risk management software in the future might use Monte Carlo simulation with the goal being a report on the projected risk performance of a particular procurement and production plan under a variety of what-if scenarios. These scenarios could be computer-generated or hand-crafted to consider demand, inventory, supplier lead-time, and capacity variations—see Nagali et al. (2008) for a procurement risk application at Hewlett-Packard.

In light of such simulation-based approaches, selling the superiority of optimization-based risk-management approaches to managers can be a challenge. One solution may be simple stochastic programming models that can work hand-in-hand with deterministic models in extended ERP/APS systems (e.g., SAP system with Advanced Planning and Optimization or i2's Supply Chain Planner) to provide risk-adjusted plans—see Sodhi and Tang (2011) for an example of this approach.

Part IV
Perspectives and Topics for Future Research

Chapter 16
Researchers' Perspectives on Supply-Chain Risk Research

Abstract Supply chain risk management is a nascent area with researchers who have approached the area from different domains and can therefore be expected to be diverse in how they perceive the scope and the appropriateness of different research tools. This chapter presents our study of this diversity among operations and supply chain management scholars that involves literature review and surveys. Our findings characterize the diversity in terms of three "gaps": a *definition gap* in how researchers understand supply-chain risk management, a *process gap* in terms of inadequate coverage of response to risk incidents, and a *methodology gap* in terms of inadequate use of empirical methods. We also list ways to close these gaps as suggested by the researchers we surveyed.

16.1 Introduction

This concluding part of the book—Part IV—focuses on future research in two different ways. First, in this chapter we report the perspectives of supply-chain researchers regarding supply-chain risk management and various "gaps" in the literature. In the next and final chapter of the book (Chapter 17), we provide our own perspective based on the preceding chapters of this book.

Supply chain risk management (SCRM) is a nascent area with researchers who have approached the area from different domains and who can therefore be expected to be diverse in how they perceive the scope of SCRM and the appropriateness of different research tools. This chapter presents our study of this diversity among operations and supply chain management scholar that we carried out in three steps: First, we reviewed the researchers' output, i.e., the recent research literature. Next, we surveyed two focused groups (members of Supply Chain Thought Leaders and International Supply Chain Risk Management groups) with open-ended questions. Finally, we surveyed operations and supply chain management researchers during the 2009 INFORMS meeting in San Diego. Our findings characterize the diversity in terms of three "gaps": a definition gap in how researchers define SCRM, a process gap in terms of inadequate coverage of response to risk incidents, and a methodology gap in terms of inadequate use of empirical methods. We also list ways to close these gaps as suggested by the researchers.

As company executives report increased concerns about the rise of supply chain risk, becomes more attractive as a research area to academics who wish to contribute to business. On the other hand, the area is still emerging and has rather unclear boundaries at this stage, leading to questions about diversity among researchers in terms of the scope of SCRM. Moreover, with researchers having different domain expertise, questions naturally arise about the diversity of research tools and their appropriateness. As such, as we seek a basis for future research in this field, we need to characterize the diversity of scope and research tools in the researchers' perception of SCRM. Such a characterization can provide a basis for collaboration among the researchers themselves and with industry as regards future research.

The fact that SCRM is still at a nascent stage makes it appealing to conduct a *field research* study of researchers in this area. Field research is a well-established research method in the management literature especially for new research areas that require exploration (Eisenhardt, 1989; Yin, 2009). In the operations management literature, Meredith (1998), Voss et al. (2002), and Seuring (2005) have argued that field research study is an appropriate approach for conduct exploratory investigations of new operations management topics that are not well defined or understood—certainly SCRM fits that description. Jehn, Northcraft and Neale (1999) have used a multi-method field study method to explore diversity in workgroups—in our case we wish to explore diversity among researchers so we too decided to use a multi-method field study.

Adapting the methodology presented by Burgess (1984) and Voss et al. (2002), we first carried out direct observations to make our perceptions more concrete, then gathered some evidence through surveys of focus groups, and finally sought confirmation and additional information through a survey. Thus, we employ the three methods—"participant observation, informant interviewing, and enumeration (sampling)"—advocated by Zelditch (1962) for field research as follows:

1. First, we obtained direct observations of diversity in the output of SCRM researchers by reviewing some recent research literature so as to formulate our own perception of diversity in scope and research tools.
2. Second, we conducted open-ended surveys of two focus groups of supply chain researchers—supply chain management researchers at the *2008 Supply Chain Thought Leaders (SCTL) Conference* in Madrid, Spain[1], and risk management scholars at the *2008 International Supply Chain Risk Management (ISCRiM) Conference* in Trondheim, Norway.[2]

---

[1] Attendees include supply chain management researchers from Columbia University, Carnegie-Mellon University, Dartmouth College, Duke University, Georgia Tech, Harvard University, IN-SEAD, Instituto de Empressa, MIT, MIT-Zaragoza Logistics Institute, Stanford University, UCLA, University of Michigan, University of North Carolina at Chapel Hill, University of Pennsylvania, Washington University in St. Louis, and Waseda University.

[2] Attendees include risk management scholars from Bowling Green State University, Cranfield University, ETH Center for Enterprise Sciences (Zurich) European Business School, Lappeenranta

3. Finally, to obtain confirmation and additional information, we surveyed a broad-based group of 200-plus researchers who attended our keynote speech on SCRM during the *2009 Institute of Operations Research and Management Science (INFORMS) National Meeting* in San Diego.[3]

Our findings characterize the diversity of scope and research tools as three research "gaps" in SCRM that can provide a general direction for future research: (1) *a definition gap*—there is no clear consensus on the definition of SCRM (because some limit the scope of SCRM to rare but large impact events while others believe that SCRM is about demand-supply uncertainties); (2) *a process gap*—there is a lack of research on an important aspect of the risk management process, namely, the response to supply chain risk incidents; and (3) *a methodology gap*—there is a shortage of empirical research in the area of SCRM. The researchers surveyed in the third step of our study also provided initial answers on how to close these gaps.

This chapter characterizes the diversity of researchers' perspectives in SCRM, thus creating a basis for researchers to collaborate with each other and with industry. However, there are limitations of our study and of our approach, in particular about not having studied moderating effects such as editorial policies or particular topical and methodological interests of research journals—see for instance, Sodhi and Tang (2008) in the context of operations research. Still, we hope this chapter will provide a basis for future research not only for researchers but also for journal editors.

The rest of this chapter is organized as follows: we provide some background on the growing interest in supply-chain risk in Section 16.2, recalling some of the points from earlier chapters in this book, and the motivation for our research in Section 16.3. Section 16.4 presents our methodology. In Sections 16.5, 16.6 and 16.7, we present our results regarding each of the three steps we undertook respectively: a study of the research output, focus groups, and formal survey. In Section 16.8, we provide suggestions made by the surveyed researchers on how to close these gaps before concluding in Section 16.9.

16.2 Growing Interest in Supply Chain Risk

Since the early 1990s, many firms have implemented various supply chain initiatives to increase revenue, to reduce costs, and/or to reduce assets. However, to meet these goals, most supply chains became more complex and consequently more vulnerable to disruptions than they were before (Craighead et al., 2007). To reduce vulnerability, there have been calls for "resilience" (cf. Sheffi, 2005 a and b) or "robustness" (cf. Tang, 2006a).

University of Technology, Loughborough University, Lund University, MIT, Norwegian University of Science and Technology, Nottingham University, Sabanci University, Swiss Federal Institute of Technology (Zurich), UCLA, University of Central Lancashire, and Western Carolina University. See http://www.uclan.ac.uk/lbs/research/iscrim.php for details.

[3] Abstract of the keynote speech is available at: http://meetings.informs.org/sandiego09/plenaries .html.

There are many examples of significant supply chain disruptions that we provided in earlier chapters. To recapitulate some of them, Mattel recalled 19 million toys due to lead paint or loose magnets in 2007, in 2006, due to a fire hazard, Dell recalled 4 million laptop computer batteries made by Sony, Ericsson reported year-end losses of $2.34 billion for the mobile phone division after its supplier's semiconductor plant caught on fire in 2000, Land Rover laid off 1400 workers after their supplier became insolvent in 2001, Dole suffered a large revenue decline after their banana plantations were destroyed after Hurricane Mitch hit South America in 1998, and Ford closed 5 plants for several days after all air traffic was suspended after September 11 in 2001 (Chopra & Sodhi, 2004; Christopher, 2004; and Sheffi, 2005 a and b).

Such disruptions can have not only long-term stock price effects but also loss of reputation and even loss of life. Based on an analysis of 827 disruption announcements made over a 10-year period, Hendricks and Singhal (2005a) found that companies suffering from the occurrence of uncertain events experienced 33–40% lower stock returns relative to their industry benchmarks over the three-year time period starting one year before and ending two years after the event announcement date.

The impact of such incidents has led to a growing interest in the area of supply chain risk and its management, as evidenced in the number of industry surveys, practitioner conferences and consultancy reports devoted to the topic, e.g., McKinsey (2006). According to a study conducted by Computer Sciences Corporation, 60% of the firms surveyed acknowledged that their supply chains are vulnerable to disruptions (CSC 2004). Supply chain executives in IBM believe that supply chain risk management (SCRM) is the second most important issue for them (IBM, 2008). Also, the research by AMR in 2007 reported that 46% of the executives believe that better SCRM is needed (Hillman and Keltz, 2007). However, few companies have taken commensurate actions (McKinsey, 2006).

Leading companies have made changes in light of man-made disasters such as 9/11 and of natural disasters such as Hurricane Katrina. Hewlett Packard formed the Procurement Risk Management Group to manage supply chain risks on the procurement side (Nagali et al., 2008), and Cisco formed the SCRM team that is responsible for ensuring supply chain resiliency. In the aftermath of Katrina, company executives from Procter & Gamble, WalMart, and SYSCO began sharing their SCRM processes (Bednarz, 2006). Ericsson implemented a new organization that is responsible for developing SCRM process after experiencing the huge financial loss caused by a small fire at a supplier's plant in March 2000, (Norrman and Jansson, 2004). After its recall of toys, Mattel formed a new division to audit, monitor and respond to supply chain risks.

At the same time, consulting firms such as Deloitte and PriceWaterhouseCoopers (PC) and insurance companies such as Zurich Insurance established SCRM as a new area of practice. Deloitte's and PWC's SCRM practices provide consulting services around assessing and mitigating supply chain risks arising from product development to outsourcing and from finance to logistics. Zurich's SCRM practice provides consulting services for reducing supply chain failures and insurance coverage including supplier defaults and supply delay so that the insurer can reduce financial risk exposure.

As a result, many researchers have been drawn to supply chain risk. Their early effort includes books on SCRM (e.g., Brindley, 2004); Zsidisin and Ritchie (2008); Wu and Blackhurst (2009); Sodhi and Tang (2011); special issues on SCRM (e.g., Ritchie and Brindley (2007), Narasimhan and Talluri (2009), Cao et al. (2009), and reviews of SCRM literature (e.g., Paulsson, 2004, and Tang, 2006b).

16.3 Motivation for This Study

Because SCRM is still a nascent area, most researchers in this area tend to come from other, more established areas. Hence, it is natural to expect a diverse set of viewpoints on what the scope of the field is and what research methodologies are appropriate at this stage. This diversity is unavoidable as we saw with supply chain management in the 1980s. It may even be beneficial in fuelling the kind of rapid growth in SCRM literature we have seen since 2005.

However, this diversity affects collaboration with other researchers and the review process in journal publication. It can also hamper research engagement with industry. . We need to establish a consensus regarding SCRM research and so in the context of industry need. A useful first step would be to characterize the diversity in scope and in the use of research methods among researchers. It is hard to escape the headline grabbing stories about major supply chain risk events—such characterization could help build a consensus regarding supply-chain risk with an eye towards industry need. Then the next step is gaining a better understanding from the research community itself regarding the scope of supply-chain risk and the appropriateness of research methodologies to various topics in the area.

16.4 Methodology

To better understand *scope diversity*, we looked at how (implicitly or explicitly) researchers have *defined* supply chain risk or SCRM and how they have addressed the *process* of SCRM: identify, assess, mitigate or respond, and communicate. To understand *research-tool diversity*, we looked at the *research methodologies* that researchers have used. To examine this diversity in scope and research tools, our multi-method field research study employs the following three steps:

Step 1: Direct observations.   We obtained direct observations of SCRM research activities by reviewing some recent research literature so as to examine how well the SCRM literature met the needs of industry in the eyes of the researchers. Our goal was to form our own perception regarding how researchers perceive the scope including the definition of SCRM and what research methodologies they employ. We reviewed their output of their efforts, i.e., academic papers that specifically mention that the focus of their investigation is risks/ uncertainties in the supply chain, taken from a wide range of peer-reviewed journals. We *excluded* papers that dealt with managing supply-demand risks within the established supply-chain management context—these included most papers using mathematical modelling as the primary methodology; see Tang (2006b) for a review of papers on risk but within the supply chain management modelling literature. Our aim was to capture the "breadth" of papers rather than conduct an exhaustive survey given our purpose to characterize diversity. This step indicated diversity among researchers in their definition of SCRM, in their addressing different aspects of the process of SCRM and in their use of different research methodologies. Ultimately, this step helped shape our perception about three "gaps" in current SCRM research; (1) a definition gap, (2) a process gap, and (3) a methodology gap—that we discuss later.

Step 2: Exploratory survey of focus groups.   To further explore researcher diversity in scope, in particular in the definition of supply chain risk and of SCRM, we surveyed two focus groups aided with a presentation and an open-ended questionnaire (Table 16.4). One group comprised supply chain management researchers at the *2008 Supply Chain Thought Leaders (SCTL) Conference* in Madrid, Spain, and the other comprised SCRM scholars at the *2008 International Supply Chain Risk Management (ISCRiM) Conference* in Trondheim, Norway. We obtained 42 responses to the open-ended questionnaire from the attendees at these two mini-conferences. Their responses helped us to further characterize the diversity of scope in terms of the definition of supply chain risk and of SCRM and also led to a starting point for scoping SCRM.

Step 3: Survey about the three gaps.   We used a presentation and a questionnaire with closed-ended as well as open-ended questions to survey a broad-based group of operations management researchers (Table 16.6). These researchers attended our keynote speech on SCRM during the *2009 Institute of Operations Research and Management Science (INFORMS) National Meeting* in San Diego. We used this survey to seek the opinion of researchers on the gaps we identified in the previous steps. We also sought views from researchers about what can or should be done to address these research gaps in SCRM. We distributed the questionnaire to approximately 200 attendees who attended the keynote speech on SCRM during the *2009 INFORMS San Diego meeting* and obtained 133 responses albeit some with incomplete responses to the open-ended questions.

16.5 Step 1 Findings: Diversity in Scope and Research Tools

We first looked into the *types* of risks identified as supply chain risks in the previous SCRM research. To do so, we reviewed only those research articles where authors specifically discussed the definition or the scope of supply chain risks and uncertainties (Table 16.1). A number of research articles on SCRM, including those we reviewed later (Tables 16.2 and 16.3), focus only on a certain aspect of risk or SCRM. However, we did not review these articles to seek diversity in definition.

Many articles categorize supply chain risks as a first step to managing these risks but they do so from widely different perspectives (Chopra and Sodhi, 2004; Christopher and Peck, 2004; Hallikas and Virolainen, 2004; Manuj and Mentzer, 2008; and Neiger et al., 2009). While the literature we surveyed reported in Table 16.1 is not exhaustive, it does indicate the following gap:

Definition gap: There is no consensus on a definition or scope for supply chain risk.

Next, we reviewed a sample of papers to understand different SCRM process elements and how these elements were covered in the literature. For instance, Jüttner et al. (2003) suggest four elements of managing supply chain risk: (1) assessing the risk sources, (2) identification of risk concepts, (3) tracking the risk drivers, and (4) mitigation risks. Likewise, Kleindorfer and Saad (2005) identify the process elements as (1) specifying sources of risks and vulnerabilities, (2) assessment and (3) mitigation.

We classified the existing SCRM literature according to four key elements for managing supply chain risks as we discussed in Chapter 1: (1) risk identification; (2) risk assessment; (3) risk mitigation; and (4) responsiveness to risk incidents, the last one sub-divided into responsiveness according to (a) operational risks (frequent risk events stemming from inherent supply-demand uncertainty); and (b) catastrophic risks (caused by natural or man-made disasters). We then indentified how articles from a broad base of journals cover these elements of the risk management process (Table 16.2).

The first SCRM element is *identifying* risks and uncertainty, covered in Chapter 2, is an initial step to manage supply chain risks according to many researchers (e.g. White, 1995; Hauser, 2003; Chopra & Sodhi, 2004; Wu et al., 2006; Hallikas et al., 2004; Manuj & Mentzer, 2008; and Neiger et al., 2009). Indeed many articles discuss this aspect of the risk management process (Table 16.2). However, with the exception of Nieger et al. (2009), researchers such as Chopra & Sodhi (2004) or Spekman & Davis (2004) cover this step only as a part of a framework of managing SCRM rather than focus on it.

The second SCRM process element is *assessment*, the subject of Chapter 3, involving the evaluation of the likelihood and of the impact (Harland et al., 2003, and Knemeyer et al., 2009). Many papers covering or mentioning assessment are conceptual papers and, as with risk identification, cover it only as a part of a broad SCRM framework. This is surprising given many researchers' stated interest in honing in on probabilities related to supply and demand uncertainty. A notable exception is the work of Hendricks and Singhal (2003, 2005 and 2007) who seek to empirically establish the impact of supply-chain disruptions on stock price, operating performance and shareholder wealth. Zsidisin et al. (2004) analyze risk assessment techniques used by purchasing organizations to mitigate risks posed by suppliers. Norrman and Jansson (2004) presented Ericsson's revised supply-chain risk assess-

Table 16.1 Diverse views of supply chain risk in articles that aim to look at SCRM comprehensively

Articles (in chronological order)	Scope of Risk
Jüttner, Peck and Christopher (2003)	Based on sources: environmental risk sources, network risk sources, and organizational risk sources
Spekman and Davis (2004)	Six dimensions of supply chain as risk sources, 1) inbound supply, 2) information flow, 3) financial flow, 4) the security of a firm's internal information system, 5) relationship with partners, and 6) corporate social responsibility
Cavinato (2004)	Based on five sub-chains/networks as risk sources, 1) physical, 2) financial, 3) informational, 4) relational, and 5) innovational.
Chopra and Sodhi (2004)	Categorize supply chain risks at a high level as disruptions or delays. These risks pertain to 1) systems, 2) forecast, 3) intellectual property, 4) receivable, 5) inventory and 6) capacity risk
Christopher and Peck (2004)	Categorize supply chain risks as 1) process, 2) control, 3) demand, 4) supply and 5) environmental
Kleindorfer and Saad (2005)	Based on the sources and vulnerabilities of risks, 1) operational contingencies, 2) natural hazards, and 3) terrorism and political instability
Bogataj and Bogataj (2007)	Categorize supply chain risks as 1) Supply risks; 2) Process risks; 3) Demand risks; and 4) Control risks
Sodhi and Lee (2007)	Categorize supply chain risks in the consumer electronics industry broadly as those requiring strategic decisions and those requiring operational decisions, in three categories: 1) Supply, 2) Demand, and 3) and Contextual risks.
Tang and Tomlin (2008)	Categorize supply chain risks as 1) supply, 2) process, and 3) demand risks, 4) intellectual property risks, 5) behavioural risks, and 6) political/social risks
Manuj and Mentzer (2008a)	Categorize supply chain risks as 1) supply, 2) operations, 3) demand and 4) other risks including security and currency risks.
	See Manuj and Mentzer (2008b) for another categorization: 1) supply 2) operational, 3) demand, 4) security, 5) macro, 6) policy, 7) competitive, and 8) resource risks.
Oke and Gopalakrishnan (2009)	Consider low-impact high-frequency and high-impact low-frequency risks in three major categories: 1) supply, 2) demand, and miscellaneous risks in the retail sector.
Rao and Goldsby (2009)	Categorize supply chain risks as 1) framework and 2) problem specific and 3) decision making risk.

Table 16.2 The elements of SCRM covered by the literature

Article	Identifi-cation	Assess-ment	Mitiga-tion	Responsiveness …	
				…to oper-ational risks	…to catas-trophic risks
Treleven and Schweikhart (1988)	X	X			
Johnson (2001)			X		
Hendricks and Singhal (2003)		X			
Chopra and Sodhi (2004)	X	X	X		
Christopher and Lee (2004)			X		
Giunipero and Eltantawy (2004)		X	X		
Norrman and Jansson (2004)	X	X	X	X	X
Spekman and Davis (2004)	X		X		
Zsidisin et al. (2004)	X	X	X		
Blackhurst et al. (2005)			X	X	
Hendricks and Singhal (2005a)		X			
Hendricks and Singhal (2005b)		X			
Kleindorfer and Saad (2005)	X	X	X		X
Brun et al. (2006)		X			
Choi and Krause (2006)			X		
Cucchiella and Gastaldi (2006)			X		
Gaudenzi and Borghesi (2006)		X			
Bogataj and Bogataj (2007)		X			
Sodhi and Lee (2007)	X		X		
Cheng and Kam (2008)	X	X	X		
Manuj and Mentzer (2008a)	X	X	X		
Tang and Tomlin (2008)			X	X	
Wagner and Bode (2008a)		X			
Braunscheidel and Suresh (2009)			X	X	
Jiang et al. (2009)			X		
Knemeyer et al. (2009)	X	X	X		X
Nieger et al. (2009)	X				
Oke and Gopalakrishnan (2009)	X		X		
Rao and Goldsby (2009)	X				
Trkman and McCormack (2009)	X	X			
Ellis et al. (2010)		X			

ment approach spanning site-specific risks to natural disaster related risks that impact many sites.

The third element of the SCRM process is ***mitigation***, i.e., reducing the likelihood of a particular risk's occurrence, reducing its potential impact or reducing both as discussed in Chapter 4. A substantial majority of papers in our sample cover the mitigation element of SCRM, although they do it as a part of broader SCRM frameworks. Still, there are papers that focus on mitigation. Braunscheidel and Suresh (2009) suggest that firm's cultural antecedents and organizational practices have significant impact on the firm' agility, which enables firms to mitigate supply chain risks better. Also, Christopher and Lee (2004) suggest that an important way to mitigate supply chain risks is improving "end-to-end" visibility of the supply chain and thereby improving confidence.

The fourth and final element of the SCRM process is ***response***, i.e., responding to an actual risk event so as to reduce the potential impact and to hasten recovery as we discussed in Chapter 5. When a company cannot prevent a risk incident from occurring, it has to figure out ways to respond quickly so as to contain the damage. In our sample, only five articles covered response this element of SCRM with only two looking at catastrophic risks (Norman & Jansson, 2004 and Kleindorfer & Saad, 2005) although Knemeyer *et al.* (2009) use the other elements for planning to respond to catastrophic risks. The rest at "operational" risks with high frequency and low per-event impact (Normand and Jansson, 2004, Braunscheidel and Suresh, 2009, and Blackhurst et al., 2005). As such, we formed a perception of a gap regarding the research literature:

Process gap: Researchers do not cover the response element of supply chain risk.

This gap is glaring in light of the well-publicised supply chain disruptions reported in the press.

Next, we first categorized SCRM articles into three groups: conceptual, quantitative empirical (statistical analysis of empirical data); and qualitative empirical (case studies). Of the papers in our sample, more than half are either conceptual or framework-type papers. Also, other than Wagner and Bode (2008b), most of the chapters in the books edited by Brindley (2004) and Zsidisin and Ritchie (2008) are conceptual. However, some simulation and mathematical models related to SCRM can be found in the book edited by Wu and Blackhurst (2009). We also found some industry studies, for instance, one on the retail sector (Oke and Gopalakrishnan, 2009) and one on the consumer electronics industry (Sodhi and Lee, 2007). Recently, Ellis et al. (2010) examined empirically about buyers' perception of supply risks in terms of the probability and the magnitude of supply disruption. Accordingly, we formed a perception of a gap from our literature review:

Methodology gap: Empirical work is not extensive in the area of SCRM even though it would be useful for such a nascent field.

Section 16.7 presents the response of researchers to a question related to this in the survey in the third part of our study.

Table 16.3 Research methodologies used in the research literature

Article	Conceptual/ Framework	Empirical (Quantitative)	Empirical (Qualitative)
Treleven and Schweikhart (1988)	X		X
Johnson (2001)			X
Hendricks and Singhal (2003)		X	
Chopra and Sodhi (2004)	X		
Christopher and Lee (2004)	X		
Giunipero and Eltantawy (2004)	X		
Norman and Jansson (2004)			X
Spekman and Davis (2004)	X		
Zsidisin et al. (2004)			X
Blackhurst et al. (2005)			X
Hendricks and Singhal (2005a)		X	
Hendricks and Singhal (2005a)		X	
Kleindorfer and Saad (2005)	X	X	
Brun et al. (2006)	X		X
Choi and Krause (2006)	X		
Cucchiella and Gastaldi (2006)	X		
Gaudenzi and Borghesi (2006)			X
Bogataj and Bogataj (2007)			
Sodhi and Lee (2007)	X		X
Cheng and Kam (2008)	X		
Manuj and Mentzer (2008a)	X		X
Tang and Tomlin (2008)	X		
Wagner and Bode (2008a)		X	
Braunscheidel and Suresh (2009)		X	
Jiang et al. (2009)		X	
Knemeyer et al. (2009)	X		
Nieger et al. (2009)	X		
Oke and Gopalakrishnan (2009)	X		X
Rao and Golsby (2009)	X		
Trkman and McCormack (2009)	X		
Ellis et al. (2010)		X	

16.6 Step 2 Findings: A Closer Look at the Definition Gap

Our literature review indicates there is much diversity in the scope including in definition of SCRM. To investigate this further, we focused on the definition of supply chain risk and of SCRM in two different focus groups. Specifically, we conducted two rounds of surveys for researchers in order to collect their opinions on what do they understand by supply chain risk and by SCRM (Table 16.6). Our findings from surveying the participants of the SCTL and the ISCRiM mini-conferences underscored our perception regarding the definition gap.

We summarized the responses of the surveyed participants to the open-ended questions into different categories. Recall that the first question (Q1) is about the respondent's definition of SCRM (Table 16.4). The tabulated results show that one-third of the respondents take a probabilistic approach and define SCRM as dealing with probabilities related to supply-demand matching (Table 16.5). About the same number take an operations view in suggesting that SCRM deals exclusively the risks stemming from supply chain operations. About 7% of the respondents believe that SCRM deals with risks arising from not only the operational aspects, but also the strategic aspects of supply chain.

Table 16.4 Questionnaire for the first survey (SCTL and ISCRiM)

No.	Questions
Q1	What is supply chain risk management (SCRM)?
Q2	How is SCRM different from supply chain management?
Q3	What is the link between SCRM and Enterprise Risk Management (ERM)?

One interesting observation is that although 19% of the respondents believe SCRM deals with rare but high impact events such as plant fires and natural disasters—this proportion could rise to nearly half if we include "dealing with the unknown" and "dealing with disruptions/disasters" as independent responses. By and large, however, research articles tend to focus on supply delays or other frequent disruptions that have low-to-moderate impacts. This phenomenon could be explained by the fact that researchers tend to cover supply chain risks that are more easily quantified with higher data availability, but it could be also be explained by researchers' fundamental understanding of SCRM within the context of supply chain management.

The second question (Q2) in our open-ended survey (Table 16.4) was to find out the link between SCRM and supply chain management. Indeed, as already speculated, about half (52.4%) participants view SCRM as a subset of supply chain management, an already established area of research and business practice. More than half of these (28.6% of the total) believe that SCRM is part of supply chain management, but with additional focus on risk elements. On the other hand, half the respondents believe SCRM includes elements outside supply chain management

Table 16.5 Response to Q1 ($N = 42$; some responses fell into more than one category)

Q1: What is supply chain risk management (SCRM)?	
Dealing with supply-demand stochastic (probability)	33.3%
Dealing with risk within supply chain operations	31.0%
Focus on low probability-high impact events	19.0%
Dealing with the unknown	14.3%
Dealing with disruptions/disasters	11.9%
Dealing with risk within supply chain strategy	7.1%
Dealing with stochastic, but need new probability-based approaches	4.8%
Dealing with financial risk	4.8%

with 16.7% of the respondents regarding SCRM as being entirely outside of supply chain management (Table 16.6).

Table 16.6 Response to Q2 ($N = 42$ respondents. Some responses fell into more than one category

Q2: How is SCRM different from supply chain management?	
SCRM is a subset of SCM	52.4%
SCRM is a subset of SCM, with additional focus on risk elements	28.6%
SCRM has something outside SCM	16.7%
SCRM is a subset on SCM but additional focus on supply sources	2.4%
SCRM overlaps with SCM and risk management, finance	2.4%

With the third question (Q3) about "the link between SCRM and enterprise risk management" (Table 16.4), we intended to find out how SCRM differs from enterprise risk management (ERM). Nearly three-fourths (74.2%) of the respondents believe SCRM to be a subset of ERM or an extension of it (Table 16.7). Also, 13.0% of these respondents underlined that the boundary of the traditional ERM tend to limit to the focal firm and the immediate surroundings but the boundary of SCRM is more extensive. Importantly, nearly a fifth the respondents believe that SCRM is separate from ERM (19.4%) while a tenth of the respondents place SCRM at the intersection of supply chain management and ERM (9.7%).

In Section 16.8.1, we speculate on a scope for SCRM based on these and other responses (Fig. 16.5).

16.7 Step 3 Findings: Three Gaps

At the 2009 INFORMS San Diego meeting, we posed three sets of questions about the definition gap, process gap, and the methodology gap. We used closed-ended

Table 16.7 Response to Q3 (Percentage was calculated out 31 respondents. Some responses fell into more than one category

Q3: What is the link between SCRM and Enterprise Risk Management (ERM)?	
SCRM is a subset of ERM	41.9%
SCRM is an extension of ERM	32.3%
SCRM is separate from ERM	19.4%
SCRM is the overlap between SCM and ERM	9.7%

questions (Table 16.8) and let the respondents know beforehand that our questions were intended for identifying the gaps in SCRM.

Table 16.8 Questionnaire for the INFORMS survey

Q1	Gap 1: There is no clear consensus on the definition of supply chain risk management. (7 point Likert-scale, strongly disagree—strongly agree)
Q2	In what terms do you think SCRM should be primarily defined (select one)? – Dealing with unknown, disruptions/disasters/low-prob, high impact events – Dealing with supply-demand stochastic (probability-based approaches) – Dealing with risk within supply chain operations – Dealing with risk within supply chain strategy – Other: (Please write)
Q3	What should we do to address this gap?
Q4	Gap 2: There is a lack of emphasis on research on response to risk incidents. (7 point Likert-scale, strongly disagree—strongly agree)
Q5	What should we do to address this gap?
Q6	Gap 3: There is a shortage of empirical research in the area of SCRM. (7 point Likert-scale, strongly disagree—strongly agree)
Q7	What should we do to address this gap?

Responding to the first question Q1 (Table 16.8), more than four-fifths of the respondents agreeing (score of 5 or more on a 7-point Likert scale) that "there is no clear consensus on the definition of supply chain management" (Fig. 16.1).

Responding to the question about the terms in which SCRM should be primarily defined, (Q2, Table 16.8), nearly half (47%) of the respondents agreed that SCRM is about dealing with low-probability and high-impact events. On the other hand, a tenth (10%) of the respondents chose to point out the risks stemming from supply-demand uncertainties. There were those who prefer to think in terms of supply chain strategy (10%) and those who emphasized supply chain operations (20%) instead. Of the remaining 13% who selected "other", more than half (7.5% of all the respondents) suggested that SCRM to encompass all of these risks (Fig. 16.2).

To confirm researchers' perception of the process gap, we posed two questions in the survey, Q4 and Q5 (Table 16.8), about the lack of emphasis on research on

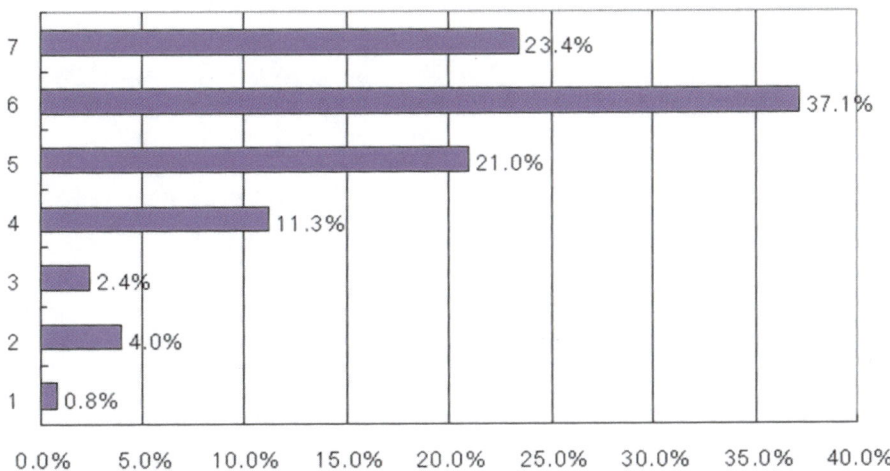

Fig. 16.1 Response to Q1: The 1st gap of SCRM research, there is no clear consensus on the definition of supply chain risk management; do you agree with this statement? (7 point Likert-scale, strongly disagree—strongly agree, $n = 124$)

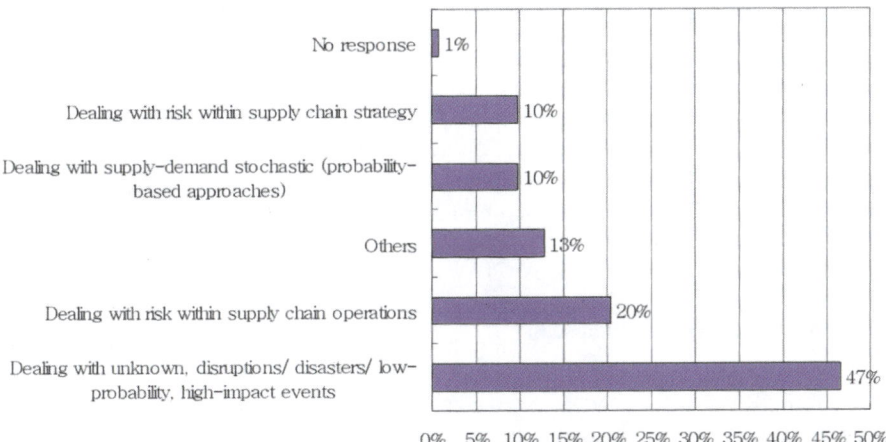

Fig. 16.2 Response to Q2: In what terms do you think SCRM should be primarily defined? ($N = 133$)

response to risk events. Nearly 70% of the respondents confirmed that there is a lack of the research on response relative to prevention and mitigation (Fig. 16.3).

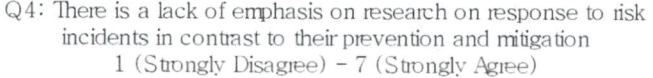
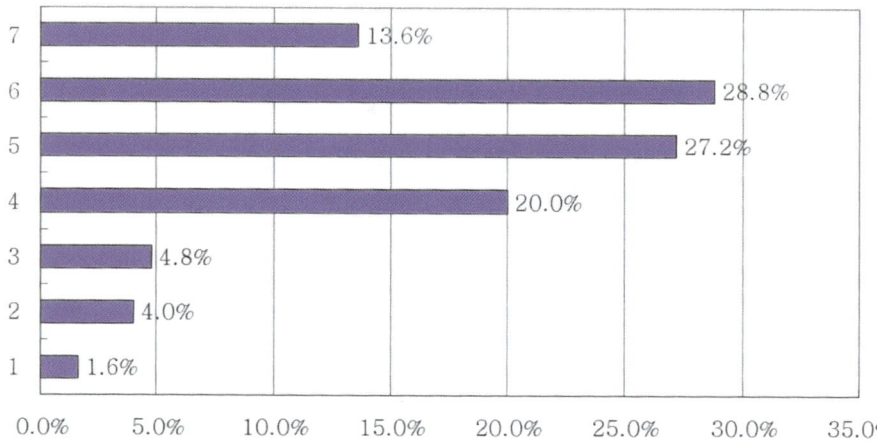

Q4: There is a lack of emphasis on research on response to risk incidents in contrast to their prevention and mitigation
1 (Strongly Disagree) – 7 (Strongly Agree)

Fig. 16.3 Response to Q4 regarding a lack of emphasis on research on response to risk incidents in contrast to their prevention and mitigation. (7 point Likert-scale, strongly disagree—strongly agree, $N = 125$)

We then sought to verify the perception of the methodology gap in our survey of operations management scholars at INFORMS with two questions, Q6 and Q7 (Table 16.8). A majority of the respondents agreed with this statement with nearly four-fifths of the respondents giving a score of 5 or higher as their response (Fig. 16.4).

16.8 Addressing the Gaps

In the INFORMS survey, we also asked three open-ended questions about how to close the three gaps using questions Q3, Q5 and Q7 in the questionnaire (Table 16.8). These responses can provide guidance to researchers and to journal editors and reviewers.

16.8.1 Closing the Definition Gap

We received many interesting suggestions from 122 respondents regarding ways to close the definition gap of SCRM. Broadly speaking, the respondents' suggestions

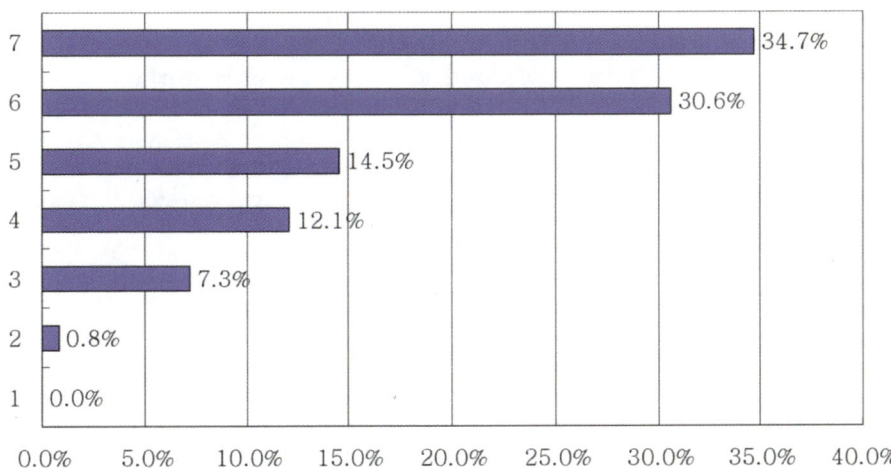

Fig. 16.4 Response to Q6. The 3rd gap of SCRM research: there is a shortage of empirical research in the area of SCRM. Do you agree with this statement? (7 point Likert-scale, strongly disagree—strongly agree, $n = 125$)

fall into five categories (Table 16.9). These categories indicate the broad range of approaches suggested.

Table 16.9 Response to Q3. What should we do to address the definition gap?

Categories of responses regarding closing the definition gap

1. As the field of SCRM mature, this gap will close itself naturally
2. More research such as survey-based papers on the definition of SCRM is needed
3. More discussions in academia as well as industry are needed to close this gap
4. There should be an official definition of SCRM by an organization such as INFORMS
5. SCRM should be limited to quantifiable risks

Although SCRM will naturally become better defined over time naturally, the process could take a long time. For instance, in the late 80s, there were different definitions of SCM ranging from supply/procurement management, logistics management, multi-echelon inventory management, etc. It took almost 20 years for the field of SCM to mature enough in order to obtain a clear definition of SCM today. With SCRM, arguably the need is much greater. Without a clear definition of SCRM, researchers would find it all the more challenging to communicate with company executives or otherwise gain access to industry to conduct applied research.

Moreover, senior managers from Cisco (McMorrow, 2009) and from Deloitte (Zhou (2009)) also perceive a definition gap of SCRM among company executives and highlight the need to develop a clear definition of SCRM.

An interesting picture emerges that could become the basis for defining SCRM while satisfying most of the respondents of the last two questions in our focus groups (Table 16.4; Q2 and Q3). This view is that SCRM has two parents: supply chain management and enterprise management. It has traits from both parents but is not a strict subset of either. Moreover, it is more than the overlap between its two parents (Fig. 16.5). We note that this view has not yet been validated or vetted and is offered here for further research and discussion.

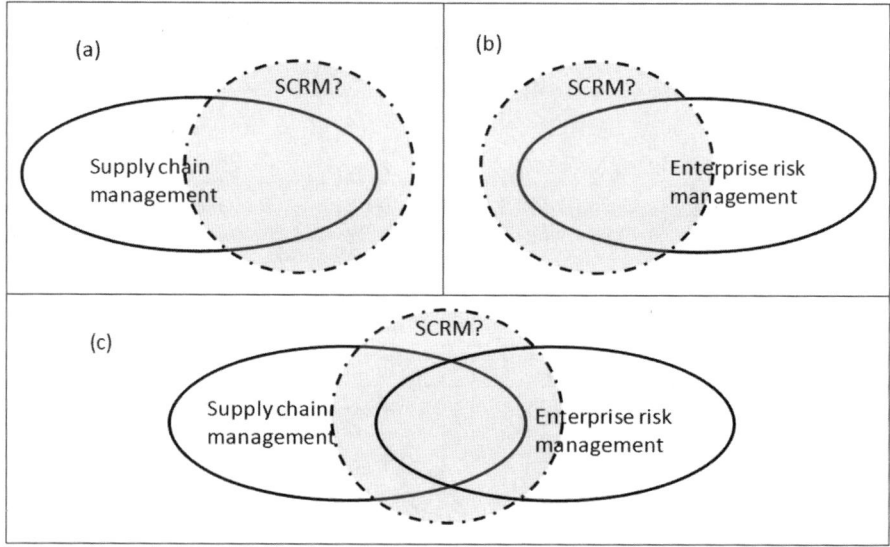

Fig. 16.5 A possible scope for SCRM combining diverse views of respondents (a) a part of supply chain management but extending it as per the responses in Table 16.6 (b) as a part of enterprise risk management but extending it as per the responses in Table 16.7, and (c) as a part of both but extending the overlap

16.8.2 Closing the Process Gap

We categorized the response to the open-ended question (Q5) (Table 16.8) regarding the process gap into three types (Table 16.10).

The first category covers what to do research on: many believe that more effort is needed for building a foundation for SCRM research. These works involve "defining the spectrum of types of supply chain risks that require responses", "building frameworks for responses in SCRM", "developing methodology and models for re-

Table 16.10 Response to Q5. What should we do to address the process gap?

Categories of responses regarding closing the process gap
1.   More effort is needed for building a foundation for SCRM research.
2.   Closer industry collaboration and case-study-based research
3.   Need better ways to publish and share research.

sponding", and "developing proper measure of scope of the risk and mathematically estimating the sequence of catastrophe". The responses in this category also include learning from the research in related areas such as "defence and military studies", "natural disasters", and "humanitarian disasters" where the focus of research is very much on response rather than on mitigation.

The second category comprises responses about how to carry out research. Many respondents suggested closer research collaboration with industry is one way to address this issue because of two major reasons: (1) many companies have experienced of responding supply chain related disruptions and delays, and (2) many companies have (or should have) detailed contingency plans for catastrophic events. Respondents also recommended "more empirical research based on case method".

The third category of the responses is about sharing and publishing SCRM research on responses. Some respondents pointed out that one way to encourage more research on this is starting a special issue on response aspect of SCRM.

As we reviewed the responses regarding ways to close the process gap, we have the following thoughts. First, to learn about response to supply chain risk first-hand, collaboration with industry is essential. There are existing industry consortia with a clear focus on SCRM that academics should consider foster collaboration. For example, the Supply Chain Risk Leadership Council (www.scrlc.com), with members such as Cisco, Boeing, Merck, etc., is a cross-industry group that is committed to learn and share best-practices about SCRM. Such consortia provide opportunities for academics to collaborate with industry and to learn of the actual responses first-hand.[4] Second, responding to supply chain disruptions shares features with response to natural disasters by means of rescue and relief efforts (Tomasini & Van Wassenhove, 2009). As such, there is an opportunity to apply the lessons learned from various humanitarian efforts to develop effective response to supply chain disruptions.

16.8.3 Closing the Methodology Gap

The response to the open-ended question (Q7) (Table 16.8) in our INFORMS survey indicates ways for closing the methodology gap. Some respondents commented

---

[4] For example, Pyke and Tang (2009) documented how Mattel handled its major toy recalls in 2007.

that the nature of SCRM, which is data on catastrophic events not easy to collect, is the major reason for the methodology gap and there is no easy solution for this. We categorized respondents' suggestions into the five categories (Table 16.11). Interestingly, many responses are cautionary in suggesting that more conceptual work is needed before we start doing empirical work. In similar vein, others suggest doing case studies on catastrophic events before embarking on similar empirical work in SCRM. Close collaboration with industry and establishing event-study-type research are deemed important as well. Finally, there are suggestions about editorial policy asking for editors to be more open to less-than-conventional research designs in the area of SCRM owing to the nature of the field and to the paucity of hard data.

Table 16.11 Categories of responses regarding closing the methodology gap as response to Q7

Q7. What should we do to address the methodology gap?
1. More conceptual work is needed before proceeding to empirical research on SCRM
2. More case studies on catastrophic events is needed before conducting empirical research on SCRM
3. Establish close collaboration with industry for data collection
4. Establish event-specific research (similar to event study in finance) as an approach for studying major catastrophic events
5. Journal reviewers should be more receptive on various research designs on SCRM where data collection is difficult

The responses highlight several major challenges for conducting empirical research on SCRM at this point. First, besides the fact data that on catastrophic events is not easy to collect, it is not clear what kind of data to collect unless there is a well defined conceptual framework. Second, as expressed by Meredith (1998), Flynn et al. (1990), and Voss et al. (2002), empirical research is perceived to be riskier than conceptual or mathematical research not only because it is time consuming but also because of researchers' perceptions about getting their papers accepted by top journals.

Still, empirical work in SCRM can pay dividends. For example, the empirical research by Hendricks and Singhal (2005a and 2005b) has received great visibility in industry. Rigorous empirical study based on a solid conceptual framework should be well received by academics and practitioners alike. SCRM being not only an emerging field but also an important one, many journals are developing special issues as of this writing. Based on our discussion with various editors, good empirical studies are in short supply. As such, there is a greater opportunity now to publish empirical results, thus providing motivation to collaborate with industry for empirical studies in SCRM.

16.9 Conclusion

SCRM is an area that has gathered much researcher interest following highly visible supply chain disruptions and surveys of managers reflecting their anxieties pertaining to supply chain risk. SCRM being an emerging area, we followed a field research approach to study the area in its nascent stage. We first carried out direct observations by studying the extant literature to form our perceptions about SCRM research, then gathered additional evidence through focus groups of two groups of supply chain researchers (SCTL 2008 participants and ISCRiM 2008 participants), and finally sought confirmation through a survey of a larger and broader audience of operations management scholars during the INFORMS Conference (2009).

We found there are three "gaps" pertinent to future research in SCRM: (1) there is no clear consensus on the definition of SCRM; (2) there is a lack of commensurate research on response to supply chain risk incidents; and (3) there is a shortage of empirical research in the area of SCRM. The INFORMS respondents also provided initial answers on how to go about closing these gaps. Taken together, their suggestions point to the need for more involvement with industry for case-study and event-study based research, while at the same time pointing out the need for more conceptual work on which to base this empirical research. Their suggestions are also aimed towards journal editors and reviewers for becoming more open-minded to research methodology for SCRM.

It would be useful to carry out a similar study of practitioner communities to determine the particular risks in their respective companies, the type of data they can provide regarding risk events and what type of collaborations they want to have with academic researchers. The present study combined with such a practitioner study would help further in creating a research agenda for SCRM researchers and for journal editors. Furthermore, in our review of the syllabi of MBA core courses in Operations Management and MBA supply chain electives at top-50 US business schools (US News and World Report 2006), we found that the topic of SCRM is rarely covered (Sodhi, Son and Tang, 2008). Given that many academic researchers also teach, we could additionally expand our study to establish a teaching agenda for SCRM keeping employers in mind.

Chapter 17
Conclusion and Future Research

Abstract In this concluding chapter, we provide areas for future research based on the content of the preceding chapters. We start by raising the question of what supply chain risk management is and, importantly for research and practice, what it is not. Taking the view of risk management as identifying, assessing, mitigating and responding to risk, we first present research questions that would lead to a better empirical understanding of how companies currently do so. We also provide research questions pertaining to the nature of the organization itself. Next we outline prescriptive research questions that are important for managers and for which empirical and quantitative modeling research could provide answers as managerial implications. Finally, we outline diverse research methods that researchers can use to address these research and managerial questions.

17.1 Introduction

It is interesting to compare supply chain management in its nascent stage in the 1980s with supply chain risk management in the late 2000s. Supply chain management was formally identified as a domain only in the 1980s even though supply chains have been around as long as there has been trade. Likewise, supply chain risk management is being identified formally as a domain in the late 2000s even though managers have had to manage supply chain risk as long as there have been supply chains. Just as formal identification of supply chain management enabled integrated supply chain solutions by way of process integration, formal identification of supply chain risk management has created a basis for practitioners and researchers to develop holistic solutions across the supply chain rather than only within the four walls of the enterprise. We must keep this in mind as we think of next steps and future research. In this chapter, we present topics for future research pertinent to to defining, assessing, mitigating and responding to supply chain risk.

M.S. Sodhi, C.S. Tang, *Managing Supply Chain Risk*,
International Series in Operations Research & Management Science 172,
DOI 10.1007/978-1-4614-3238-8_17, © Springer Science+Business Media, LLC 2012

17.2 Identifying and Assessing Supply Chain Risk

What is supply chain risk and what is supply chain risk management? The survey of different focus groups reported in Chapter 16 suggested that supply chain risk management could centre on the intersection of supply chain management and enterprise risk management. In addition, crisis management or business continuity approaches could provide a way to handle disruptions when they do occur.

We can sharpen this view by defining supply chain risk management as focused on *supply chain solutions that ensure supply continues to meet demand in case of a disruption or soon after the occurrence of such a disruption*. Supply chain risk would then to be tied to potential incidents that could create a significant and prolonged mismatch between demand and supply; such a mismatch usually arises from a disruption in supply or from a sudden surge or plunge in demand. We must also understand what is not supply chain risk management. Clearly, a disruption may require a companys efforts that go well beyond restoring the supply-demand match, for instance to restore its reputation. Such efforts may be squarely within the domain of enterprise risk management. For instance, insuring against losses due to supply chain disruption is not a supply chain solution and therefore not part of supply chain risk management.

Identifying supply chain risks can be challenging because a risk incident can have a different impact on the different entities in the supply chain. For example, depending on the number of suppliers that a company has, a fire at a suppliers plant is a risk to the supplier itself but may not a supply chain risk to the company if it can rely on other suppliers instead. However, when a company relies on a single supplier, it is exposed to the same supply chain risk that its supplier faces. For instance, after an earthquake and tsunami hit Japan in March 2011, many Toyotas direct suppliers were severely damaged. Without redundant suppliers, Toyota's production output in Japan and elsewhere plummeted due to supply shortages. As late as August 2011, Toyota predicted that the normal production output would not be restored until end of the year.[1] As Toyota has been unable to replenish its dealers inventories, the auto company has suffered a significant drop in sales since the quake. To reduce the supply risk in the future, Toyota is considering developing more foreign suppliers.

In Chapter 2, we described different risk factors. However, there is no consistent terminology in practice to facilitate communication and coordination among the different entities in the supply chain. Hence, there is a research opportunity to develop a standard glossary of supply chain risk terminology that would enable researchers and practitioners to share ideas. However, before developing such a classification, we must understand what companies are currently doing to classify supply chain risks and why they are doing so. Therefore, one research question is:

– How do companies identify and classify supply chain risks? (Do they take into account the network nature of the supply chain when classifying supply chain risks?)

---

[1] Takahashi, Y., "Toyota, Honda output grinds to a halt in April after quake," Wall Street Journal, May 27, 2011.

Assessing such risk means thinking in terms of the threat of a significant and pro-longed supply-demand mismatch and only that, even though such disruptions may result in lives lost or the resignation of a senior manager that should be assessed as enterprise risk rather than as supply chain risk. Eventually, the purpose is to develop a shared sense of urgency about a subset of identified risks and catalyze risk miti-gation effort, e.g., through investment in inventory and capacity—what, where and how much and what type (e.g., fixed versus flexible capacity)—and in processes such as those needed to respond to a risk incident in different parts of the supply chain. For each identified risk factor (e.g., fire hazards, system security), we pre-sented a survey-based approach in Chapter 3 to conduct risk assessment within a company.

We also discussed how and why such assessment is difficult. First, without a clear definition and terminology about risk events, risk factors, and consequences, it is fairly common for the participants to get confused among these elements. Sec-ond, because most assessment exercises are conducted without quantitative data, most participants find it difficult to evaluate the likelihood and to estimate the con-sequences associated with a risk event. These two major challenges motivate us to propose the following research question as an empirical starting point:

– *How do companies assess supply chain risks? (Do they take into account the nature of risk, the paucity of (hard) information and the self-interest of the different entities along the supply chain in terms of willingness to disclose in-formation?)*

17.3 Mitigating and Responding to Supply Chain Risk

Once the underlying risk factors are identified and accessed, it is natural to seek ways to reduce the likelihood of the occurrence and/or limit the consequences of the associated risk events. In Chapters 4 and 5, we described different ways to pre-vent/avoid certain risks and to reduce certain risks that are based on redundancy and flexibility. Also, we presented related strategic approaches—flexibility, redundancy, etc.—in Chapter 7 for mitigating certain supply chain risks. Furthermore, in Chap-ters 7 and 8 we discussed various tactical approaches—dynamic pricing, financial and operational hedging—for reducing certain risks. In Chapters 9, 10, 11 and 12, we showed how different mitigation approaches can be applied in different con-texts (long-term capacity planning, outsourcing, supply management, and product recall). Finally, in Chapters 13, 14 and 15, we reviewed and presented quantitative models and their corresponding solution techniques that are intended to reduce the consequence associated with the occurrence of different risk events.

Chapter 7 presented a few research opportunities for developing quantitative models for robust strategies:

– *How should one measure the effectiveness of a robust strategy and what are the conditions for one robust strategy to dominate another?*

– How can we model the benefits of supply-chain visibility provided by use of pertinent information technology that may help reduce supply-chain risk and thus increase robustness?

While operational mechanisms are intended to mitigate medium- and long-term risks, the financial instruments can hedge against short-term risks. This observation raises the following questions:

– Why isn't supply chain finance so popular? Why isn't hedging supply chain risks as popular in practice as it is appealing in theory?

Most mitigation strategies are implemented "before" the occurrence of different risk events. However, while certain mitigation strategies such as contingency plans can be put in place in advance, the deployment of different contingency plans can only be implemented "after" the occurrence of risk events. When a risk event occurs, we can reduce the time to restore supply chain operations and thus reduce the impact caused by the risk event as we discussed in Chapter 5. The consequence associated with a risk event can be reduced dramatically if a supply chain can respond to the risk event quickly. Regarding response, the natural questions that arise are: How do companies use "quick response" as a strategy to manage supply chain risks?

– How do companies improve their preparedness for and responsiveness to risk events?

We used examples and some natural disruption models (epidemic model and fire model) in Chapter 5 to argue that firms can use the time-based risk management concept to focus on reducing response lead time, which will in turn reduce recovery lead time and the impact of a disruption. However, the impact models and the concept require further validation in practice and further research:

– How should we quantitatively model the relationship between response time and the impact of a risk event?

We can use such models to understand whether it is the case that (1) the recovery lead time and cost can be significantly higher if the deployment of a recovery plan is delayed; and (2) the execution of a recovery plan can become much more difficult if its deployment is delayed.

17.4 The Nature of the Organization

To develop and execute any supply chain risk management strategy, it is essential to understand the link between the nature of the organization and its exposure to supply chain risk or its approach to supply chain risk management. For instance, we could ask why companies assess risk, or why any individual manager assesses risk. Answers to this question may shed light on why we see apprehension without action' (Chapter 1) whereby companies report awareness of supply chain risk but at the same time inadequate mitigation efforts and preparation:

– Do managers view their own supply chain risk management efforts commensurate to their perception of supply chain risk their organizations face? If not, then why not?

Reasons for inadequate action may include incentives and performance metrics, local responsibility (enterprise-focused) versus global impact (across the supply chain), roles and responsibilities of various managers in own and in other supply chain entities.

Recall from Chapters 2 and 6 that we attempted to classify supply-chain risk according to a 2×2 matrix for local/global causes and local/global impact. Because a local cause can have global impact on the supply chain, we need to examine the following questions:

– Do companies perceive higher risk when its supply chain partners are spread across the globe?
– How do we model the probability distribution of the loss per incidence for network risks? (Can we learn from networks like those for telephone or electricity grids?)
– What metrics are appropriate for measuring risk exposure and risk-adjusted performance in the different quadrants of the 2×2 matrix presented in Chapters 2 and 6?

Boeing's Dreamliner program described in Chapter 11 involved dramatic shifts in supply chain strategy from traditional methods used in the aerospace industry. Organization-related questions that arise are:

– What rationale do companies use to form leadership teams in charge of developing new supply chains? What sort of internal partnership, if at all, is built between management and labour for new supply chains?

In Chapter 13 we presented some research on managerial attitudes towards risk in general and towards incentives for managing supply-chain disruptions. As such, it is of interest to investigate the following questions in the near future:

– How do companies incentivize managers for managing identified supply-chain risks (if at all)?
– How do managers perceive supply chain risk in terms of trading off risks (and costs) and rewards (long-term and short-term)?

17.5 Managerial Impact

We presented a number of research questions in the two preceding sections in terms of what is'. However, from managers' perspective, questions of interest would be in terms of what should be'. As such, we can recast some of the high-level questions presented in the previous section for prescriptive use for managers. For instance, we can start with:

– How **should** *companies identify and classify supply chain risks?*
– How **should** *companies assess supply chain risks?*

The answers to these questions may depend on different factors that include the characteristics of the supply chain and the nature of the products. As noted in Chapter 3, there is no known standard approach for classifying or assessing supply chain risks. Therefore, there is an opportunity to develop a conceptual framework building on research on the questions in the preceding sections.

Due to the self-interest of the different entities in the supply chain, the incentive for reducing the impact of a risk event on the entire supply chain is rarely considered by all the different entities. Consequently, unless each party has a clear understanding about how his/her action would affect other parties, a clear understanding of his/her role and responsibility, and a clear performance metric for each party, it is difficult to motivate an entity to act in the interest of other parties or of the entire supply chain. As such, a question of interests for managers is:

– How **should** *companies change performance metrics and define roles and responsibilities of different entities to align each entity's interests in managing supply chain risk?*

With so many related risks and risk-mitigation approaches to consider, managers must do two things: First, they must create a shared, organization-wide understanding of supply chain risk. Then they must determine how to adapt general risk-mitigation approaches to the circumstances of their particular company. Managers can achieve the first through stress testing and the second through tailoring their reserves of excess inventory, excess and/or flexible capacity, ability to respond, and redundant suppliers.

As noted in Chapter 6, there are many reasons why most companies are not well prepared to respond to risk events including poor communication/coordination and no contingent plan. Clearly, to improve preparedness and responsiveness, one should consider some of the practical approaches as described in Chapter 6. However, to gain a deeper understanding into why companies are not prepared to respond to risk events and how company should respond to risk events, managers would be interested in answers to:

– How **should** *companies improve their preparedness and responsiveness to risk events?*

17.6 Research Methods

Traditionally, the research literature on risk emphasizes quantitative modelling. However, given the paucity of our knowledge in the nascent area of supply chain risk management, we should consider employing other research methods before delving too deeply into quantitative modelling.

Case study analysis.   In Chapter 6, we explored a risk framework for studying risk management practices in companies. To build a theory about supply chain risk management, case study (or field-based) research is well established in the management literature especially for new research areas that require exploration. Case study research has been advocated in the operations and supply chain management literature (Voss, Tsikriktsis and Frohlich 2002; Seuring 2005) for exploration, theory building, theory testing and theory extension/refinement. Among the case studies presented in this book, we analyzed and evaluated the way Mattel managed its recalls in 2007.

Multi-case analysis.   In the same chapter (Chapter 6), we discussed how Samsung Electronics developed its risk management processes at the local (plant or region) level to prevent or otherwise mitigate losses at the local level as well as formal risk management processes for managing risks faced by the company as a whole. However, we cannot simply take these processes as best practice' for all companies despite Samsung Electronic being a leading company. Because of the early stage of the research presented here, much is needed for future research. A multi-case study with other companies in the same and other sectors is very much needed to understand risk management processes and the organization. Eventually, we should be able to design generic risk management strategies for each of the four quadrants including quadrant-specific risk metrics.

Behavioural research.   Such methods can help us answer questions about human behaviour, something of value for shedding light on how managers deal with risk given their incentives as presented in Chapter 13. For instance, to develop performance metrics and roles and responsibilities that would entice each party to act in the interest of the entire supply chain, one way could be to conduct behavioural experiments to gain a deeper understanding about the relationships among performance metrics and supply chain risks. Behavioural experiments can yield testable hypotheses and can help us to develop quantitative models.

Event study.   As we reported in Chapter 16, the bulk of the quantitative models used mathematical analysis to examine ways to reduce various supply chain risks rather than respond to events (although there are models to prepare optimally for response to events such as earthquakes or humanitarian crises). In general, there is a shortage of quantitative empirical analysis including event-study for examining the effectiveness of different mitigation strategies that are intended to reduce certain identified supply chain risks or to respond to certain risk events.

Quantitative modelling.   In Chapter 13, we discussed that even for such modelling, there are many questions that arise, which include: (1) How to model a disruption process? (2) How to develop objective function that would align the incentives of different supply chain partners? (3) How to use dynamic pricing / revenue management model to manage demand? The finance literature deals with hedging decisions and the operations literature focuses on production planning under uncertainty so we may consider developing quantitative models by exploiting the existing methodologies develop in finance and operations such as the use of stochastic programming (c.f., Sodhi and Tang, 2008; 2011) as described below.

Stochastic programming.   In Chapter 15, we presented a stochastic programming formulation for a simple supply chain planning model that involves the purchase of inputs from a supplier, their conversion to finished goods at a single plant, and the eventual stocking and selling of these goods from a single warehouse facing uncertain demand. We have also compared the models in these two domains, drawing out similarities and differences. We have argued that the asset-liability management literature leads the supply-chain planning literature in its use of stochastic programming and is therefore a useful guide for researchers and practitioners wishing to develop stochastic programming models for supply-chain planning. One conclusion from this survey is that no matter how fast computers become or how much solver technology improves over time, stochastic programming will remain as much of an art as it is a science. As such, modelling choices and solution choices have to be made carefully, matching these choices to the objectives and being able to compromise efficiently between diverse multiple objectives.

Simulation.   While simulation models can enable us to evaluate the effectiveness of different strategies for mitigating different supply chain risks, there is also a cautionary lesson we need to draw from the use of quantitative modelling in practice such as that for asset-liability management where there is much emphasis on Monte Carlo simulation to 'stress test' a portfolio. Regardless of the sophistication of scenarios, the simulation-only approach is simplistic compared to dynamic stochastic optimization. Still, simulation is more accessible than stochastic optimization. For models for supply chain risk management, it is conceivable to optimize a production or procurement plan under a variety of simulated what-if scenarios to consider demand, inventory, supplier lead-time, and capacity variations.

Scenario planning.   Finally, the joint use of scenario planning, whereby managers devise plausible scenarios based on their readings of trends and forces at plan, and optimization models, whereby modellers propose the best decision or course of action for each of the scenarios, may be better than either approach by itself for supply chain risk management especially for disruptions (see Chapter 9).

17.7 Final Thoughts

Supply chain risk management is an area that has gathered much researcher interest following highly visible supply chain disruptions and surveys of managers reflecting their anxieties pertaining to supply chain risk. While supply chain risk management is still a nascent area as of this writing, it has drawn interests from practitioners and researchers, and has great potential for collaboration between industry and academia. We hope this book will bring together researchers and practitioners to develop efficient and effective ways for managing supply chain risk.

References

Abery J, Stark E (2008) Global chief procurement officer survey 2008. Capgemini Consulting report

Agrawal N, Nahmias S (1998) Rationalization of the Supplier Base in the Presence of Yield Uncertainty. In: Lee HL, Ng SM (eds) Global supply chain and technology management. Production Oper Management Society Publishers, Florida

Akella R, Rajagopalan S, Singh M (1992) Part dispatch in random yield multi-stage flexible test systems for printed circuit boards. Oper Res 40:776–789

Allmendinger G, Lombreglia R (2005) Four strategies for the age of smart services. Harvard Business Rev (October):1–10,

Alonso-Ayuso A, Escudero LF, Garin A, Ortuno MT, Perez G (2003) An approach for strategic supply chain planning under uncertainty based on stochastic 0-1 programming. J Global Optim 26:97–124

AMR (2006) AMR research report on managing supply chain risk. AMR Research, Inc

Anderson C (2002) Hazards and vulnerability analysis. Report, University of North Texas, http://training.fema.gov/EMIWeb/downloads/UNTServiceLearning.pdf

Anthony R, Loveman G (1996) Laura Ashley and Federal Express strategic alliance. Harvard Business School, Case 9-693-050

Anupindi R (1993) Supply management under uncertainty. PhD thesis, Graduate School of Industrial Administration, Carnegie Mellon University

Anupindi R, Akella R (1993) Diversification under supply uncertainty. Management Sci 39(8):944–963

Anupindi R, Akella R (1997) An inventory model with commitments. Working paper, Kellogg School of Management, Northwestern University

Anupindi R, Bassok Y (1998) Supply contracts with quantity commitments and stochastic demand. In: Tayur S et al (eds) Quantitative models for supply chain management. Kluwer

Anupindi R, Bassok Y (1999) Centralization of stocks: Retailers vs. manufacturer. Management Sci 45:178–191

Arntzen B, Brown G, Harrison T, Trafton L (1995) Global supply chain management at digital equipment corporation. Interfaces 25:69–93

Aron R, Singh J (2005) Getting offshoring right. Harvard Business Rev (December):135–143

Arora VK, Boer GJ (2005) Fire as an interactive component of dynamic vegetation models. J Geophys Res 110:1-20

Atah A, Lee H, Ozer O (2005) If the inventory manager knew: Value of RFID under imperfect inventory information. Working paper, Graduate School of Business, Stanford University

Atallah M, Elmongui H, Deshpande V, Schwarz L (2004) Secure supply chain collaboration. Working paper, Krannert School of Management, Purdue University

Auguste B, Hao Y, Singer M, Wiegand M (2002) The other side of outsourcing. The McKinsey Quarterly 1:53–63

Austin RD, Nolan RL (2000) IBM Corp. Turnaround. Harvard Business School, Case 9-600-098

Automotive News (2006) Supplier turmoil: It ain't over till it's over. Automotive News 80:54

Aviv Y (2001) The effect of collaborative forecasting on supply chain performance. Management Sci 47:1326–1443

Aviv Y (2002) Gaining benefits from joint forecasting and replenishment processes: The case of auto-correlated demand. Manufacturing & Service Oper Management 4:55–74

Aviv Y (2004) Collaborative forecasting and its impact on supply chain performance. In: Simchi-Levi D, Wu D, Zhen Z (eds) Handbook of quantitative supply chain analysis. Kluwer

Aviv Y (2005) On the benefits of collaborative forecasting partnerships between retailers and manufacturers. Working paper, Olin School of Management, Washington University, St. Louis

Aviv Y, Federgruen A (1998a) The operational benefits of information sharing and vendor managed inventory programs. Working paper, Olin School of Business, Washington University, St. Louis

Aviv Y, Federgruen A (1998b) The benefits of design for postponement. In: Tayur S et al (eds) Quantitative models for supply chain management. Kluwer

Aviv Y, Federgruen A (2001a) Design for postponement: A comprehensive characterization of its benefits under unknown distribution. Oper Res 49:578–598

Aviv Y, Federgruen A (2001b) Capacitated multi-item inventory with random and seasonal fluctuating demands: Implications for postponement strategies. Management Sci 47:512–531

Babich V (2008) Independence of capacity ordering and financial subsidies to risky suppliers. Working paper, Department of Industrial and Operations Engineering, University of Michigan

Babich V, Burnestas AN, Ritchken PH (2007) Competition and diversification effects in supply chains with supplier default risk. Manufacturing & Service Oper Management 9(2):123–146

Babich V, Burnestas A, Ritchken P (2004) Competition and diversification effects in supply chains with supplier default risk. Working paper, Department of Industrial and Operations Engineering, University of Michigan

Baccarach SB (1989) Organisational theories: Some criteria for evaluation. AMR 14(4):496-515

Badinelli R (2000) An optimal, dynamic policy for hotel yield management. Eur J Oper Res 121:476–503

Bagahana MP, Cohen M (1998) The stabilizing effect of inventory in supply chains. Oper Res 46:572–583

Bailey D (2009) Half of US auto suppliers face bankruptcy: Study. Reuters News, March 19

Bajaj V (2008) Surprises in a closer look at credit-default swaps. New York Times, November 5

Barnes-Schuster D, Bassok Y, Anupindi R (2002) Coordination and flexibility in supply contracts with options. Manufacturing & Service Oper Management 4:171–207

Bassok Y, Akella R (1991) Ordering and production decisions with supply quality and demand uncertainty. Management Sci 37:1556–1574

Bassok Y, Anupindi R (1997) Analysis of supply contracts with total minimum commitment. IIE Transactions 29:373–381

Bazaraa MS, Sherali HD, Shetty CM (1993) Nonlinear programming: Theory and algorithms. John Wiley & Sons, New York

Bednarz A (2006) Supply chain execs share disaster-planning techniques. Computer world. http://www.computerworld.com/s/article/9000810/Supply_chain_execs_share_disaster_planning_techniques

Bell DE, Schleifer A Jr (1995) Risk management. Coursetechnology, Cambridge, MA

Benders JF (1962) Partitioning procedures for solving mixed variables programming problems. Numerische Mathematik 4:238–252

Bengtsson L, Berggren C (2004) Rethinking outsourcing in manufacturing: A tale of two telecom firms. Eur Management J 22(2):211–223

Berman B (1999) Planning for the inevitable product recall. Business Horizon 69–78

Bhalla A, Sodhi MS, Son BG (2008) Is more IT offshoring better? An exploratory study of western companies offshoring to South East Asia. J Operations Management 26(2):322–335

Billington C, Lee H, Tang C (1998) Successful strategies for product rollovers. Sloan Management Rev 39(3)(Spring):23–30

Billington C, Johnson B (2002) A real options perspective on supply chain management in high technology. J Applied Corporate Finance (Summer):20–28

Birge JR (1982) Decomposition and partitioning methods for multi-stage stochastic linear programs. Oper Res 33(5):989–1007

Birge JR (1985) Aggregation bounds in stochastic linear programming. Math Programming 31:25–41

Birge J (2000) Option methods for incorporating risk into linear capacity models. Manufacturing & Service Oper Management 2(1):19–31

Birge JR, Louveaux F (1997) Introduction to stochastic programming. Springer Verlag, New York

Bitran G, Dasu S (1992) Ordering policies in an environment of stochastic yields and substitutable products.' Oper Res 40:999–1017

Bitran G, Gilbert S (1996) Managing hotel reservations with uncertain arrivals. Oper Res 44:35–49

Black F, Derman E, Toy W (1990) A one-factor model of interest rates and its application to treasury bond options. Financial Analysts Journal (January–February):33–39

Blackburn J (1990) Time-based competition: The next battleground in american manufacturing. McGraw-Hill

Blackhurst J, Craighead CW, Elkins D, Handfield RB (2005) An empirically derived agenda of critical research issues for managing supply-chain disruptions. Int J Production Res 43(19):4067–4081

Boeing (2008) www.boeing.com (accessed Nov. 18)

Boer L, Labro E, Morlacchi P (2001) A review of methods supporting supplier selection. Eur J Purchasing Supply Management 7:75–89

Bogataj D, Bogataj M (2007) Measuring the supply chain risk and vulnerability in frequency space. Int J Production Economics 108(1-2):291–301

Boone Y, Ganeshan R, Stenger A (2002) The benefits of information sharing in a supply chain: An exploratory simulation study. In: Geunes J, Pardalos P, Romeijn E (eds) Supply chain management models, applications and research directions. Kluwer

Bovet D (2005) Balancing global risk and return. Supply Chain Strategy 1(6):8–10

Bradley SP, Crane DB (1972) A dynamic model for bond portfolio management. Management Sci 19(2):139–151

Braunscheidel MJ, Suresh MC (2009) The organizational antecedents of a firm's supply chain agility for risk mitigation and response. J Oper Management 27(2):119–140

Breitman RL, Lucas JM (1987) Planets: a modeling system for business planning. Interfaces 17(1):94–106

Bresnahan TF, Reiss PC (1985) Dealer and manufacturer margins. Rand J Economics 16:253–268

Brindley C (ed) (2004) Supply chain risk. Ashgate Publications, Aldershot, England

Brown AO, Lee HL (1997) Optimal pay to delay capacity reservation with applications to the semiconductor industry. Working paper, Stanford University

Brown AO, Tang CS (2005) The impact of alternative performance measures on single-period inventory policy. J Industrial Management Optimization 2(3):297–318.

Brown AO, Lee HL, Petrakian R (2000) Xilinx improves its semiconductor supply chain using product and process postponement. Interfaces 30(July-August):65–80

Brown AO, Chou M, Tang CS (2005) The implications of pooled returns policies. Working paper, UCLA Anderson School

Brown D, (2004) How U.S. got down to two makers of flu vaccine. Washington Post (October 16)

Brun A, Caridi M, Fahmy Salama K, Ravelli I (2006) Value and risk assessment of supply chain management improvement projects. Int J Production Economics 99(1–2):186–201

Burgess RG (1984) In the field: An introduction to field research. Allen and Unwin Publishers, UK

Burnestas AN, Ritchken PH (2005) Options pricing with downward-sloping demand curves: the case of supply chain options. Management Sci 51(4):566–580

Butter F, Linse KA (2008) Rethinking procurement in the era of globalization. Sloan Management Rev (Fall):76–80

Cachon G (2003) Supply chain coordination with contracts. In: De Kok AG, Graves S (eds) Handbooks in Operations research and management science. Elsevier

Cachon G (2004) The allocation of inventory risk and advanced purchase discount in a supply chain. Working paper, The Wharton School, University of Pennsylvania

Cachon G, Fisher M (1997) Campbell soup's continuous replenishment program: Evaluation and enhanced inventory decision rules. Production Oper Management 6:266–276

Cachon G, Fisher M (2000) Supply chain inventory management and the value of shared information. Management Sci 46:1032–1048

Cachon G, Lariviere M (2005) Supply chain coordination with revenue sharing contracts: Strengths and limitations. Management Sci 51:30–44

Camm J, Chorman T, Dull F, Evans J, Sweeney D, Wegryn G (1997) Blending OR/MS, judgment, and GIS: Restructuring P&G's supply chain. Interfaces 27(1):128–142

Cao D, Tang O, Nakashima K (2009) Call for papers: Special issue on "Supply Chain Risk Management." Int J Production Economics
http://www.elsevierscitech.com/pdfs/PROECO_CfP_Cao_SupplyChainRiskManagement.pdf

Cariño DR, Ziemba WT (1998a) Formulation of the Russell-Yasuda Kasai financial planning model. Oper Res 46(4):433–449

Cariño DR, Myers DH, Ziemba WT (1998b) Concepts, technical issues, and uses of the Russell-Yasuda Kasai financial planning model. Oper Res 46(4):450–462

Cariño DR, Kent T, Myers DH, Stacy C, Sylvanus M, Turner AL, Watanabe K, Ziemba WT (1994) The Russell-Yasuda Kasai model: An asset/liability model for a Japanese insurance company using multistage stochastic programming. Interfaces 24(1):24–49

Caro F, Gallien J (2005) Dynamic assortment with demand learning for seasonal consumer goods. Working paper, UCLA Anderson School of Management

Carr MJ, Konda SL, Monarch I, Ulrich FC, Walker CF (1993) Taxonomy based risk identification. Research Report # SEI-93-TR-6, Carnegie Mellon University

Carr S, Lovejoy W (2000) The inverse newsvendor problem: Choosing an optimal demand portfolio for capacitated resources. Management Sci 47:912–927

Carter JR, Vickery SK (1989) Currency exchange rates: Their impact on global sourcing. J Purchasing Materials Management (Fall):19–25

Casey N, Pasztor A (2007) Safety agency, mattel clash over disclosures. Wall Street Journal (September 4):A.1

Cavinato JL (2004) Supply chain logistics risks: From the back room to the board room. Int J Physical Distribution & Logistics Management 34(5):383–387

Cetinkaya S, Lee CY (2000) Stock replenishment and shipment scheduling for vendor managed inventory. Management Sci 46:217–232

Chandler C, Fung A (2006) Not exactly counterfeit. Fortune (May 1):108–116

Chapman P, Christopher M, Juttner U, Peck H, Wilding R (2002) Identification and managing supply chain vulnerability. Logistics Transport Focus 4(4):59–64

Chen F, Ryan A, Simchi-Levi D (2000a) The impact of exponential smoothing forecasts on the bullwhip effect. Naval Res Logistics 47:269—286

Chen F, Ryan A, Simchi-Levi D (2000b) Quantifying the bullwhip effect in a simple supply chain: The impact of forecasting, lead times and information. Management Sci 46:436–443

Chen F, Drezner A, Ryan J, Simchi-Levi D (1998) The bullwhip effect: Managerial insights on the impact of forecasting and information on variability in a supply chain. In: Tayur S et al (eds) Quantitative models for supply chain management. Kluwer

Cheng SK, Kam BH (2008) A conceptual framework for analyzing risk in supply networks. J Enterprise Information Management 22(4):345–360

Cheng TCE, Wu YN (2005) The impact of information sharing in a two-level supply chain with multiple retailers. J Operational Research Society 56:1159–1165

Chesbrough H (2003) Open innovation: The new imperative for creating and profiting from technology. Harvard Business School Press

Chod J, Rudi N (2005) Resource flexibility with responsive pricing. Oper Res 53:532–548

Chod J, Rudi N, Van Mieghem J (2009) Operational flexibility and financial hedging: Complements or substitutes? Working paper, Boston College'

Choi T, Hartley J (1996) An exploration of supplier selection practices across the supply chain. J Oper Management 14(4):333–343

Choi TY, Krause DR (2006) The supply base and its complexity: Implications for transaction costs, risks, responsiveness, and innovation. J Oper Management 24(5):637–652

Choi T, Dooley K, Rungtusanatham MJ (2001) Supply networks and complex adaptive systems: Control versus emergence. J Oper Management 19:351–366

Chong JK, Ho T, Tang CS (2001) A modeling framework for category assortment planning. Manufacturing & Service Oper Management 3(3):191–210

Chong JK, Ho TH, Tang CS (2004) Demand modeling in product line trimming: substitutability and variability. In: Chakravarty AK, Eliashberg J (eds) Managing business interfaces: Marketing, engineering, and manufacturing perspectives, Kluwer Academic Publishers

Chopra S, Meindl P (2004) Supply-chain management: Strategy, planning, operations, 2nd ed, Prentice Hall

Chopra S, Sodhi M (2004) Managing risk to avoid supply chain breakdown. Sloan Management Rev 46(1):53–61

Chowdhry B, Howe J (1999) Corporate risk management for multinational corporations: Financial and operational hedging policies. Eur Finance Rev 2:229–246

Christopher M (2004) Creating resilient supply chains. Logistics Europe (February):14–21

Christopher M, Lee H (2004) Mitigating supply chain risk through improved confidence. Int J Physical Distribution & Logistics Management 34(5):388–396

Christopher M, Peck H (2004) Building the resilient supply chain. Int J Logistics Management 15(2):1–14

Ciarallo F, Akella R, Morton T (1994) A periodic review production planning model with uncertain capacity and uncertain demand. Management Sci 40:320–332

Clark A, Scarf H (1960) Optimal policies for a multi-echelon inventory problem. Management Sci 6:475–490

Clark TH and Hammond JH (1997), Reengineering channel reordering ?processes to improve total supply-chain performance. Production and ?operations management, 6:248–265

Closs D, McGarrell E (2004) Enhancing security through the supply chain. IBM Center for the Business of Government Special Report Series (April)

Cohen MA, Agrawal N (1999) An analytical comparison of long and short term contracts. IIE Transactions 31(8):783–796

Cohen M, Lee H (1988) Strategic analysis of integrated production-distribution systems: Models and methods. Oper Res 36(2):216–228

Consigli G, Dempster MAH (1998) Dynamic stochastic programming for asset-liability management. Annals of OR 81:131–161

Cook T (1998) Sabre soars. http://lionhrtpub.com/orms/orms-6-98/sabre.html. OR/MS Today (June)

Corbett C (2001) Stochastic inventory systems in a supply chain with asymmetric information: Cycle stocks, safety stocks, and consignment stocks. Oper Res 49:487–500

Corbett C, de Groote X (2000) A supplier's optimal quantity discount policy under asymmetric information. Management Sci 46:444–450

Corbett C, Tang CS (1998) Designing supply contracts: contract type and information asymmetric information. In: Tayur S et al (eds) Quantitative models for supply chain management. Kluwer

Corsten D, Saraf B (2008) Nestlé Russia LLC: Supplier finance programme. Instituto de Empressa business school case, Madrid

Cottrill K (2006) A game plan for disaster recovery. MIT Supply Chain Strategy 2(6, June):4–9

Craighead CW. Blackhurst J, Rungtusanatham MJ, Handfiel RB (2007) The severity of supply chain disruptions: design characteristics and mitigation capabilities. Decision Sci 38(1):131–156

Craighead C, Blackhurt J, Rungtusanatham MJ, Handfield RB (2007) The severity of supply chain disruptions: design characteristics and mitigation capabilities. Decision Sci 38(1):131–156

Crew M, Kleindorfer P (1986) The economics of public utility regulation. MIT Press, Cambridge, MA

Crown J (2008) Will Boeing pay for delays? Spiegel Online International (April 4) http://www.spiegel.de/international/business/0,1518,545365,00.html, (accessed May 2009)

Cucchiella F, Gastaldi M (2006) Risk management in supply chain: A real option approach. J Manufacturing Technology Management 17(6):700–720

Current JR, Weber CA (1994) Application of facility location modeling constructs to vendor selection problems. Eur J Oper Res 76:387–392

Dahel N (2003) Vendor selection and order quantity allocation in volume discount environments. Supply Chain Management: An International Journal 8:335–342

Dana J, Spier K (2001) Revenue sharing and vertical control in the video rental industry. J Industrial Economics 59:223–245

Dana J (1998) Advance-purchase discounts and price discrimination in competitive markets. J Political Economy 166:395–422

Dana J (1999) Using yield management to shift demand when the peak time is unknown. Rand J Economics 30:456–474

Dapiran P (1992) Benetton: Global logistics in action. Asian Pacific Int J Business Strategy 5:7–11

Dasu S, Li L (1997) Optimal operating policies in the presence of exchange rate variability. Management Sci 43:705–722

Dattatreya RE, Fabozzi FJ (1995) Active total return management of fixed-income portfolios. Irwin, Burr Ridge, IL

Dawson A, Hassenpflug S, Sloan J (1998) California agriculture trade: Combating the med fly menace. Case study, Center for Trade and Commercial Diplomacy, Monterey Institute of International Studies, Monterey, California

de Geus A (1997) The Living Company. Harvard Business School Press, Boston: 69.

Deleris LA, Elkins D, Paté-Cornell ME (2004) Analyzing losses from hazard exposure: A conservative probabilistic estimate using supply chain risk simulation. In: Ingalls RG, Rossetti MD, Smith JS, Peters BA (eds) Proceedings of the 2004 simulation conference 1384–1391

Demeester L, Tang CS (1996) Reducing cycle time at an IBM wafer fabrication facility. Interfaces 26(2, March–April):34–49

Denardo EV (2002) The science of decision making: A problem-based approach using Excel. Wiley, New York

Denardo E, Lee T (1996) Managing uncertainty in a serial production line. Oper Res 44:382–392

Denardo E, Tang CS (1992) Pulling a Markov production system. Oper Res 40:259–278

Denardo E, Tang CS (1997) Control of a stochastic production system with estimated parameters. Management Sci 43:1296–1307

Dert CL (1999) A dynamic model for asset liability management for defined benefit pension funds. In: Ziemba WT, Mulvey JM (eds) Worldwide asset and liability modeling. Cambridge University Press, Cambridge (reprinted)

De Waart D (2006) Getting SMART about risk management. Supply Chain Management Rev (November):26–33

Ding Q, Dong L, Kouvelis P (2007) On the integration of production and financial hedging decisions in global markets. Oper Res 55(3):470–489

Disney SM, Towill DR (2003) The effect of vendor managed inventory (VMI) dynamics on the bullwhip effect in supply chains. Int J Production Economics 85:199–215

Doherty NA (2000) Integrated risk management. McGraw-Hill, New York, NY

Doig SJ, Ritter RC, Speckhals K, Woolson D (2001) Has outsourcing gone too far?" The McKinsey Quarterly 4:25–37

Donohue K (2000) Efficient supply contracts for fashion goods with forecast updating and two production modes. Management Sci 46(11):1397–1411

Dormer A, Vazacopoulos A, Verma N, Tipi H (2005) Modeling and solving stochastic programming problems in supply chain management using Xpress-SP. In: Geunes J, Pardalos P (eds) Supply chain optimization. Springer, New York

Dornier P, Ernst R, Fender M, Kouvelis P, (1998) Global operations and logistics: Text and cases. Wiley, New Jersey

Drijver SJ, Haneveld WKK, van der Vlerk MH (2000) Asset liability management modeling using multistage mixed-integer stochastic programming. Working paper, SOM, University of Groningen, www.som.rug.nl

Dyer J (1996) How Chrysler created an American keiretsu. Harvard Business Rev (August):42–56

Dyer J, Ouchi W (1993) Japanese style business partnerships: Giving companies a competitive edge. Sloan Management Rev 35(1):51–63

Eckert R (2007) In defence of Mattel. Wall Street J (September 11):A19

Eisenhardt K (1989) Building theories from case study research. Acad Management Rev 14(4):532–550

Eliashberg J, Steinberg R (1993) Marketing-production joint decisions. In: Eliashberg J, Lilien G (eds) Handbook of operations research and management science: Marketing. Elsevier, Amsterdam, The Netherlands

Ellis SC, Henry RM, Shockley J (2010) Buyer perceptions of supply disruptions risk: A behavioral view and empirical assessment. Journal of Operations Management 28(1):34–46

Elmaghraby W, Keskinocak P (2003) Dynamic pricing in the presence of inventory considerations: Research overview, current practices, and future directions. Management Sci 49:1287–1309

Emmons H, Gilbert S (1998) Returns policies in pricing and inventory decisions for catalogue goods. Management Sci 44:276–283

Eppen G, Iyer A (1997a) Backup agreements in fashion buying: The value of upstream flexibility. Management Sci 43(11):1469–1484

Eppen G, Iyer A (1997b) Improved fashion buying with Bayesian updates. Oper Res 45:805–819

Eppen G, Schrage L (1981) Centralized ordering policies in a multi-warehouse system with lead times and random demand. In: Schwarz LB (ed) Multi-level production/inventory control systems: Theory and practice. North-Holland, Amsterdam

Eppen GD, Martin RK, Schrage L (1989) A scenario approach to capacity planning. Oper Res 37(4):517–524

Ernst R, Kouvelis P (1999) The effects of selling packaged goods on inventory decisions. Management Sci 45:1142–1155

Eskew M (2004) Mitigating supply chain risk. CEO (April):25–26

Esterl M (2009) Southwest Airlines reports loss. Wall Street J (April 17):B4

Federgruen A (1998) Comments on: Variability reduction through operations reversal in supply chain re-engineering. Working paper, Columbia University

Feitzinger E, Lee H (1997) Mass customization at Hewlett Packard: The power of postponement. Harvard Business Rev 75:116–121

Ferdows K, Lewis M, Machuca J (2004) Rapid fire fulfillment. Harvard Business Rev (November):104–117

Fischhoff B, Lichtenstein S, Slovic P, Derby S, Keeney R (1981) Acceptable risk. Cambridge University Press, New York

Fisher M (1997) What is the right supply chain for your product? Harvard Business Rev 75(2):105–116

Fisher ML, Hammond J, Obermeyer WR, Raman A (1994) Making supply meet demand in an uncertai world. Harvard Business Rev (March-April):106–116

Fisher M, Raman A (1996) Reducing the cost of demand uncertainty through accurate response. Operations Res 44:87–99

Flynn BB, Sado S, Schroeder RG, Bates KA, Flynn EJ (1990) Empirical research methods in operations management. J OperManagement 9(2):250–284

Fukuda Y (1964) Optimal policies for the inventory problem with negotiable lead time. Management Sci 10:690–708

Gaivoronski AA, de Lange PE (2000). An asset liability management for casualty insurers: complexity reduction versus parameterized decision rules. Eur. J. of Op. Research 99, 227–250.

Gan X, Sethi S, Yan H (2004) Coordination of supply chains with risk-averse agents. Prod Oper Management 13(2):135–149

Gan X, Sethi S, Yan H (2005) Channel coordination with a risk neutral supplier and a downside-risk-averse retailer. Prod Oper Management 14(1):80–89

Garg A, Lee HL (1996) Effective postponement through standardization and process sequencing. IBM Research Report RC 20726, IBM TJ Watson Research Center, Yorktown Heights, NY 10598

Garg A, Lee HL (1998) Managing product variety: An operations perspective. In: Tayur S et al (eds) Quantitative models for supply chain management. Kluwer

Garg A, Tang CS (1997) Postponement strategies for product families with multiple points of differentiation. IIE Transactions 29:641–650

Gates D (2000) Boeing strike ends; Machinists back on the job Sunday. Seattle Times (Nov 2) http://seattletimes.nwsource.com/html/localnews/2008340022_webmachinists02m.html (accessed April 2009)

Gaudenzi B, Borghesi A (2006) Managing risks in the supply chain using the AHP method. Int J Logistics Management 17(1):114–136

Gaur V, Giloni A, Seshadri S (2005) Information sharing in a supply chain under ARMA demand. Management Sci 51:961–969

Gavirneni S, Kapuscinski R, Tayur S (1999) Value of information in capacitated supply chains. Management Sci 45:16–24

Geoffrion A (1976) Better distribution planning with computer models. Harvard Business Rev (July–August):92–99

Geoffrion A, Powers R (1995) Twenty years of strategic distribution system design: An evolutionary perspective. Interfaces 25(5):105–127

Gerchak Y, Vickson R, Parlar M (1988) Periodic review production models with variable yield and uncertain demand. IIE Transactions 20(2):144–150

Ghemawat P, Nueno J (2003) Zara: Fast Fashion. Harvard Business School, Case # 9-703-497

Giddy IH (2009) The corporate hedging process. Online article, Stern School of Business, New York University. http://people.stern.nyu.edu/igiddy/corphdg.htm (accessed April 2009)

Gilbert K (21005) An ARIMA supply chain model. Management Sci 51:305–310

Gilbert S, Ballou R (1999) Supply chain benefits from advanced customer commitments. J Oper Management 18:61–73

Giunipero LC, Eltantawy RA (2004) Securing the upstream supply chain: a risk management approach. Int J Physical Distribution & Logistics Management 34(9):698–713

Gondzio J, Kouwenberg R (2001) High-performance computing for asset-liability management. Oper Res 49(6):879–891

Gong L, Matsuo H (1997) Control policy for a manufacturing system with random yield and rework. J Optimiz Theory Appl 95(1):149–175

Gottfredson M, Puryear R, Phillips S (2005) Strategic sourcing: From periphery to the core. Harvard Business Rev (February) 83(2):132–139, 150.

Graves SC, Tomlin BT (2003) Process flexibility in supply chains. Management Sci 49(7):907–919

Gumbel P (2006) Trying to untangle wires. Time Europe Magazine (October) http://www.time.com/time/europe/magazine/printout/0,13155,1543879,00.html)

Gunasekaran A, Ngai E (2005) Build-to-order supply chain management: A literature review and framework for development. J Oper Management 23:423–451

Gunsalus J (2007) Boeing sticks to revised 787 Dreamliner schedule. Bloomberg Press (Dec 11) http://www.bloomberg.com/apps/news?pid=20601103&sid=aqgUCtUslurM&refer=us (accessed April 2009)

Gupta D, Benjaafar S (2004) Make-to-order, make-to-stock, or delay product differentiation? A common framework for modeling and analysis. IIE Transactions 36:529–546

Gurnani H, Tang CS (1999) Note: Optimal ordering decisions with uncertain cost and demand forecast updating. Management Sci 45:1456–1462

Ha A (2001) Supplier-buyer contracting: Asymmetric cost information and cut-off level policy for buyer participation. Naval Res Logistics 48:41–64

Hallikas J, Virolainen VM (2004) Risk management in supplier relationships and networks. Burlington, USA, Ashgate Pub Ltd

Hallikas J, Karvonen I, Urho P, Veli-Matti V, Markku T (2004) Risk management processes in supplier networks. Int J Production Economics 90(1):47–58

Hammond J, Raman A (1995) Sport Obermeyer Ltd. Harvard Business School, Case # 695022

Handfield R, McCormack KP (eds) (2008) Supply chain risk management: Minimizing disruptions in global sourcing. Auerbach Publications (Taylor and Francis Group), Boca Raton FL

Harland CM, Brenchley R, Walker H (2003) Risk in supply networks. J Purchasing Supply Management 9(2):51–62

Harrington L (2004) Speeding global shipments. Inboundlogistics.com (November)

Hartwigsen K (2006) Integrated crisis management: Building a framework for unified response. Nike internal presentation

Haskett J, Signorelli S (1989) Benetton (A). Harvard Business School, Case 9-685-014

Hauser LM (2003) Risk-adjusted supply chain management. Supply Chain Management Rev 7(6):64–71

Hawk J (2005) The Boeing 787-Dreamliner: More than an airplane. (May) http://www.aiaa.org/events/aners/Presentations/ANERS-Hawk.pdf (accessed April 2009)

Headley JS, Tufano (1994) Dell Computer Corporation. Harvard Business School, Case 9-294-051

Heath D, Jarrow R, Morton A (1990) Bond pricing and the term structure of interest rates: A discrete time approximation. J Finan Quantitative Anal 25(4):419–440

Heinrich C (2005) RFID and beyond. Wiley

Helper S (1991) How much has really changed between U.S. automakers and their suppliers. Sloan Management Rev 32(4):15–28

Hendricks KB, Singhal VR (2003) The effect of supply chain glitches on shareholder wealth. J Oper Management 21(5):501–522

Hendricks KB, Singhal VR (2005a) An empirical analysis of the effect of supply chain disruptions on long-run stock price performance and equity risk of the firm. Production Oper Management 14(1):35–52

Hendricks KB, Singhal VR (2005b) Association between supply chain glitches and operating performance. Management Sci 51(5):695–711

Herbst M (2008) Hedging against $200 oil. Bus Week (May 7)

Henig M, Gerchak Y (1990) The structure of periodic review policies in the presence of variable yield. Oper Res 38:634–643

Hicks M (2002) When supply chain snaps. eWeek (Feb 18)

Hillman M, Keltz H (2007) Managing Risk in the Supply Chain: A Quantitative Study, AMR Research Report.

Ho T, Lee S (1986) Term structure movements and pricing interest rate contingent claims. J Finance 41:1011–1029

Ho T, Tang CS (eds) (1998) Product variety management: Research advances. Kluwer, Massachusetts

Holland W, Sodhi M (2008) A simple method for regulators to cross-check operational risk loss models for banks. In: Christodoulakis G, Satchell S (eds) The analytics of risk model validation. Butterworth and Heinemann/Elsevier

Hopkin P (2010) Fundamentals of risk management: Understanding, evaluating and implementing effective risk management. Kogan Page, London

Hopkins K (2005) Value opportunity three: Improving the ability to fulfill demand. Bus Week (January 13)

Hoyland K, Wallace S (2001) Generating scenario trees for multistage problems. Management Sci 47:295–307

Hsu A, Bassok Y (1999) Random yield and random demand in a production system with downward substitution. Oper Res 47:277–290

Huchzermeier A, Loch CH (2001) Project management under risk: Using the real options approach to evaluate flexibility in R&D. Management Sci 47(1):85–101

Huchzermeier A, Cohen MA (1996) Valuing operational flexibility under exchange rate uncertainty. Oper Res 44(1):100–113

Hucko C (2007) Airbus A380 vs. Boeing 787: Poll reveals that passengers prefer a smaller plane. Suite101 (April 28) http://airplanes.suite101.com/article.cfm/airbus_a380_vs_boeing_787 (accessed May 2009)

Hull JC (2006) Options, futures, and other derivatives. 6th ed. Prentice Hall, New Jersey

Hunt B (2000) Issue of the moment: The rise and the rise of risk management. Finan Times (June 27)

Iassinovski S, Artiba A, Bachelet V, Riane F (2003) Integration of simulation and optimization for solving complex decision making problems. Int J Prod Economics 85(1):3–10

IBM (2008) Supply chain risk management: A delicate balancing act. White paper, IBM Global Business Services, New York

Iyer A (1998) Modeling the impact of information on inventories. In: Tayur S et al (eds) Quantitative models for supply chain management. Kluwer

Iyer A, Bergen ME (1997) Quick response in manufacturing-retailer channels. Management Sci 43:559–570

Iyer A, Deshpande V, Wu Z (2003) A postponement model for demand management. Management Sci 49:983–1002

Jain N, Paul A (2001) A generalized model of operations reversal for fashion goods. Management Sci 47:595–600

Janssen F, de Kok T (1999) A two-supplier inventory model. Int J Prod Economics 59:395–403

Janssens M (2000) Introduction to mathematical fire modeling. 2nd edition, CRC Press

Jehn KA, Northcraft GB, Neale MA (1999) Why differences make a difference: A field study of diversity, conflict, and performance in workgroups. Admin Sci Quarterly 44(4):741–763

Jiang B, Baker RC, Frazier GV (2009) An analysis of job dissatisfaction and turnover to reduce global supply chain risk: Evidence from China. J Oper Management 27(2):169–184

Johnson EM (2001) Lessons in managing supply-chain risk from the toy industry. California Management Rev (Spring)

Johnson ME (2001) Learning from toys: Lessons in managing supply chain risk from the toy industry. California Management Rev 43(3):106–124

Johnson ME, Davis T, Walker M (1999) Vendor managed inventory in the retail supply chain. J Bus Logistics 20:183–203

Jordan W, Graves SC (1995) Principles on the benefits of manufacturing process flexibility. Management Sci 41(4):577–594

Jüttner U, Peck H, Christopher M (2003) Supply chain risk management: Outlining an agenda for future research. Int J Logistics, Research & Applications 6(4):197–210

Kallberg JG, Ziemba WT (1983) Comparison of alternative utility functions in the portfolio selection problems. Management Sci 29(11):1257–1276

Kandybin A, Florham P (2004) Raising your return on innovation investment, strategy + business. Booz & Company Publications 35:1–12

Kapuscinski R, Tayur S (1999) Variance vs. standard deviation—note: On variability reduction through operations reversal in supply chain re-engineering. Management Sci 45:21–23

Kazaz B, Dada M, Moskowitz H (2005) Global production planning under exchange-rate uncertainty. Management Sci 51:1101–1119

Kersten W and Blecker T (2006) Managing risks in supply chains: How to build reliable collaboration in logistics. Erich Schmidt Verlag, Berlin.

Klaassen P (1997) Discretized reality and spurious profits in stochastic programming models for asset/liabilty management. Eur J Oper Res 101(2):374–392

Klaassen P (1998) Financial asset-pricing theory and stochastic programming models for asset/liabilty management: A synthesis. Management Sci 44(1):31–48

Kleindorfer PR, Saad GH (2005) Managing disruption risks in supply chains. Production & Oper Management 14(1):53–68

Kleindorfer PR, Wu DJ (2003) Integrating long-term and short-term contracting via business-to-business exchanges for capital-intensive industries. Management Sci 49(11):1597–1615

Knemeyer AM, Zinn W, Eroglu C (2009) Proactive planning for catastrophic events in supply chains. J Oper Management 27(2):141–153

Kogut B (1985) Designing global strategies: Profiting from operational flexibility. Sloan Management Rev (Fall):27–38

Kogut B, Kulatilaka N (1994) Operating flexibility, global manufacturing, and the option value of a multinational network. Management Sci 40(1):123–139

Kopczak L, Lee H (1993) Hewlett-Packard: Deskjet printer supply chain. Stanford Graduate School of Business case

Kotha S, Kotha S, Olesen D (2005) Boeing 787: The Dreamliner. Harvard Business School Case 5-305-104

Kouvelis P (1998) Global sourcing strategies under exchange rate uncertainty. In: Tayur S et al (eds) Quantitative models for supply chain management. Kluwer

Kouvelis P (2006) Interfaces of operational and financial decisions. SCTL workshop presentation. Olin School of Business, Washington University in St. Louis

Kouvelis P, Gutierrez G (1997) The newsvendor problem in a global market: Optimal centralized and decentralized control policies for a two-market stochastic inventory system. Management Sci 43(5):571–585

Kouvelis P, Rosenblatt M (2002) A mathematical programming model for global supply chain management: Conceptual approach and managerial insights. In: Geunes J et al (eds) Supply chain management: Models, applications, and research directions. Kluwer

Kouwenberg R (2001) Scenario generation and stochastic programming models for asset liability management. Eur J Oper Res 134:279–292

Krazit T (2004) Trouble in east fishkill? IBM chip group struggles. InfoWorld.com (April 21)

Krishnan M, Kekre S, Mukhopadhyay T, Srinivasan K (1998) Managing variety in software industry. In: Ho TH, Tang CS (eds) Product variety management: Research advances. Kluwer

Kunreuther H (1976) Limited knowledge and insurance protection. Public Policy 24:227–261

Kusy MI, Ziemba WT (1986) A bank asset and liability management model. Oper Res 34(3):356–376

Kwon D, Lippman SA, McCardle K, Tang CS (2009) Time based constracts with delayed payments. Working paper, UCLA Anderson School

Lariviere M (1998) Supply chain contracting and co-ordination with stochastic demand. In: Tayur S et al (eds) Quantitative models for supply chain management. Kluwer

Lariviere M, Porteus E (2001) Selling to the newsvendor: An analysis of price-only contracts. Manufacturing & Service Oper Management 3:293–305

Larsen E (2000) What Is Scenario Planning?. Teaching note, Cass Business School

Lederer PJ, Mehta T (2005) Economic evaluation of scale-dependent and irreversible technology investments. Production Oper Management 14(1): 21–34

Lee HL (1996) Effective management of inventory and service through product and process redesign. Oper Res 44:151–159

Lee HL (2004) The triple-A supply chain. Harvard Business Rev (October):102–112

Lee H, Tang CS (1996) Managing supply chains with contract manufacturing. Asian J Bus Information Systems 1(1):11–22

Lee H, Tang CS (1997) Modeling the costs and benefits of delayed product differentiation. Management Sci 43(1):40–53

Lee H, Tang CS (1998a) Variability reduction through operations reversal. Management Sci 44(2):162–172

Lee HL, Tang CS (1998b) Managing supply chains with contract manufacturing. In: Lee HL, Ng SM (eds) Global supply chain and technology management. Production Oper Management Society Publishers, Florida

Lee H, Whang S (1998) Value of postponement. In: Ho TH, Tang CS (eds) Produce variety management: Research advances. Kluwer

Lee H, Whang S (2006) Seven-Eleven Japan. Stanford Graduate School of Business Case #GS-18

Lee H, Wolfe M (2003) Supply chain security without tears. Supply Chain Management Rev 7(1):12–20

Lee HL, Padmanabhan P, Whang S (1997a) Information distortion in a supply chain: The bullwhip Effect. Management Sci 43:546–558

Lee HL, Padmanabhan V, Whang S (1997b) The bullwhip effect in supply chains. Sloan Management Rev 38:93–102

Lee HL, So KC, Tang CS (2000) The value of information sharing in a two-level supply chain. Management Sci 46:626–643

Leeham Co (2005) 787 is not meeting 24hour-engine change promo, lessor says. Leeham Co. LLC (July 18)
http://www.leeham.net/filelib/SCOTTSCOLUMN071805.pdf (accessed May 2009)

Leung SCH, Wu Y, Lai KK (2006) A stochastic programming approach for multi-site aggregate production planning. J Oper Res Soc 57:123–132

Levy D (1995) International sourcing and supply chain stability. J Int Bus Studies 26:343–360

Li CL, Kouvelis P (1999) Flexible and risk-sharing supply contracts under price uncertainty. Management Sci 45:1378–1398

Li G, Wang S, Yan H, Yu G (2005) Information transformation in a supply chain: A simulation study. Computers Oper Res 32:707–725

Lilien G, Kotler P, Moorthy S (1992) Marketing models. Prentice Hall, New Jersey

Lim WS, Tang CS (2005) Optimal product rollover strategies. Eur J Oper Res 174(2): 905–922

Luenberger D (1998) Investment science. Oxford University Press, New York

Lunsford J (2007a) Boeing replaces head of 787 Dreamliner program. Wall Street J (Oct 17)

Lunsford J (2007b) Boeing, in embarrassing setback, says 787 Dreamliner will be delayed. Wall Street J (Oct 11)

Lynch G (2009) Single point of failure: The 10 essential laws of supply chain risk management. John Wiley & Sons, Hoboken, NJ

MacDuffie JP, Sethuraman K, Fisher M (1996) Product variety and manufacturing performance: Evidence from the international automotive assembly plant study. Management Sci 42(3): 350–369

Mandal A, Deshmukh SG (1996) Vendor selection using interpretive structural modeling. Int J Oper Production Management 14(6):52–59

Mandel TF, Wilson I (1993) How Companies Use Scenarios: Practices and Prescriptions. SRI International, report no. 822.

Manuj I, Mentzer JT (2008a) Global supply chain risk management strategies. Int J Physical Distribution & Logistics Management 38(3):192–223

Manuj I, Mentzer JT (2008b) Global supply chain risk management. J Bus Logistics 29(1):133–155

March J, Shapira Z (1987) Managerial perspectives on risk and risk taking. Management Sci 33:1404–1418

Martha J, Subbakrishna S (2002) Targeting a just-in-case supply chain for the inevitable next disaster. Supply Chain Management Rev 6(5):18–23

Martin M, Hausman W, Ishii K (1998) Design for variety. In: Ho TH, Tang CS (eds) Product variety management: Research advances. Kluwer

Martinez-de-Albeniz V, Simchi-Levi D (2005) A portfolio approach to procurement contracts. Production Oper Management 14(1): 90–114

McCardle K, Rajaram K, Tang CS (2004) Advance booking discount programs under retail competition. Management Sci 50(5):701–708

McCardle K, Rajaram K, Tang CS (2005) Bundling retail products: Models and analysis. Eur J Oper Research 177(2):1197–1217

McCue A (2006) Sainsbury's saves big by insourcing. Bus Week (Nov 21)

McKinsey (2006) Understanding supply chain risk: A McKinsey global survey. The McKinsey Quarterly

McKinsey Quarterly Report (2008) Managing Global Supply Chains

McMorrow J (2009) Private Communication. Cisco Systems

Meredith J (1998) Building operations management theory through case and field research. J Oper Management 16:441–454

Miller LT, Park CS (2005) A learning real options framework for process design and capacity planning. Production Oper Managament 14(1): 5–20

Milne R (2009) Early warnings in the supply chain. Financial Times (March 24):10

Minner S (2003) Multiple-supplier inventory models in supply chain management: A review. Int J Production Economics 281:265–279

Mintzberg H (1976) Planning on the left side and managing on the right. Harvard Business Review 54(July-August): 49–59.

Mintzberg H (1994) The Rise and Fall of Strategic Planning. Prentice Hall International, London.

Mishina K, Flaherty T (1988) Intercon Japan. Harvard Business School case #688-056

Moinzadeh K, Nahmias S (1988) A continuous review model for an inventory system with two supply modes Management Sci 34:761–773

Mollison D (2003) Epidemic models: Their structure and relation to data. Cambridge University Press

Moon Y (2003) The birth of the swatch. Harvard Business School case 9-504-096

Mortimer J (2004) Vertical contracts in the video rental industry. Working paper, Department of Economics, Harvard University

Mula J, Poler R, Garcia-Sabtel JP, Lario FC (2006) Models for production planning under uncertainty: A review. Int J Prod Economics 103(1):271–285

Mulvey JM (1996) Generating scenarios for the Towers Perrin investment system. Interfaces 26(2):1–15

Mulvey JM, Ruszczynski A (1995) A new scenario decomposition method for large scale stochastic optimization. Oper Res 43:477–490

Mulvey JM, Shetty B (2004) Financial planning via multi-stage stochastic optimization. Computers and Oper Res 31:1–20

Mulvey JM, Thorlacius AE (1999) The Towers Perrin global capital market scenario generation system. In: Ziemba WT, Mulvey JM (eds) Worldwide asset and liability modeling. Cambridge University Press, Cambridge, UK

Mulvey JM, Vladimirou H (1992) Stochastic network programming for financial planning problems. Management Sci 38(11):1642–1664

Murray C (2007) Boeing 787 Dreamliner rolls out smoother ride with gust suppression. Design News (June 4). http://www.designnews.com/article/439-Boeing_787_Dreamliner_Rolls_Out_Smoother_Ride_with_Gust_Suppression.php (accessed April 2009)

Nagali V, Hwang J, Sanghera D, Gaskins M, Pridgen M, Thurston T, Mackenroth P, Branvold D, Scholler P, Shoemaker G (2008) Procurement risk management (PRM) at Hewlett-Packard Company. Interfaces 38(1):51–60

Nagurney A, Curz J, Dong J, Zhang D (2005) Supply chain networks, electronic commerce, and supply side and demand side risk. Eur J Oper Res 164:120–142

Narasimhan R, Talluri S (2009) Perspectives on risk management in supply chains. J Oper Management 27(2):114–118

Narayanan VG, Raman A (2004) Aligning incentives in supply chains. Harvard Business Rev (November):94–103

Neiger D, Rotaru K, Churilov L (2009) Supply chain risk identification with value-focused process engineering. J Oper Management 27(2):154–168

Nielsen SS, Zenios SA (1993) A massively parallel algorithm for nonlinear stochastic network problems. Oper Res 41:319–337

Norman A, Jansson U (2004) Ericsson's proactive supply chain risk management approach after a serious sub-supplier accident. Int J Physical Distribution & Logistics Management 34(5):434–456

Norris G (2009) 787s move along, weight problems persist. Aviation Daily (May 6). http://www.aviationweek.com/aw/generic/story_generic.jsp?channel=aviationdaily&id= news/787NEW05059.xml (accessed May 2009)

Norrman A, Jansson U (2004) Ericsson's proactive supply chain risk management approach after a serious sub-supplier accident. Int J Physical Distribution & Logistics Management 34(5):434–456

Oh S (2004) Hansol paper improves labour relationship by improving working condition. Korean Economic Daily (June 22)

Oke A, Gopalakrishnan M (2009) Managing disruptions in supply chains: A case study of a retail supply chain. International Journal of Production Economics 118(1): 168-174

Padmanabhan V, Png I (1997) Manufacturer's returns policy and retail competition. Marketing Sci 16:81–94

Parlar M, Goyal S (1984) Optimal ordering decisions for two substitutable products with stochastic demand. OPSEARCH 21:1–15

Parlar M, Perry D (1996) Inventory models of future supply uncertainty with single and multiple suppliers. Naval Res Logistics 43:191–210

Pasternack B (1985) Optimal pricing and return policies for perishable commodities. Marketing Sci 4:166–176

Pasternack B (2002) Using revenue sharing to achieve channel coordination for a newsboy type inventory model. In: Geunes J, Pardalos P, Romeijn E (eds) Supply chain management models, applications and research directions. Kluwer

Paulsson U (2004) Supply chain risk management. In: Brindley C (ed) Supply chain risk. Ashgate, Aldershot, UK

Paulsson U, Nilsson CH (2008) Potential risk handling alternatives for supply chain disruptions in liquid food production: The case of V&S Vin & Sprit AB, the Sundsvall site. Research report 1015, Lund University, Sweden

Perry L (2001) Oursourcing industry leaders. Pharmaceutical Technology (January):68–73

Petruzzi N, Dada M (1999) Pricing and the newsvendor problem: A review with extensions. Oper Res 47:183–194

Petruzzi N, Dada M (2001) Information and inventory recourse for a two-market, price-setting retailers. Manufacturing & Service Oper Management 3:242–263

Pflug GC, Swietanowski A, Dockner E, Moritsch H (2000) The AURORA financial management system: Model and parallel implementation design. Annals Oper Res 99:189–206

Philips M (2008) The monster that ate Wall Street. Newsweek (October 6)

Poirier C, Quinn F (2003) Calibration Supply Chain Management. Computer Sciences Corporation Report

Porter M (1985) Competitive advantage. The Free Press, New York

Porteus E (2002) Foundations of stochastic inventory theory. Stanford University Press

PriceWaterhouseCoopers (2008) Global sourcing: Shifting strategies. Survey report (June)

Pyke D, Tang CS (2008) How to mitigate product safety risks proactively? Process, challenges and opportunities. Working paper, UCLA Anderson School

Pyke D, Tang CS (2009) Product recall management: A continuous improvement framework. Working paper, UCLA Anderson School

Quek T (2009) China firms turn to home market. Straits Times (February 28)

Quelch J (2007) Mattel: Getting a recall right. Harvard Business Online (August 27), http://hbswk.hbs.edu/item/5755.html

Quelch J, Kenney D (1994) Extend profits, not the product lines. Harvard Business Rev 72(September–October):153–160

Raghunathan S (2001) Information sharing in a supply chain: A note on its value when demand is non-stationary. Management Sci 47:605–610

Rajaram K, Tang CS (2001) The impact of product substitution on retail mechandising. Eur J Oper Res 135:582–601

Raju J, Sethuraman R, Dhar S (1995) The introduction and performance of store brands. Management Sci 41(6): 957–978

Raman A (1998) Managing inventories for fashion products. In: Tayur et al (eds) Quantitative models for supply chain management. Kluwer

Raman A, Schmidt C, Gaul V (2008) Airbus A380—turbulence ahead. Harvard Bus School Case Number N9-609-041

Ramasesh R, Ord J, Hayya J, Pan A (1991) Sole versus dual sourcing in stochastic lead-time (s, Q) inventory models. Management Sci 37:428–443

Ramdas K (2003) Managing product variety: An integrative review and research directions. Prod Oper Management 12:79–101

Ramstad E (2007) Samsung faces growing price pressure. The Wall Street Journal (European edition) (January 15):6

Ray S (2008) Boeing buys Vought Venture to stem delays on 787. Bloomberg Press (March 28), http://www.bloomberg.com/apps/news?pid=20601103&sid= aw4dIEC3nhGs&refer=news (accessed April 2009)

Rao S, Goldsby TJ (2009) Supply chain risks: A review and typology. Int J Logistics Management 20(1):97–123

Repenning N, Sterman J (2001) Nobody ever gets credit for fixing problems that never happened. California Management Rev (43, Summer):64–88

Reynolds P (2005) Multiple failures caused relief crisis. BBC News (September 7) (Article available at: http://news.bbc.co.uk/1/hi/world/americas/4216508.stm)

Rice B, Caniato F (2003) Building a secure and resilient supply-chain. Supply Chain Management Rev 7(5):22–30

Richards GD (1995) A general mathematical framework for modelling two dimensional wildland fire spread. Int J Wildland Fire 5:63–72

Rigby D (2001) Management tools and techniques: A survey. California Management Rev 43:139–159

Rigby B (2008) Spirit Aero to get early payments from Boeing. Reuters (April 10) http://uk.reuters.com/article/basicIndustries/idUKN0946545920080410 (accessed April 2009)

Ritchie B, Brindley C (2007) An emergent framework for supply chain risk management and performance measures. J Oper Res Soc 58:1398–1411

Roberts D (2006) How rising wages are changing the game in China. Business Week (March 27)

Rockafeller RT, Wets RJ-B (1991) Scenarios and policy aggregation in optimization under uncertainty. Mathematics Oper Res 16(1):119–147

Roth AV, Tsay AA, Pullman M, Gray JV (2007) Unraveling the food supply chain: Strategic insights from China and the 2007 recalls. Unpublished working paper, Clemson University, Clemson, SC

Ruwitch J (2009) China factory gloom plays into state plan. Reuters News (March 9)

Sahin F, Robinson EP Jr (2005) Information sharing and coordination in make-to-order supply chains. J Oper Management 23:579–598

Sarin R (2001) A social decision analysis of the earthquake safety problem. In: Kleindorfer PR and Sertel MR (eds) Mitigation and financing seismic risks. Kluwer, The Netherlands

Scheller-Wolf A, Tayur S (1999) Managing supply chains in emerging markets. In: Tayur et al (eds) Quantitative models for supply chain management. Kluwer, 703–735

Schoemaker PJH (1993) Multiple Scenario Development: Its Conceptual and Behavioral Foundation. Strategic Management Journal 14:193–213

Schoemaker PJH (1995) Scenario Planning: A Tool for Strategic Thinking. Sloan Management Review 36:25–40.

Schwarz P (1991) The Art of the Long View. New York: Doubleday

Sedarage D, Fujiwara O, Luong H (1999) Determining optimal order splitting and reorder level for n-supplier inventory systems. Eur J Oper Res 116:389–404

Sen S (2001) Stochastic programming: Computational issues and challenges. In: Gass S, Harris C (eds) Encyclopedia of operations research and management science. Springer, New York

Serel DA, Dada M, Moskowitz H (2001) Sourcing decisions with capacity reservation contracts. Eur J Oper Res 131:635–648

Seshadri S, Subrahmanyam M (2005) Introduction. Production Oper Management 14(1):1–5

Seshadri S, Khanna A, Harche F, Wyle R (1998) A method for strategic asset-liability management at the Federal Home Loan Bank of New York. Oper Res 47(3):345–360

Sethi A, Sethi S (1998) Flexibility in manufacturing: A survey. Int J Flexible Manufacturing Systems 2:289–328

Seuring S (2005) Case study research in supply chains—An outline and three examples. In: Kotzab H, Seuring S, Muller M, Reiner G (eds) Research methodologies in supply chain management. Physica-Verlag, Heidelberg

Shapira Z (1986) Risk in managerial decision making. Unpublished manuscript, Hebrew University

Sheffi Y (2001) Supply chain management under the threat of international terrorism. Int J Logistics Management 12(2):1–11

Sheffi Y (2005a) The resilient enterprise. MIT Press, Cambridge, MA

Sheffi Y (2005b) Creating demand-responsive supply chains. Harvard Business Rev (April):3–5

Shin H, Collier D, Wilson D (2000) Supply management orientation and supplier/buyer performance. J Oper Management 18:317–333

Signorelli S, Heskett JL (1984) Benetton. Harvard Business School Case # 685020

Slovic P (1987) Risk Perception Science, New Series, 236(4799): 280–285

Smith C, Gilbert S, Burnestas A (2002) Partial quick response policies in a supply chain. In: Geunes J, Pardalos P, Romeijn E (eds) Supply chain management models, applications and research directions. Kluwer

Smith NC, Thomas RJ, Quelch JA (1996) A strategic approach to management product recalls. Harvard Business Rev (September–October):102–112

So K, Tang CS, Zavala R (2003) Model for improving team productivity at the Federal Reserve Bank. Interfaces 33(2):25–36

Sodhi M (2000) Getting the most from planning technologies. Supply Chain Management Rev (Winter supplement):19–23

Sodhi M (2001) Applications and opportunities for operations research in internet-enabled supply chains and electronic marketplaces. Interfaces 31(2):56–69

Sodhi M (2003) How to do strategic supply-chain planning. Sloan Management Rev 45(1):69–75

Sodhi M (2005a) Tactical planning under demand risk for a global electronics company. Production Oper Management 14(1):69–79

Sodhi M (2005b) LP modeling for asset-liability management: A survey of choices and simplifications. Oper Res 53(2):181–196

Sodhi M, Lee S (2007) An analysis of sources of risk in the consumer electronics industry. J Oper Res Soc 58(11):1430–1439

Sodhi M, Tang CS (2008). The OR/MS ecosystem: strengths, weaknesses, opportunities, and threats. Oper Res 56(2):267–277

Sodhi M, Tang CS (2009a) Time-based disruption management: A framework for managing supply chain disruption. In: Blackhurst J, Wu T (eds) Managing supply chain risk and vulnerability: Tools and methods for supply chain decision makers. Springer, New York

Sodhi M, Tang CS (2009b) Modeling supply chain planning under demand uncertainty using stochastic programming: A survey motivated by asset-liability management. Int J Production Economics (forthcoming). SSRN download http://ssrn.com/abstract=910579

Sodhi M, Tang CS (2009c) Supply chain risk management. In: Cochran JJ, Anthony CL, Keskinocak P, Kharoufeh JP, and Smith JC (eds) Encyclopedia of operations research and management science. Wiley

Sodhi M, Tang CS (2011) An exact quantification of the bullwhip effect. European Journal of Operational Research 215(2):374–382

Sodhi M, Son BG, Tang CS (2008) What employers demand from applicant for MBA-level supply chain jobs and the coverage of supply chain topics in MBA courses. Interfaces 38(6):469–484

Spekman RE, Davis EW (2004) Risky business: Expanding the discussion on risk and the extended enterprise. Int J Physical Distribution & Logistics Management 34(5):414–433

Spinler S, Huchzermeier A (2006) The valuation of options on capacity with cost and demand uncertainty. Eur J Oper Res 171:915–934

St. George A (1998) Li and Fung: Beyond 'Filling in the mosaic.' Harvard Business School case 9-398-092

Stalk G, Hout T (1991) Competing against time: How time-based competition is reshaping global markets. Free Press, New York

Sterman JD (1989) Modeling managerial behavior: Misperceptions of feedback in a dynamic decision making experiment. Management Sci 35:321–339

Stremersch S, Tellis GJ (2002) Strategic bundling of products and prices: A new synthesis for marketing. J Marketing 66:55–72

Su J, Chang YL, Ferguson M (2005) Evaluation of postponement structures to accommodate mass customization. J Oper Management 23:305–318

Sunday Times (2003) Can suppliers bring down your firm? (November 23)

Swaminathan J, Tayur S (1998a) Stochastic programming models for managing product variety. In: Tayur S et al (eds) Quantitative models for supply chain management. Kluwer

Swaminathan J, Tayur S (1998b) Managing broader product lines through delayed differentiation using vanilla boxes. Management Sci 44:S161–S172

Swaminathan J, Tayur S (1999a) Managing design of assembly sequences for product lines that delay product differentiation. IIE transactions 31:1015–1026

Swaminathan J, Tayur S (1999b) Stochastic programming models for postponement. In: Tayur S, Ganeshan R, Magazine M (eds) Quantitative models for supply chain management, revised edition. Kluwer, Norwell, MA

Swinney R, Netessine S (2009) Long-term contracts under the threat of supplier default. Management Sci 11(1):109–127

Tagaras G, Lee H (1996) Economic models for vendor evaluation with quality cost analysis. Management Sci 42(11):1531–1543

Talluri K, Van Ryzin G (2005) The theory and practice of revenue management. Kluwer

Tang CS (1990) The impact of uncertainty on a production line. Management Sci 36(12):1518–1531

Tang CS (1999) Supplier relationship map. Int J Logistics Research Appl 2(1):39–56

Tang CS (2006a) Robust strategies for mitigating supply chain disruptions. Int J Logistics Research Appl 9(3):33–45

Tang CS (2006b) Perspectives in supply chain risk management. Int J Production Economics 103(4):451–488

Tang CS (2007) Boeing's 787 supply chain: a dream or a nightmare? Unpublished paper, UCLA Anderson School

Tang CS, Deo S (2005) Rental price and rental duration under retail competition. Working paper, UCLA Anderson School

Tang CS, Tomlin B (2008) The power of flexibility for mitigating supply chain risks. Int J Production Economics 116(1):12–27

Tang CS, Rajaram K, Alptekinoglu A, Ou J (2004) The benefits of advance booking discount programs: Models and analysis. Management Sci 50(4):465–478

Tang K (1988) An economic model for vendor selection. J Quality Technology 20:81–89

Tarim S, Kingsman BG (2004) The stochastic dynamic production/inventory lot-sizing problems with service-level constraints. Int J Production Economics 88(1):105–119

Tayur S, Ganeshan R, Magazine M (1998) Quantitative models for supply chain management. Kluwer

Terwiesch C, Ren ZJ, Ho TH, Cohen MA (2005) An empirical analysis of forecast sharing in the semiconductor equipment supply chain. Management Sci 51:208–220

Tighe C (2004) Samsung to quit UK site in quest for cheaper labour. Financial Times (January 16)

Tirole J (1988) The theory of industrial organization. The MIT Press, Cambridge, MA

Tomasini R, Van Wassenhove L (2009) Humanitarian logistics. McMillan Publishers, UK

Treleven M, Bergman Schweikhart S (1988) A risk/benefit analysis of sourcing strategies: Single vs. multiple sourcing. J Oper Management 7(3-4): 93–114

Trkman P, McCormack K (2009) Supply chain risk in turbulent environments—A conceptual model for managing supply chain network risk. Int J Production Economics 119(2):247–258

Tsay A, Lovejoy W (1999) Quantity-flexibility contracts and supply chain performance. Manufacturing & Service Oper Management 1(2):89–111

Tsay AA, Nahmias S, Agrawal N (1998) Modeling supply chain contracts: A review. In Tayur et al (eds) Quantitative models for supply chain management. Kluwer

Ulrich K, Randall T, Fisher M, Reibstein D (1998) Managing product variety. In: Ho TH, Tang CS (eds) Product variety management: Research advances. Kluwer

Van de Ven AH (1989) Nothing is so practical as a good theory. AMR 14(4):486–489

Van de Ven AH, Johnson PE (2006) Knowledge for theory and practice. AMR 31(4):802–821

Van Mieghem J, Dada M (2001) Price versus production postponement: Capacity and competition. Management Sci 45:1631–1649

Van Wassenhove L (2006) Humanitarian aid logistics: Supply chain management in high gear. J Oper Res Soc 57:475–489

Veverka M (1999) A DRAM shame. Barron's (October 15):15

Vlachos D, Tagaras G (2001) An inventory system with two supply modes and capacity constraints. Int J Production Economics 72:41–58

Vokurka RJ, Choobineh J, Vadi L (1996) A prototype expert system for the evaluation and selection of potential suppliers. Int J Oper Production Management 16(12):106–127

Voss C, Tsikritsis N, Frohlich M (2002) Case research in operations management. Int J Oper Production Management 22(2):195–219

Wagner SM, Bode C (2008a) An empirical examination of supply chain performance along several dimensions of risk. J Business Logistics 29(1):307–325

Wagner SM, Bode C (2008b) Dominant risks and risk management practices in supply chains. In: Zsidisin G, Ritchie B (eds) Supply chain risk: A handbook of assessment, management, and performance. Springer, New York.

Wagner SM, Bode C, Koziol P (2009) Supplier default dependencies: Empirical evidence from the automotive industry. Eur J Oper Res (forthcoming)

Wallace J (2006a) Aerospace notebook: Lightning a weighty issue for the 787. Seattle P-I (July 12) http://www.seattlepi.com/business/277220_air12.html (accessed April 2009)

Wallace J (2006b) Airbus unveils widebody, says A350 XWB will top 787 and 777. SeattlePI.com (July 18) http://www.seattlepi.com/business/277877_airshow18.html (accessed April 2009)

Wang Y, Gerchak Y (1996) Periodic review production models with variable capacity, random yield, and uncertain demand. Management Sci 42(1):130–137

Wang H, Barney JB, Reuer JJ (2003) Stimulating firm-specific investment through risk management. Long Range Planning 36(1):49–59

Waters D (2007) Supply chain risk management: Vulnerability and resilience in logistics. Kogan Page, London

Weatherford LR, Bodily S (1992) A taxonomy and research overview of perishable-asset revenue management: Yield management, overbooking and pricing. Oper Res 40:841–844

Weber CA, Current JR (1993) Multi-objective analysis of vendor selection. Eur J Oper Res 68:173–184

Weber CA, Current JR, Desai A (2000) An optimisation approach to determining the number of vendors to employ. Supply Chain Management: An International Journal 5:90–98

Weng K, Parlar M (1999) Integrating early sales with production decisions: Analysis and insights. IIE Transactions 31:1051–1060

Weng K, Parlar M (2005) Managing build-to-order short life-cycle products: Benefits of pre-season price incentives with standardization. J Oper Management 23:482–495

West K (2007) Boeing 787 program not out of woods. MSNBC (Dec 12)

Wets RJ-B (1983) Solving stochastic problems with simple resource. Stochastics 10(3-4):219–242

Whang S, Lee HL (1998) Value of postponement. In: Ho T, Tang CS (eds) Product variety management: Research advances. Kluwer

White D (1995) Application of system thinking to risk management: A review of the literature. Management Decision 33(10):35–45

Worzel KJ, Vassiadou-Zenio C, Zenios SA (1994) Integrated simulation and optimization models for tracking indices of fixed-income securities. Operations Res 42(2):223–233

Wright SE (1994) Primal-dual aggregation and disaggregation for stochastic linear programs. Math Oper Research 19(4):893–908

Wu T, Blackhurst J (eds) (2009) Managing supply chain risk and vulnerability: Tools and methods for supply chain decision makers. Springer, London

Xie J, Shugan S (2001) Electronic tickets, smart cards, and online prepayments: When and how to advance sell. Marketing Sci 20:219–243

Xu X, Birge J (2004) Joint production and financing decisions: Modeling and analysis¿ Working paper, School of Business, University of Chicago

Yang B, Burns N, Backhouse C (2004) Postponement: A review and an integrated framework. Int J Oper Production Management 24:468–487

Yano C, Gilbert S (2004) Coordinated pricing and production/procurement decisions: A review. In: Chakravarty AK, Eliashberg J (eds) Managing business interfaces: Marketing, engineering, and manufacturing perspectives. Kluwer Academic Publishers

Yano C, Lee HL (1993) Lot sizing with random yields: A review. Oper Res 43:311–334

Yeh C, Yang C (2003) A cost model for determining dyeing postponement in garment supply chain. Int J Advanced Manufacturing Technology 22:134–140

Yu G, Arguello M, Song G, McCowan S, White A (2003) A new era for crew recovery at Continental Airlines. Interfaces 33(1):5–22

Zeldith M Jr (1962) Some methodological problems of field studies. Amer J Sociology 67(5):566–576

Zhou YC (2009) Private Communication. Deloitte Consulting

Zenios SA, Ziemba WT (eds) (2004) Handbook of asset-liability management. Handbooks in finance series. North Holland, Amsterdam, The Netherlands

Zetter K (2008) FAA: Boeing's new 787 may be vulnerable to hacker attack. Wired (Jan 8) http://www.wired.com/politics/security/news/2008/01/dreamliner_security. (accessed May 2009)

Zhang VL (1996) Ordering policies for an inventory system with three supply modes. Naval Research Logistics 43:691–708

Zhang X (2004) The impact of forecasting methods on the bullwhip effect. Int J Production Economics 88:15–27

Zhao X, Xie J, Leung J (2002) The impact of forecasting model selection on the value of information sharing in a supply chain Eur J Oper Res 142:321–344

Zhou YC (2009) Private Communication. Deloitte Consulting

Ziemba WT, Mulvey JM (eds) (1999) Worldwide asset and liability modeling. Cambridge University Press, Cambridge, UK (first printed 1998)

Zipkin P (1980a) Bounds on the effect of aggregating variables in linear programs. Oper Res 28(2):403–418

Zipkin P (1980b) Bounds for row-aggregation in linear programming. Oper Res 28(4):903-916

Zipkin P (2000) Foundation of inventory management. McGraw Hill, New York

Zsidisin G, Melnyk S, Ragatz G (2005) An institutional theory perspective of business continuity planning for purchasing and supply management. Int J Production Res 43(16):3401–3420

Zsidisin G, Panelli A, Upton R (2001) Purchasing organization involvement in risk assessments, contingency plans, and risk management: An exploratory study. Supply Chain Management: An International Journal 5(4): 187–197

Zsidisin G, Ragatz G, Melnyk S (2004a) Effective practices and tools for ensuring supply continuity. In: Brindley C (ed) Supply chain risk. Ashgate Publishers, Hampshire, England

Zsidisin GA, Ritchie B (2008) Supply chain risk: A handbook of assessment, management, and performance. Int Series in Operations Research & Management Science), Springer, New York

Zsidisin G, Ellram LM, Cater JR, Cavinato JL (2004) An analysis of supply risk assessment techniques. Int J Physical Distribution & Logistics Management 34(5):397–413

Printed by Printforce, the Netherlands